IBM WebSphere and Lotus

Implementing Collaborative Solutions

IBM Press Series—Information Management

On Demand Computing Books

On Demand Computing
Fellenstein

Grid Computing
Joseph and Fellenstein

Autonomic Computing
Murch

Business Intelligence for the Enterprise
Biere

DB2 Books

The Official Introduction to DB2 for z/OS
Sloan

High Availability Guide to DB2
Eaton and Cialini

DB2 Universal Database v8.1 Certification Exam 700 Study Guide
Sanders

DB2 Universal Database v8.1 Certification Exams 701 and 706 Study Guide
Sanders

DB2 for Solaris: The Official Guide
Bauch and Wilding

DB2 Universal Database v8 for Linux, UNIX, and Windows Database Administration Certification Guide, Fifth Edition
Baklarz and Wong

Advanced DBA Certification Guide and Reference for DB2 Universal Database v8 for Linux, UNIX, and Windows
Snow and Phan

DB2 Universal Database v8 Application Development Certification Guide, Second Edition
Martineau, Sanyal, Gashyna, and Kyprianou

DB2 Version 8: The Official Guide
Zikopoulos, Baklarz, deRoos, and Melnyk

Teach Yourself DB2 Universal Database in 21 Days
Visser and Wong

DB2 UDB for OS/390 v7.1 Application Certification Guide
Lawson

DB2 SQL Procedural Language for Linux, UNIX, and Windows
Yip, Bradstock, Curtis, Gao, Janmohamed, Liu, and McArthur

DB2 Universal Database v8 Handbook for Windows, UNIX, and Linux
Gunning

Integrated Solutions with DB2
Cutlip and Medicke

DB2 Universal Database for OS/390 Version 7.1 Certification Guide
Lawson and Yevich

More Books from IBM Press

Developing Quality Technical Information: A Handbook for Writers and Editors
Hargis, Carey, Hernandez, Hughes, Longo, Rouiller, and Wilde

Enterprise Messaging Using JMS and IBM WebSphere
Yusuf

Enterprise Java Programming with IBM WebSphere, Second Edition
Brown, Craig, Hester, Stinehour, Pitt, Weitzel, Amsden, Jakab, and Berg

IBM WebSphere and Lotus

Implementing Collaborative Solutions

IBM Press

WebSphere

John Lamb, Michael Laskey, and Gopal Indurkhya

PRENTICE HALL PTR

An Imprint of PEARSON EDUCATION
Upper Saddle River, NJ • New York • San Francisco
Toronto • London • Munich • Paris • Madrid
Capetown • Sydney • Tokyo • Singapore • Mexico City
www.phptr.com

Library of Congress Cataloging-in-Publication Data
A CIP catalog record for this book can be obtained from the Library of Congress
LOC Number: 2004107328

Acquisitions Editor: John Neidhart
Marketing Manager: Robin O'Brien
International Marketing Manager: Tim Galligan
Managing Editor: Gina Kanouse
Senior Project Editor: Sarah Kearns
Copy Editor: Chrissy Andry
Indexer: Angie Bess
Proofreader: Andrew Wahnsiedler
Composition: Jake McFarland
Cover Design: Nina Scuderi
Manufacturing Manager: Dan Uhrig

Prentice Hall PTR offers excellent discounts on this book when ordered in quantity for bulk purchases or special sales. For more information, please contact: U.S. Corporate and Government Sales: 1-800-382-3419, corpsales@pearsontechgroup.com. For sales outside the U.S., please contact: International Sales, international@pearsoned.com.

ISBN: 0-13-144330-5
Text printed on recycled paper
2 3 4 5 6 7 8 9 10 IBT 08 07 06
Second printing: February 2006

Pearson Education Ltd.
Pearson Education Australia Pty., Limited
Pearson Education South Asia Pte. Ltd.
Pearson Education Asia Ltd.
Pearson Education Canada, Ltd.
Pearson Educacion de Mexico, S.A. de C.V.
Pearson Education--Japan
Pearson Malaysia S.D.N. B.H.D.

To my wife Penny and our son Jack.
—JL

To my wife Laura and our daughter Lauramarie.
—ML

To my wife Vandana and our sons Sagar and Aakash.
—GI

Acknowledgments

Many of the ideas and details presented in this book are based on the WebSphere and Domino installations designed by IBM for its internal and commercial users. Therefore, we would like to thank all those involved with the design of these IBM implementations.

Richard Wood of IBM Global Services was our most dedicated reviewer. "Woody" looked over almost all the revisions and gave us many valuable comments. Caleb Barlow and Brian White of Lotus reviewed the manuscript from the Domino and Lotus Collaboration perspective to make sure we were not missing anything major in the manuscript. We would like to thank our IBM management chains for the support needed to make an undertaking like this happen. Our first-line managers Boris Vishnevsky, Woody Wood, and Donald Hooker; our second-line management Beth Hazuka; and our third-line managers (directors or principals in IBM) Susan Gryzbowski and Anil Cheriyan.

Paul Dantzig, manager of the High-Volume Web Serving group at IBM Research in Hawthorne, New York, graciously allowed us to include much of his work on the IBM Web sites for the Olympic Games, Grand Slam Tennis events, and other sports events in Chapter 15, "Implementing High-Volume Web Sites."

Michael Haley, IBM Distinguished Engineer and Global Executive Grid Computing, provided expert advice on the sections in Chapter 15 relating to Grid Computing and the design aspects of On Demand Web systems. Rohit Sahasrabudhe of IBM Austin, Texas, provided much of the material for Appendix F on installing Lotus Workplace 1.1. Carol Zimmet, Louis Bradbard, and Martha Moreau of Lotus created some of the figures that help describe Lotus collaboration products and function. Mark Feinman and Chris Pepin of IBM contributed to the information in Chapter 15 on IBM's internal implementation of Lotus Web Conferencing and

Instant Messaging, respectively. Tony Payne of Lotus Sametime Development helped mightily with the implementation of our Lotus Web conferencing commercial offering described in Chapter 15. Christopher Ahrendt helped with the design of the WAS MQ cluster diagram given at the end of Chapter 7.

Tony Cusato of IBM Global Services contributed to Chapter 15 on the use of Domino for the Web sites used for the Olympics. Kevin Finnegan of IBM AMS (Application Management Systems) contributed to the sections on IBM's WebSphere Portal technology. Charles Lin and Melva Jeter of Groupware Support at IBM Boulder provided review from the aspect of a hosted solution provider. Charles installed WebSphere Portal Server 5.01 on a lab server at IBM Boulder and was able to verify all the installation steps and make screenshots for Appendix E on the WebSphere Portal Server. He made his tests on a Redhat Linux Advanced Server 2.1. Charles also installed Lotus Workplace 1.1 at the Boulder lab and was able to give us information for Appendix F on Lotus Workplace 1.1. Charles did point out that since the Lotus Workplace 1.1 installation included the configuration for the LDAP directory and DB2, that installation was more automated.

Much of the information on Lotus products was taken with permission from the Lotus technical Web site at `http://www-136.ibm.com/developerworks/lotus` (also accessible via its former name of `http://www.notes.net`). Leigh Ann Fowler of IBM/Lotus Software Group Media Relations served as our "window" for help in obtaining a Lotus executive for both an author of the Foreword for our book and for an IBM "Executive Champion" for this IBM Press book.

Finally, we would like to thank Steve Robinson, Vice President of IBM Software Services for Lotus, who not only wrote the Foreword for our book, but also acted as the IBM "Executive Champion" for our book—a new post required for books published under the IBM Press logo.

Table of Contents

About the Authors

John P. Lamb is currently a Senior Technical Staff Member for IBM in White Plains, New York. His work includes architecture and design for IBM Web services for several companies including the New York Stock Exchange and several large banks. Service offerings to corporations for collaboration services using products from Lotus, such as Web conferencing, instant messaging, and Web teamrooms, have recently become another focal area. John holds a B.A. from the University of Notre Dame, M.S. degrees from the University of Utah and Pace University, and a Ph.D. in Engineering Science from the University of California at Berkeley. He is a senior member of the IEEE and ASME engineering societies. He has published more than 40 technical papers and has previously co-authored three books, *Lotus Notes Network Design* (published by McGraw-Hill, 1996), *Lotus Notes and Domino Network Design* (published by McGraw-Hill, 1997), and *Lotus Notes and Domino 5 Scalable Network Design* (published by McGraw-Hill, 1999). John is a Principal Certified Lotus Professional (PCLP) in Systems Administration and he is a Certified Senior Consulting I/T Architect. In addition, John is an avid airplane pilot who holds a commercial pilot's license with glider rating.

Michael Laskey is a Senior Software Engineer for IBM Global Services in Tampa, Florida. Mike has over 25 years experience as a software developer and architect and is currently responsible for requirement analysis, architecture, design, and development of e-Commerce solutions for IBM EDI service offerings and specific customer projects. These solutions involve a wide range of Internet-related product software, technologies, and development tools. Michael has authored several technical articles on Domino and WebSphere integration and is a co-author of the first edition of the IBM Redbook, *Domino and WebSphere Together*. Michael enjoys time with his wife and daughter, exploring west Florida by kayak, golf, tennis, and tutoring mathematics.

Gopal Indurkhya is currently a Certified Consulting I/T Architect for IBM in Charlotte, North Carolina. He manages development and develops solution architectures for organizations to assist them in meeting their business requirements. Gopal holds M.S. degrees from IIT, Kanpur, India, and the University of Nebraska, Lincoln. He has published several technical papers in leading journals and has co-authored an IBM Redbook, "Servlet/JSP/EJB design and Implementation Guide for WebSphere Application Server." He also teaches and assisted in developing IBM's "Architectural Thinking" class.

Abstract

Lotus collaboration products, mostly based on Lotus Notes and Domino, are the premier offerings in the field of collaboration. WebSphere is IBM's premier entry in the area of tools for building Web sites. Thus, for companies that have a Web application that requires the strengths of Lotus collaboration offerings (e.g., Teamrooms, Knowledge Bases, Corporate Portals, Help Desks, Sales Force Automation, Web Meetings that include "chat," Project Tracking, Intranets, built-in messaging, etc.) and the strengths of WebSphere (e-commerce, Brokerages, B2B Exchanges, Supply Chain Integration, Online Banking, Web Account Access, Web Portals, Extranets, etc.), a combined Lotus Domino and WebSphere system would be ideal. In addition to details on how to have WebSphere and Domino work together, this book discusses how to use the new Lotus Workplace products and WebSphere Portal as ways to implement collaboration tools with WebSphere. There are several ways to implement Lotus collaborative solutions with WebSphere, and this book discusses the options.

One of the significant growth areas for many companies has been commerce between companies, or business-to-business (B2B) commerce. B2B commerce requires much greater need for relationships and collaborative technologies (such as Web conferencing and "chat") to facilitate purchasing for large, complex transactions between teams of people. Also, sites with just transactional capabilities are realizing that they could increase the amount and frequency of purchasing by assisting transactions. If a company is building Web applications that focus on collaboration, Domino is an excellent choice since Domino excels when an application requires sophisticated collaborative capabilities, such as workflow and distributed content authoring, or deals with a lot of unstructured data or knowledge. Likewise, if a company is building a Web site that needs to handle a lot of transactions or requires the use of Enterprise Java components, then WebSphere is the good choice since WebSphere excels when an application requires industrial-strength transaction management or massive scalability, or where business logic is wholly encapsulated in distributed components such as servlets or Enterprise JavaBeans (EJBs). If a company needs to build a site that requires both of these capabilities, then Domino and WebSphere can work together over the same or separate HTTP stacks to deliver the best of both worlds.

The main idea behind the use of collaboration tools for corporations should be to help in making business processes more efficient and effective. This book addresses the needs of those IT architects, network designers, and Web system developers who must help implement Web systems to meet requirements for both collaboration and Web services.

Foreword

Businesses are challenging IT professionals to help them achieve a greater degree of productivity—organizational productivity. Technology can no longer be just a series of stand-alone applications. Solutions need to integrate people, information, and business processes in order to satisfy the need for on demand business that is more efficient, flexible, and responsive to changing market conditions.

Organizations want to increase their productivity with dynamic work environments where role-based workplaces can be created with ease. And they want open standards-based solutions that integrate across their entire product portfolio.

IBM's technology portfolio provides end-to-end solutions for our customers. When real-time collaborative solutions from Lotus are coupled with a leading web services platform such as WebSphere, the combination extends collaboration to a broader range of users, business partners, and customers. The rich capabilities utilizing components can be embedded in any client, allowing for the creation of dynamic, role-based workplaces. Together, Lotus and WebSphere are fundamentally changing the way people and organizations work.

IBM WebSphere and Lotus: Implementing Collaborative Solutions provides a concise step-by-step guide on the best methods of implementing a merged Lotus and WebSphere strategy. It is a technical guide that will be of great use to developers and administrators, particularly those looking to design or implement large, enterprise-wide deployments of WebSphere and Lotus solutions.

As corporations evolve in the on demand world, IT professionals must deliver practical and cost-effective solutions that address the ever-changing business environment. Future success requires focusing on growth, dedication to core competencies, accelerating the time to market, reducing costs, and earning a prominent position in the global marketplace. IBM Lotus and WebSphere are responding by providing IT professionals with innovative solutions that harness the power of open standards while helping businesses reach their maximum potential through organizational productivity.

Steve Robinson
Vice President, IBM Software Services for Lotus
IBM Lotus Software

Preface

This book focuses on the design of the WebSphere and Lotus Collaboration implementations used within IBM and in IBM's commercial offerings. As such, it is geared mostly toward the design of large, enterprise-wide deployment of WebSphere, Domino, and other Lotus Collaboration tools. The basic concepts for the design of WebSphere and Lotus Collaboration implementations, however, apply to all sizes of deployments, so the book should prove useful to your WebSphere and Lotus Collaboration implementations—no matter what the size.

Cutting-edge collaboration can streamline business processes and speed response times. Success usually depends on convergence and consolidation of your processes. If you can place everything from security to content repositories on a common platform, you can help break down the barriers between your employees, partners, and customers. Convergence and consolidation with the help of collaboration tools means fewer devices to manage and maintain, which translates into lower administrative costs and higher return on investment (ROI).

Productivity is another payoff. Convergence lets employees move seamlessly from application to application in real-time, whether instant messaging, Web conferencing, or document management. With collaboration tools, you can find information on the fly and save, modify, or send as you go. The use of collaboration tools with your Web-based systems can be much more efficient than following email strings or sorting through voice messages.

It should be noted that the IBM/Lotus Company is continually looking at the role of the WebSphere and Domino platforms. For example, Lotus has been looking at the use of WebSphere and DB2 for many of its future messaging and collaboration products. In 2003, Lotus introduced Workplace, a set of browser-based collaboration solutions that are built on WebSphere and DB2. It's a new server with a back-end data repository that runs on DB2 and uses WebSphere as the Web-serving engine for the browser-based user interface. The server authenticates users through an LDAP directory. WebSphere and DB2 provide stable foundations for the product and give developers more options for J2EE-based application integration. The first Lotus Workplace component introduced was Workplace messaging. The messaging market is diverging. There exists messaging that is tied closely to groupware and offers collaboration capabilities (as Domino does); but with Workplace messaging, IBM has developed a portal approach to messaging where messaging becomes a service or feature that's integrated into a portal. This approach is significant for IBM and its ongoing portal, wireless, and On Demand Business strategy.

Domino, by way of its collaborative capabilities, also has that association between applications and email. Domino is still the best platform for users who require high-end email and collaboration functionality. But the purely WebSphere solution has advantages because of its J2EE standards and ability to easily integrate with other standards-based solutions such as portals. The next generation of most Lotus and IBM products will offer closer orientation to a portal. A full-fledged portal can take over much of what more traditional groupware used to do and often can do it faster, cheaper, and better. By the way, we've all been using portals for some time because, in essence, a portal is an internet Web site that allows people to search for information from a number of different internet sources, as well as the portal Web site itself. In effect, a portal acts as gateway to the World Wide Web. Products such as WebSphere Portal Server and Lotus Workplace have now made it easy to set up and modify portals.

We should all value the application of technology to solve serious business problems. Computing is no longer seen as a collection of piece parts. Today, computing is viewed holistically—as a technical architecture, an end-to-end system, and a computing model. On Demand Business has to do with how trends in technology and application of technology to the needs of business are coming together in a new era, and it's that era that is being called "On Demand Business." An On Demand Business enjoys enormous competitive advantages. These companies are able to

- Be extremely responsive to external threats, marketplace opportunities, and the needs of their customers, employees, and partners.
- Convert fixed costs into variable costs.
- Greatly reduce their inventories.

To achieve these goals, a business must integrate its processes throughout the organization and with their suppliers and customers. Once a company's various processes are optimized and pulled together into an integrated system, they'll enjoy greater productivity and will be able to meet their own customers' escalating demand and the changing dynamics of their own industries and the global economy. Combining collaboration and Web server software is one important step toward integrated systems. The intent of this book is to help you with the different options in integrating Lotus Collaboration and WebSphere—including the significant option of using the WebSphere Portal Server to help make that integration happen.

How to Use This Book

This book is organized largely around the two basic reader groups for a book of this type: 1) Lotus Domino and Lotus Collaboration designers and developers and 2) IBM WebSphere designers and developers. If you're currently a WebSphere developer, then you would probably want to skip Chapter 3 and Appendix A on WebSphere Application Server 5. Likewise, if you're currently a Domino/Lotus Collaboration designer, then you may want to skip Chapter 4 and Appendix B on Domino 6.

Chapters 1 through 5 are used to give us a background on WebSphere, Domino, and other Lotus Collaboration tools.

Chapters 6 through 10 give details and hints on the best ways to combine WebSphere and Domino to optimize the foundation for using Lotus Collaboration tools with a web-based infrastructure.

Chapters 11 though 15 give details and implementation tips for implementing important aspects of a combined WebSphere and Domino/Lotus Collaboration system. Topics such as security and Single Sign-On, WebSphere and Domino Clustering, performance and high availability, and high-volume Web sites are covered in detail. Within these chapters are real-world case studies of how IBM has implemented these features both for our internal Web sites and for commercial offerings and for our customers.

Chapter 16 summarizes the book and looks to the future. Appendices A through F give further details on the areas in the chapters and include details and case studies on the use of the WebSphere Portal Server and the new Lotus Workplace product as very convenient ways to implement collaboration tools with your web sites.

This book was written to allow you to go to a chapter or section of interest and begin reading without being required to have first read the previous chapters. Details on IBM's own large-scale internal use of Lotus Collaboration tools to improve efficiency and foster teamwork are given in several places throughout the book. Almost all of IBM's 340,000 employees use Lotus Instant Messaging on a daily basis. IBM's internal use of Lotus Web Conferencing and use of the IBM Portal Server to enhance teamwork and improve workplace efficiency is another theme, with detailed statistics on IBM's collaboration tools usage growth given in Chapter 15 and Appendix E.

Enjoy this book. We had an enjoyable time writing it.

John Lamb, Mike Laskey, and Gopal Indurkhya
June, 2004

Chapter 1

Introduction

> **"Apollo 13 depended upon everyone being 'part of the solution.' A groupware tool like Lotus Notes and Domino would have been ideal."**
>
> **—Jim Lovell, Apollo 13 spacecraft commander, at LotuSphere 1999 in Orlando, Florida**

> **"According to Gartner Dataquest's annual report on market share, IBM is riding a champion steed named WebSphere. The software brand took a trio of titles in 2002, garnering it a Triple Crown for application servers, business portals, and integration software."**
>
> **—*e-Pro Magazine*, May, 2003**

WebSphere and Domino (together with all the Lotus Collaboration tools based around Domino) each offer unique capabilities in helping you build robust Web sites. The Lotus Domino collaboration product and the Web platform WebSphere product working together can bring significant new functions to a Web site. One of the significant growth areas in IBM, and for most customers, has been commerce between companies or business-to-business (B2B) commerce. B2B commerce requires a much greater need for relationships and collaborative technologies (such as Web conferencing and "chat") to facilitate purchasing for large, complex transactions between teams of people. In addition, sites with just transactional capabilities are realizing that they could increase the amount and frequency of purchasing by assisting transactions.

Why Implement Both WebSphere and Domino?

If a company is building Web applications that focus on collaboration, Domino has many benefits. Domino excels when an application requires sophisticated collaborative capabilities, such as workflow, distributed content authoring, or deals with a lot of unstructured data or knowledge. Likewise, if a company is building a Web site that needs to handle many transactions or requires the use of Enterprise Java components, then WebSphere provides significant benefits. WebSphere excels when an application requires industrial-strength transaction management, massive scalability, or where business logic is wholly encapsulated in distributed components, such as servlets or Enterprise JavaBeans (EJBs). Many companies are now finding that they need Web sites that require both of these capabilities. Domino and WebSphere can work together over the same or separate Hypertext Transfer Protocol (HTTP) stacks to deliver the best of both worlds. For example, one customer has built an application that uses a Domino workflow infrastructure to implement content management for its Web site catalog. Domino-based workflow and collaboration software is used to manage the content generation and content approval process for catalog entries and prices. The highly scalable transaction services of WebSphere are used in tandem to implement the actual catalog site. Domino messaging is used to send confirmation information to the customer and maintain the dialog with the customer.

This book addresses strategies of merging Domino and WebSphere application functions to develop powerful Web sites. Development along J2EE lines is stressed. This book was written to have a primarily role as a developer's reference—and as such, the use of tools such as WebSphere Studio is emphasized. However, Web infrastructure, including network bandwidth considerations, high availability, clustering, performance, and scalability considerations, all have a significant place in the text. The book includes the following topics:

- What to put on WebSphere Application Servers (WAS—pronounced "waz") vs. Domino (for example, JavaServer Pages or JSPs vs. Domino forms, servlets vs. agents, or WAS vs. Domino servlet engine)
- Accessing J2EE elements from Domino (for example, Java from LotusScript, JSP tag library for Domino elements, or XML)
- Accessing Domino from J2EE (Domino classes from Java or Web services access to Domino, for example)
- Development tools (WSAD, Eclipse plug-ins for Domino development, Domino Designer features for J2EE development)
- Use of the WebSphere Studio Development Platform
- Issues related to WAS and Domino together: Single Sign-on (SSO), clustering, performance, high availability, and so on
- The role of the new Lotus Workplace components in implementing WebSphere and Lotus Collaboration solutions

- The role of the WebSphere Portal Server in implementing WebSphere and Lotus Collaboration solutions

A real-world case study that uses WebSphere, Domino, and several Lotus Collaboration tools is discussed in Chapters 13 and 15. That real-world example is IBM's On Demand Service (ODS) offerings for Lotus Collaboration Applications, which are good examples of IBM's hosted offerings in the on demand era. These offerings are hosted at IBM On Demand Center (ODC) server farms, such as the IBM ODC in Boulder, Colorado. The On Demand Lotus Collaboration applications include Web Conferencing, Instant Messaging, Web Mail, and Lotus TeamRoom.

These On Demand Service offerings are hosted offerings designed to allow multiple customers to securely share hardware resources. Security access controls restrict customers to specific virtual resources. The On Demand Lotus Collaboration offerings are based on a set of IBM Software Group software products including Tivoli Monitoring, Tivoli Storage Management (TSM), IBM Directory Server (used as an LDAP directory), WebSphere Application Server (WAS), IBM/Lotus Web Conferencing 6.5.1 with the Enterprise Meeting Server (EMS) option, Lotus Instant Messaging, Lotus Workplace Messaging, Lotus TeamRoom, WebSphere Everyplace Subscription Manager (WESM), DB/2, WebSphere MQ server, and Domino 6. These offerings utilize IBM's new Universal Management Infrastructure (UMI) at the IBM Boulder On Demand Center. The billing model is based on connection time and/or maximum concurrent users. The design includes Gigabit speeds to the Internet and the use of cluster technology for all components that allow applications to scale to a very large number of customers. IBM is using hardware SSL Accelerators and Load Balancers to maximize performance. The IBM design team worked closely with the team that designed Lotus Web conferencing for IBM internal use and many "lessons learned" were shared. Although the IBM internal implementation of Lotus Web Conferencing was designed primarily for internal use, we also invite many customers to Web conferences on this system. Since all users must be registered with userid and password, the IBM LDAP directory to authenticate for this Web conference collaboration system has over 2 million names (including 340,000 IBMers), so it definitely falls into the high-volume, high-capacity collaboration system based on WebSphere, Domino, and Lotus Collaboration tools. The case studies in Chapters 13 and 15 discuss the details of the design and architecture for all of these high-volume, high-availability, high-performance, on demand collaboration offerings.

The WebSphere "Umbrella"

The IBM WebSphere brand encompasses a variety of products, from those focused on e-business applications to integration and portals. In fact, the WebSphere brand has become an "umbrella" for many different IBM products. IBM describes WebSphere as Internet infrastructure software, or middleware, that enables companies to develop, deploy, and integrate next-generation e-business applications. WebSphere supports business applications from simple Web publishing through enterprise-scale transaction processing.

IBM identifies WebSphere products as fitting into one of three areas: foundation and tools, business portals, or business integration.

1. **Foundation and Tools.** The foundation and tools provide customers with the ability to build, deploy, and manage the creation of J2EE-based Web applications. The products include WebSphere Application Server, WebSphere Application Server Express, and WebSphere Studio.
2. **Business Portals.** Business portals let organizations extend applications to users in a single, customized environment and also extend the applications to a variety of electronic devices. The products include WebSphere Commerce, WebSphere Portal, WebSphere Portal Express, WebSphere Everyplace, and WebSphere Personalization.
3. **Business Integration.** Business Integration products provide for integration between new and existing applications in a network. The products include WebSphere MQ, WebSphere Business Integration, IBM CrossWorlds solutions, and adapters and connectors.

Within each area, some products overlap in what they can do for an organization, but they may still provide varying degrees of functions and capabilities. The title of this book only mentions the WebSphere brand, but the WebSphere Application Server product is actually the focus of this book – however all areas under the WebSphere umbrella are discussed to a degree. An important aspect of discussing the use of WebSphere Application Server and Domino together falls in the WebSphere area of Business Portals.

IBM's Portal Strategy

IBM's continuing enhancements to the WebSphere Portal are very relevant to the direction of merging Domino and WebSphere application development along J2EE lines. Over the past few years, the Lotus range of products has expanded to embrace technologies such as instant messaging or chat. IBM has recognized that these new technologies of presence offer the possibility of incremental collaboration. Use of the WebSphere Portal can be the payoff. It incorporates the Lotus-derived Collaboration Center, which offers "portlets" integrated with Lotus technology for messaging and team workplaces. Placing parts of Lotus collaboration under the WebSphere umbrella will help enhance the scalability of Lotus Collaboration for the large enterprise. In addition, the Portal strategy is a good way to allow IBM and Lotus to introduce the WebSphere community to the joys of collaboration.

The move is also significant because of what it says about IBM's portal strategy. Some analysts divide portal servers into application portal suites (APS) and smart enterprise suites (SES). An APS is essentially an application server with limited collaborative facilities, and an SES is a high-powered collaborative workspace unattached to any specific application. With the enhancements to WebSphere Portal, IBM is moving in the direction of an APS/SES hybrid, adding industrial-strength collaboration and workflow facilities to the tried-and-tested WebSphere application-based setup.

With Lotus embracing DB2 and J2EE, much of the communication the Lotus and WebSphere communities has been one way. Now with WebSphere embracing collaborative techniques, the communication is becoming "two way."

Lotus Workplace—A Portal-Based Collaboration Platform

Lotus Workplace, made available toward the end of 2003, is a portal-based collaboration platform that lets users interact with colleagues, customers, and partners, providing immediate access to the applications and information they need to provide superior service. Because it is estimated that 80 percent of a business' knowledge resides with its people, the potential productivity gains are very significant. With Lotus Workplace there are several products that allow customers to add modular collaborative capabilities to their business applications. For example, Lotus Team Collaboration lets users instantly access people and information, minimizing the cycles associated with e-mail and voicemail "tag" and complex searches for timely information.

Lotus Workplace ties together key collaborative applications and functions based on an employee's role and delivers it through a single desktop interface. It lets a user keep an eye on important events and projects outside of the current area of focus, allowing for easy multitasking. Lotus Workplace also simplifies the user experience by enabling access to critical business information anywhere, anytime from whatever device the user chooses.

Over the years, businesses have purchased a variety of products to improve collaboration between individuals, teams, and workplace communities. The result is a disjointed user experience with multiple interfaces, each with a different sign-on. This actually impedes a user's ability to collaborate and share information, slowing productivity and response time.

The Lotus Workplace platform contains all the Lotus Collaboration capabilities—messaging, calendaring, scheduling, awareness, contextual collaboration, documents, e-learning, and workflow. Each area of Lotus Workplace is a portlet that includes Lotus collaborative capabilities or an application based on those capabilities. Portal-based collaboration will continue to make it easier to integrate the capabilities of WebSphere and Domino/Collaboration-based applications.

Portals provide a single point of access to information. They allow businesses to operate faster and more efficiently by letting people access information and communicate across different systems in real-time via a Web browser. Lotus Workplace features components for e-mail, collaboration, content management, and learning.

IBM Lotus Workplace Web Content Management allows non-technical users to publish material within the Workplace environment, on a portal, or on an application server. The application is the result of the acquisition of Australian content-management provider, Aptrix. In addition, IBM has added features that give users the ability to import and export Microsoft Excel, PowerPoint, and Word files, as well as IBM Lotus 1-2-3 and Lotus Freelance files.

Additional updates to the Lotus Workplace Web Content Management tool include enhanced support for Oracle applications via the WebSphere Portal Application Integrator tool. The product extends customers' ability to access enterprise applications. For example, because the tool extends access to Oracle's human resources, finance, and manufacturing applications, a business can bring Oracle access to its employees in a centralized and integrated environment. This is a tool for the Human Resources (HR) manager who understands the HR process very well but not HTML programming.

WebSphere Portal's productivity tools, which allow users to view, create, convert, and edit basic text files, spreadsheets, and presentations without having to leave the portal environment, have also seen recent improvements. The WebSphere Portal Catalog has over 1,200 solutions and services that allow users to build and deploy portlets that provide easier access to corporate applications such as enterprise resource planning, customer relationship management, and supply chain management.

Software companies offering services for WebSphere Portal are able to make them available worldwide via the WebSphere Portal Catalog. By being able to access services via the catalog, customers have access to technology expertise, such as consulting, designing, developing, and integrating portal solutions.

What Is IBM's Direction for Domino and WebSphere?

With the move of so many products in the WebSphere direction, it is natural to ask, "What will happen to Domino in the long run?" Ambuj Goyal, the head of Lotus, recently answered this way, "We have a huge user base on Notes and Domino. They would like us to move them forward. They would like Notes to become a portal client, in addition to being an e-mail, instant messaging, and collaboration client. For Domino applications, we are saying we could deliver applications in any rich client or Web browser—not only through Notes. I can move (the base) forward by continuing to build on client-server technology. I would rather move them forward by building on open standards that every browser and device is using. I'd rather I didn't have to invest in developing the infrastructure element over and over again."

Because Domino continues to excel as a communication, content, and collaboration solution, and WebSphere continues to excel for Web sites needing high transactions, massive scalability and the power of J2EE, the advantages of combining the strengths of these two products will increase and not decrease over the next few years. The best ways to combine the strengths of Lotus Domino along with the other Lotus Collaboration products and the WebSphere products are described throughout this book.

CHAPTER 2

The Architecture of WebSphere and Lotus Collaboration Together

"Ah, to build, to build! That is the noblest of all the arts."

—Henry Wadsworth Longfellow, from the poem, "Michael Angelo" (pt. I, II, l.54)

This chapter describes the architectures for the different ways to combine WebSphere with Lotus Collaboration capabilities. This chapter specifies which software pieces provide which function when the WebSphere and Domino platforms are used together—along with recommendations as to which software to use when there are alternatives. Given that both WebSphere and Domino can act independently as fully functional Web servers, there are some choices to be made—but that's good. The flexibility in combining WebSphere with Lotus Collaboration (whether through the use of Domino, WebSphere Portal, Lotus Workplace, Lotus Collaboration Center, or any combination of these elements) will allow you to tailor a system best suited for your company's needs.

The WebSphere Portal can play an important role in combining WebSphere and Lotus Collaboration tools. A portal is the next-generation desktop, the tool that provides you the access you need to get your job done, in one place and from any client or device. Essentially, the idea is to aggregate both the data and applications needed by specific workers in a unified interface to improve productivity. Portlets are the components that enable this aggregation. They are independent, reusable components that are the keys to providing access to the data and applications; but even more, they provide the value to portals by making useful integration an expected part of the portal interface. Another requisite feature of portals is collaboration. Collaboration is how you build communities using your portals, and it's one way your users get more value out of the portal than they would out of simply getting very good with the use of the Alt+Tab keys. The success of collaborative features in portals is considered one of the keys to the success of portal adoption itself. To achieve success here, you must provide both formal and informal collaboration. Formal collaboration is usually document-centric—that is, mostly e-mail. Informal collaboration includes features such as expertise location, chat and awareness, and discussions. The "awareness" aspect of collaboration tools is very significant. The more formal term, "awareness of presence," is also used. Your Instant Messaging "buddy list" indicates who is online and hence makes you "aware" of the colleagues whom you could "chat" with online. Many Lotus Collaboration tools are including awareness as a part of the tool or as an option.

If your organization already has some kind of investment in collaborative technology, it's most likely in an integrated collaborative environment (ICE), defined by IDC and others to include such products as Lotus Domino/Notes and Microsoft Exchange/Outlook and such features as e-mail, awareness, chat, and Web conferencing. In the past few years, Meta and IDC coined the term "contextual collaboration" to describe the process of embedding these collaborative features into business applications. That process and that term finally finds its home when you embed collaboration tools into WebSphere Portal. There is both interest in and resistance to WebSphere Portal from both Lotus and WebSphere professionals. Lotus professionals tend to feel that they can create all the features a portal offers within Domino. In truth, many Lotus professionals have created portals over the years using existing Lotus technologies. To integrate those collaborative technologies fully with the other business applications in use in the environment, however, Lotus professionals need to step outside a strictly Domino environment to create the most value for their users.

Some WebSphere professionals, on the other hand, have a resistance not to the portal technology itself, necessarily, but to the collaborative features provided by the Lotus technologies. WebSphere pros need to recognize that collaboration, both on a group level and on an organizational level, has been the Lotus business since the company came into being. It makes sense for us to take advantage of their expertise in the collaborative arena and use the components included with WebSphere Portal. Those components include available portlets, such as Sametime, QuickPlace, and Domino portlets, and the Collaboration Center, which comprises a team workplace portlet that integrates with QuickPlace, a Lotus Web Conferencing portlet, and the People Finder. And, of course, there are many other ways to integrate existing Lotus collaborative applications.

Pros and Cons for the WebSphere/Domino Option

Some of the advantages of using Domino in conjunction with WAS are that Domino provides mature, extensive implementations of certain functions that are only provided as APIs in J2EE, such as mail, directory, calendaring, workflow, document handling, and so on. WebSphere Application Server provides a robust e-business application deployment environment with a complete set of application services including capabilities for transaction management, security, clustering, performance, availability, connectivity, and scalability. WAS version 5 offers full

J2EE specification support, which includes servlet 2.3, JSP 1.2, and EJB 2.0 as well as a variety of extensions. The following sections give details on the unique strengths of Domino and the unique strengths of WebSphere.

Domino Replication

Replication has always been the "Crown Jewel" of the Lotus Notes and Domino product. Domino cluster replication technology allows a group of Domino servers to work together to appear to the user as if they were a single Domino server. The cluster is used to provide high availability of data and services, to balance workload, and to significantly increase the number of users your system can support (scalability). Servers in a Domino cluster use cluster replication (real-time replication) to keep all servers in the cluster the same. If one server in a cluster goes down, users are transparently switched to another server in the cluster (failover). Cluster workload balancing ensures that heavily used servers are able to pass requests to other cluster servers and that work is evenly distributed across the cluster. Additional servers can easily be added to a cluster to allow the Domino service, with workload balancing, to scale to a very large number of users.

Domino 6 Streaming Replication

Streaming replication involves a single server request, which then pulls in all the data (Notes documents and their attachments) into the database. This feature dramatically reduces replication time and works in all Domino 6 client/server scenarios. Streaming replication also means that when using the Notes client, you no longer have to wait until the replication is over before seeing replicated documents in folders. They appear individually as soon as they are pulled into the system, and you can begin to work on them before the database has finished replicating. In addition, documents are replicated in ascending order by size.

To use streaming replication, both parties involved in the replication must be running Notes/Domino 6. Streaming replication is PULL only. This feature uses a single streamed RPC (Remote Procedure Call) to read a number of documents and attachments. This is an improvement over the non-streaming method of requesting and acknowledging one database note at a time.

Streaming replication saves time and consumes less CPU and network activity to replicate the same number of documents. Network latency is reduced, especially on slower links. Documents are also replicated in ascending size order (smallest first). This allows the client to abort a long replication but still receive some documents (for example, mail). Partial replication is also supported, allowing an aborted replication to be reinitiated and to continue where it left off. Views in folders are also updated incrementally while a replication is in progress (as opposed to the end of replication). This allows users to begin responding to initial documents (mail messages, for example,) while larger documents continue downloading.

End-user benefits of streaming replication include shorter replication times, ability to start viewing and working on documents before replication is completed, and replication of smallest documents first. If replication is cancelled, the target system still has documents (especially smaller ones) that were replicated before the replication session ended. Incrementally updated folders means you no longer have to wait until the replication is over before seeing replicated documents in folders.

Advantages for IT departments (for example, server to server replication) include faster replication, partial replication, potentially less network traffic due to single streamed RPC, and reduction of ACK (Acknowledgment) TCP/IP responses. Lotus found that databases with 10,000 and 25,000 documents replicated 15% faster with streaming replication on. A database with 50,000 documents replicated 60% faster with streaming replication on.

Domino 6 Network Compression

Network compression is an important performance feature offered in Notes/Domino 6. When you enable network compression, data is automatically compressed before it is sent over the network. This improves network performance, especially over slower line speeds. With network compression enabled on both the sending and the receiving computers, the Notes client and/or the Domino server attempts to compress any data (for example, mail or replicated documents) before transmitting it over the network. Network compression works with mail routing, replication, or any other data sent via NRPC (Notes Remote Procedure Call). This includes design elements in new or updated replica databases.

Notes/Domino 6 network compression offers a number of immediate benefits. For example, by reducing the amount of data being transmitted, you can improve the performance of your routing and replicating hub servers, especially if they are currently laboring under heavy workloads. You also can enable it by default so that all your users with Notes 6 clients can take advantage of this functionality without having to select it themselves. Because network compression is a standard, out-of-the-box feature, it doesn't require any additional code, which helps simplify administration. CPU and memory resources used on the system increase by a small amount. In many of the environments where it is very expensive or not even even an option to upgrade the existing network infrastructure, this is a welcome trade-off.

Lotus also enhanced attachment compression in Domino 6. Attachment compression can now take advantage of the LZ1 (Lempel-Ziv) algorithm (the same algorithm used in network compression). In R5, attachment compression used the Huffman algorithm. To use LZ1 compression in Domino 6, you must enable it in the Database Properties box. (By default, databases still use the Huffman algorithm to maintain backward compatibility.) The LZ1 attachment compression typically reduces attachment file size by 5% to 40% over Huffman attachment compression. The percent reduction in size depends upon the type and complexity of the file attachment.

HTTP Server Plug-In

Both WebSphere and Domino have HTTP server capability. Starting from WAS 4.0, the HTTP server is also embedded in WAS itself. The HTTP handling capability embedded within the application server supports a connection from an HTTP client to the Application Server. An HTTP client can connect to a Web server, and the HTTP plug-in can forward the request to the Application Server.

Built-In Domino Functions

The Domino platform has the capability to provide many key groupware and collaboration functions "out of the box." These key built-in functions include e-mail or messaging, directory services, collaboration, document management, and security services.

Mail, Address Book, and Calendaring

Domino has long provided excellent support for mail, address book (directory services), and calendaring. Mail and Directory Services are discussed further in the paragraphs that follow. Please see Appendix B for additional discussion on these Domino 6 features.

Collaboration and Document Management

Providing collaboration support has always been the main focus of the Lotus Notes and Domino products. In addition to built-in collaboration features, Domino provides the platform for the Lotus Instant Messaging, Lotus Web Conferencing, and Lotus Team WorkPlace (QuickPlace 3) collaboration products. IBM Lotus Domino Document Manager (formerly called Lotus Domino.Doc) is a ready-built solution for organizing documents for effective shared access by work teams, managing versions so that each team member has the latest and automating document-driven processes such as review and approval, assembly and publishing, archiving, and records management.

Mail

Domino functions as an e-mail server with support for several Internet open standards: POP3, SMTP, and IMAP, in addition to the Lotus Notes e-mail protocol. These e-mail protocols have been provided by Domino almost since its inception and have been continually improved and made more robust.

In Domino 6 the IMAP server has been redesigned to improve performance throughout by use of multi-threading and data streaming and by enhancing the Domino database format to support IMAP data elements directly. Other Domino 6 features enable better detection and control of unsolicited e-mail ("spam"). The Domino product has the capability to provide many key groupware and collaboration functions out of the box. These key built-in functions include e-mail or messaging, directory services, collaboration, document management, and security services.

Directory Services

The Domino directory is central to the function of a Domino server environment. Information about users, servers, and processes are maintained by the Domino directory. The key feature of the Domino directory relative to a combined Domino/WebSphere environment is its support of the Lightweight Directory Access Protocol (LDAP). In many ways, LDAP can be considered the directory access protocol of the Internet. WebSphere and the J2EE platform also enable LDAP access to directory information especially for naming and authentication, which can be configured to access a Domino directory. Domino 6 provides enhancements to LDAP support including improving the performance of LDAP directory access, providing a new directory schema database, and better change tracking.

Security

Domino has provided strong security features since its first release. It pioneered the use of Public Key Infrastructure (PKI) mechanisms for user authentication, document encryption, and digital signature functions. Although the PKI implementations within Domino have been largely proprietary, they are being made more open, along the lines of Internet defined PKI standards. As they become more open, the Domino security mechanisms will better interoperate with Internet and J2EE security mechanisms. An example of this direction is the recent support for "single sign-on" between Domino and WebSphere-based Web applications.

Also along this direction are the Domino 6 enhancements of its Certificate Authority (CA) function. With Domino 6, administrators will have a unified mechanism to generate both Notes and Internet user certificates and to register notes and Internet keys.

WAS J2EE Functions—Servlets, JSPs, and EJBs

The IBM WebSphere Application Server product at its core implements the Java 2 Enterprise Edition (J2EE) technologies of servlets, Java Server Pages (JSPs), and Enterprise Java Beans (EJBs). These three principal J2EE technologies are the foundation upon which large-scale distributed and high-performance Web-enabled applications can be built. With the addition of Web services technology, WAS stands as a premier platform for providing Web-based applications.

> **NOTE**
> In WebSphere Application Server, EJBs are executed by the EJB (Enterprise Java Beans) container, and servlets and JavaServer Pages files are executed by the Web container.

Servlets

J2EE servlets can be viewed as small Java programs running on the Web server side of a Web browser to server interaction. Servlets handle a browser request, formulate the response, and convey the response back to the browser. More technically, servlets consist of Java objects invoked by the Web application server. (Servlets execute as threads under the Java Virtual Machine established for the application server, referred to as the "servlet container.") The servlet object accepts and validates an HTTP request and formulates an HTTP response message, which it writes back to the application server that, in turn, returns it to the browser. The Java classes and interfaces used in this HTTP request and response processing are defined by the J2EE servlet specification (and thus provided in the same way by any J2EE-compliant application server).

It is important to keep in mind that although servlets are Java programs and have the full range of J2EE function available to them, they are meant to run in the context of a server response to a browser request. Servlets are appropriate for functions such as generating Web pages with dynamic content, handling HTML form submissions (HTTP POST requests), triggering more complex server-side processing, and so on. Servlets are not the best choice for implementing persistent data transactions, complex application logic, or computation intensive processing.

For readers already familiar with Domino server processing, the J2EE servlets are similar to Domino server-side database agents. The fact that Domino agents can be coded in Java makes it tempting to think of servlets and agents as equivalent; however, there are significant differences. Chapter 7 gives more details on this in the section, "Servlets vs. Web Agents."

Java Server Pages

JSPs can be viewed as a specific implementation of the servlet model geared toward providing dynamic HTML content. A JSP consists of HTML tags, some of which contain JavaScript code that is executed by the application server when the JSP is requested. A JSP is executed by the application server in the same way as a servlet is and produces a response containing HTML. JSPs are the J2EE way to implement powerful dynamic HTML functions.

You will see in later chapters that IBM/Lotus provides extensive support for JSPs and JavaScript tags to access documents in Domino databases. The Domino-provided JavaScript tag libraries open up the full scope of Domino document databases to the J2EE application developer and are a key point of integration between Domino and WebSphere.

Enterprise JavaBeans

Enterprise JavaBeans (EJBs) are the Java component that can be combined with other resources to create J2EE applications. They combine Java server-side components with distributed object technologies: CORBA + RMI. EJBs are always distributed, which make them fundamentally different from standard JavaBeans. EJBs are not concerned with presentation issues and therefore must be used in conjunction with display functions, such as servlets or JSPs for HTML clients or Java applications that use technologies such as AWT or Swing.

WebSphere Application Server provides an implementation of the EJB defined component models. This implementation is referred to as the "EJB container." EJBs can be used in a broad scope of applications beyond Web applications, but they are often used to provide the transaction processing of persistent data (often called the "business logic") of a Web application. It is often EJBs that are invoked to update bank account or reservation records. WAS provides the full range of functions required to make EJB use robust, secure, and scalable.

There are three types of enterprise beans: entity beans, session beans, and message-driven beans. All types of EJBs reside in EJB containers, which provide an interface between the beans and the application server on which they reside. Entity EJBs store permanent data. Entity beans with container-managed persistence (CMP) require connections to a form of persistent storage.

Bean-managed persistence implies that the task of managing the lifecycle of the entity within the datastore is performed by the bean itself. The bean developer has to develop both the business logic as well as the logic to persist the bean to the datastore. The disadvantage is that the persistence code has to be hardcoded within the bean itself. As a result, the bean is targeted for a single datastore and becomes difficult to reuse.

Entity beans differ from session beans in several ways. Entity beans are persistent, allow shared access, have primary keys, and may participate in relationships with other entity beans. Session beans do not require database access, although they can obtain it indirectly as needed through entity beans. Session beans also can obtain direct access to databases (and other resources) through the use of resource references. Session beans can be either stateful or stateless. Use a stateless session bean to act as the entry point for business logic. Message-driven beans (MDB) are new in the EJB specification version 2.0., which enable asynchronous message servicing.

The EJB container and a Java Message Service (JMS) provider work together to process messages. When a message arrives from another application component through JMS, the EJB container forwards it through an onMessage() call to a message-driven bean instance, which then processes the message. In other respects, message-driven beans are similar to stateless session beans. A typical e-business application with JSP, servlet, and EJBs is depicted in Figure 2-1.

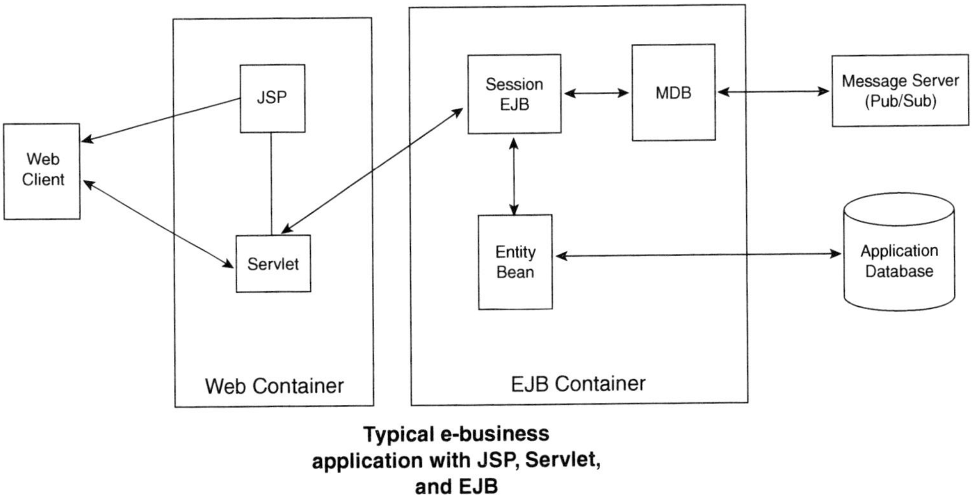

Figure 2-1 Diagram of the EJB model.

Use the following best practices when designing and developing the enterprise beans: In an Enterprise JavaBeans (EJB) Version 2.0 environment, use local interfaces to improve communication between enterprise beans in the same JVM. Local calls avoid the overhead of RMI/IIOP and use pass-by-reference semantics instead of pass-by-value. For each call, the caller and callee beans share the state of arguments. EJB 2.0 beans can have both a local and remote interface but more typically have one or the other. For communicating with remote clients, provide remote and remote home interfaces. For communicating with local clients such as servlets, entity beans, and message-driven beans, provide local and local home interfaces.

Method-Level Object Security with EJB

Until the advent of enterprise beans, there was no way to restrict access to an object or method by a particular user. Enterprise beans now, however, allow method-level security on any enterprise bean or method. Users and user groups can be created, which can be granted or denied execution rights to any EJB or method. In WebSphere, these same user groups can be granted or denied access to Web resources (servlets, JSPs, and HTML pages), and the user IDs can be seamlessly passed from the Web resources to the EJBs by the underlying security framework.

As Domino is not oriented toward transaction processing, there is no analogous function to EJBs in Domino. EJBs are most often invoked from Domino components, such as from form documents or from agents. It is also possible to invoke Domino functions from within EJBs and there are some technical issues in doing so. In this book, we'll focus on the issues with invoking EJBs managed by WAS from Domino and vice versa, rather than those involved with using EJBs themselves.

WebSphere and Domino Role in On Demand Business

On demand operating environments, with their collective computing resources, provide a more systemic type of business value. Rather than a hodgepodge of proprietary interfaces and inefficient links, these environments give users "virtual" control over disparate systems and applications, effectively fusing them into an integrated and coherent whole.

The open standards on which an on demand operating environment is built enable different systems to work together, so a business's various processes can be integrated and transactions can be done faster—or even automatically.

Benefits include significantly improved efficiency, cost-effectiveness, and built-in flexibility. Free from the need to see or directly manage the complexity of the infrastructure, users will be able to respond to business conditions with unprecedented speed and agility. Given the power to dynamically reshape their IT resources, users can implement their core business goals, capitalizing on changing marketplace conditions rather than dreading them.

Unlike some previous phases of computing, developing an on demand operating environment does not mean companies have to rip out their existing computer systems and start from scratch. Existing systems and processes can be integrated in some cases with the basic first step of server consolidation.

As systems are consolidated and integrated into internal or external computing grids, the collected capacity can be used more efficiently, so existing systems work harder and deliver more output. And often, the savings generated by simply running systems more efficiently can be invested to help fund further on demand development.

The new offerings provide integration, virtualization, automation, and flexibility—critical elements of on demand solutions. They include the following:

- TotalStorage virtualization family, data as a single pool
- WebSphere virtualization for grids, managing disparate apps
- Variable Open Infrastructure Offering, flexible financing
- Stand-by capacity on demand, blades on call

Currently, customers only use about 20 percent of the server capacity they have installed and paid for. IBM's On Demand Business initiative will help drive utilization rates up to 40 percent or more, delivering significant cost savings. Storage efficiency is equally startling: average storage utilization stands at only 48 percent and can be improved by 20 percent or more.

Here are some specifics on how the new products and services help customers build their on demand operating environments.

Driving Integration

An on demand operating environment drives the integration of people, business process, and information anywhere, any time, from any device. Through the advanced capabilities found in the latest WebSphere and DAB offerings including IBM Business Integration, IBM Collaboration Portal, and IBM Information Integration, capabilities from across IBM are being packaged, tested, and priced to simplify and support customers' business priorities.

Enabling Automation

A typical infrastructure includes a variety of systems running different applications to support the overall e-business. At any given time, one system could slow down the entire operation if it becomes overloaded. By using IBM's automated Availability Management, Web Server Provisioning, or IBM eServer zSeries Optimization tools to balance the workloads on their infrastructure, customers can prevent this. IBM User Management is also available to automate system security. By enabling the systems to virtually manage themselves, customers can focus their efforts and resources on other projects.

Employing Virtualization

Virtualization technology helps customers see the resources they have and how they are being used and enables them to assign additional workloads to any servers, storage, or distributed systems that are not being fully utilized. By leveraging virtualization to pool the resources throughout the customer's entire network, IBM is helping them overcome the business silos that have developed through years of IT expansion. Depending on the customer's size and needs, an IBM eServer zSeries Linux consolidation; iSeries Linux, NT, and UNIX consolidation; or a virtual Linux server solution could be employed today to boost the customer's productivity. IBM also offers pSeries, xSeries, and Enterprise Storage Server consolidation opportunities that leverage virtualization technology.

Offering Flexibility

The idea of on demand is to offer customers exactly what they need, when they need it. Today, on demand has been expanded to include stand-by capacity on demand, which allows TotalStorage and IBM eServer BladeCenter customers to install more systems than they need and only pay for them after they are used. IBM is also offering customers its new Variable Open Infrastructure on demand, which makes it easier for customers to pay for and expand their infrastructure as business needs dictate.

C H A P T E R 3

WebSphere Application Server (WAS) 5

"WebSphere Application Server 5.0 awarded 'Best Application Server' for its tight integration with tools, smooth application deployment, and granular management."

—*InfoWorld Magazine*, December, 2003

As IBM's flagship J2EE application server product, WebSphere Application Server Version 5 (WAS V5) contains more than five year's worth of stress-tested code providing a comprehensive set of functions for building high-performance, highly scalable Web applications. Its strengths are clearly its wide range of support for J2EE, HTTP, and database components, its flexibility in deployment and clustering options, its scalability, and its performance. It excels as the base for high-volume, interactive, transaction-oriented Web applications often required by many enterprises. This chapter describes the features of WAS V5. Appendix A gives further WAS V5 details; Appendix D discusses some of the new features of WAS V6. At the time this book was written, WAS V5 was the latest version available.

Strengths and Weaknesses

Some of the strengths of the WAS product are also its liabilities. With all of its flexibility and scope of functions, WAS is certainly complex and can be daunting to application developers and system support staff. Exercising its full range of functions can lead to difficulties in system administration and application maintenance. The performance of certain J2EE components, such as EJBs, is sensitive to the way in which WAS is configured and deployed, and it may take extensive application tuning to achieve good performance.

WAS, primarily a J2EE component platform by nature, lacks certain functions required by some applications, such as e-mail, workflow, and collaboration. Fortunately, these are precisely the strengths of the Domino product.

J2EE 1.3 Support

WAS V5 provides support for the J2EE Version 1.3 platform. The following table (Table 3-1) shows the specific versions of the individual J2EE components supported.

Table 3-1 Levels of J2EE Components Provided by WAS V5

Technology	J2EE 1.3
Java 2 Platform, Enterprise Edition (J2EE)	1.3
Enterprise Java Beans (EJB)	2.0
Java Server Pages (JSP)	1.2
Java Servlets	2.3
Java Authentication and Authorization Service (JAAS)	1.0
Java Activation Framework	1.0
Java Mail	1.2
Java API for XML Parsing (JAXP)	1.1
J2EE Connector Architecture	1.0
JDBC Standard Extension (JDBC)	2.0
Java Message Service (JMS)	1.0
Java Naming and Directory Interface (JNDI)	part of J2SE 1.3
Java Transaction API (JTA)	1.0
RMI/IIOP	part of J2SE 1.3

Java Servlet 2.3 Support

Filters

A Servlet filter intercepts the servlet request and response data so that this data can be modified or used in some way. A filter can be defined to apply to all servlet (JSP) requests and responses and thus are a means to achieve better servlet code modularization and reuse. There are many uses for servlet filters such as response data transformation (e.g., HTML to WML), request authentication and authorization, data validation, and data localization.

Lifecycle Events

The new servlet lifecycle events are generated by the servlet container in response to a change in a servlet's context or session. An application can register listeners for events signaling the creation or destruction of the servletContext or HTTPSession objects or changes in attributes of these objects.

Internationalization (I18N) Support

With the new internationalization support, a servlet can now specify which character set encoding is used to parse the request or set the response. This feature makes it easier for servlets to handle forms with parameters or content encoded in various ways other than the default ISO 8859-1 encoding.

Java Server Pages 1.2 Support

XML Syntax

JSP source can now be specified in XML. Although it can be simply stated, this feature has far reaching implications. The use of XML for JSP source allows the JSPs to be validated against an XML DTD or schema, and it allows for easier manipulation or transformation of the JSP source via XML transformation mechanisms, such as XSLT. A JSP written in the new XML format is referred to as a **JSP document**.

Tag Library Improvements

JSP tag libraries provide a way for developers to reuse and share common JSP tags. Improvements to tag libraries with the JSP 1.2 specification include a clearer interface definition and additional elements for author documentation. A set of tag libraries are provided by Domino for using JSPs to display Domino database documents. We'll explore the use of these tag libraries in Chapter 10.

JavaMail 1.2 Support

The JavaMail component of J2EE supports the creation and transfer of mail messages from a client to a mail server. Implementations of JavaMail provide support for the common Internet open mail exchange protocols: POP3, IMAP, and SMTP. JavaMail also provides interfaces for the creation of MIME message content. MIME is an open and widely used mail message format on the Internet, supporting multiple content types within the same mail message such as text and images.

In the context of the WAS and Domino products, JavaMail is a natural interface to the mail handling functions of Domino. The JavaMail classes and interfaces can be invoked from a Web application running on a WAS node to send and receive mail via a Domino server. We'll also explore the use of JavaMail in a combined WAS and Domino application server environment in later chapters.

JAXP 1.1

The Java API for XML Processing (JAXP) provides a standard interface to XML processing functions within the J2EE component framework. A more recent 1.2 version of the JAXP specification has been defined; however, JAXP 1.1, along with DOM Version 2 and SAX Version 2 XML parsers, are required by the J2EE 1.3 version. The later versions of JAXP, DOM, and SAX are not supported by J2EE 1.3. Applications written to other XML processing APIs (such as the XML API for Java) must be changed to support only the APIs defined in JAXP 1.1.

JAAS 1.0

The Java Authentication and Authorization Services component is a key extension to the security model defined by the Java 2 platform. In the Java 2 security model, control of access to Java runtime objects is based solely on the location of the code and its signer. JAAS 1.0 extends this model by enabling access control based on authentication of the user for whom the code is being run and authentication of this user (often referred to as the **principal**).

WAS V5 fully utilizes the JAAS functions to perform user authentication and access control to the servlet, JSP, EJB, and other resources within the application server. In addition, WAS V5 provides its own extensions to the JAAS function, primarily for providing role-based authorization of server resources.

Enterprise JavaBeans 2.0 Support

EJB 2.0 introduces new features for EJB invocation and management. The following key EJB 2.0 features are provided by WAS V5.

Message Driven Beans (MDB)

A Message Driven Bean is a third type of EJB defined by the EJB 2.0 specification. In addition to session and entity EJBs, MDBs provide a means for a process to invoke the function of an EJB running in a different process and most likely on a different server. Unlike session or entity EJBs, MDBs are invoked by sending messages to the MDBs; there is no "home" or "remote" interface associated with an MDB. The Java Message Service (JMS) API is used to accomplish the MDB invocation. A client can direct a JMS message to an EJB 2.0 node where it is processed by an MDB. In WAS V5, the JMS API is implemented by IBM's mature MQSeries (now the WebSphere/MQ) product that is provided as part of all WAS V5 versions—except for WAS Express.

Improved Container Managed Persistence (CMP) Features

The EJB 2.0 specification added several new features to the CMP functions that an EJB container can provide. WAS V5 supports these and some additional container features aimed at further improving EJB performance.

Container Managed Relationships (CMR) for CMP Entity Beans

The WAS EJB container can manage relationships between EJB entity beans. This feature significantly reduces the programming required to construct sets of interrelated EJBs, such as those often found in sophisticated Web applications.

EJB Query Language (EJBQL)

IBM has provided an SQL-like query language for EJBs in past versions of WAS. A good portion of this query language is now part of the EJB 2.0 specification. WAS V5 supports the 2.0 version of the EJBQL and provides several additional commands and operators.

Local Interfaces

The local interfaces feature is also new with the EJB 2.0 specification. It applies to the situation where the home and remote (client and server) parts of an EJB reside on the same server and improves the performance of these types of EJBs by creating a direct link between the home and remote interfaces.

In addition to the new EJB 2.0 features, WAS V5 provides several new EJB features that will have additional performance and development benefits.

Entity Bean Inheritance

With WAS V5 it is possible to write EJB superclasses that contain shared members and methods useful to derived subclasses. This feature allows common object-oriented design practices to be used with EJBs. The WSAD product also supports this feature.

Access Intent Policies

With access intent policies, application developers can specify to the application server the intended use (or not) of EJBs during a transaction, thus allowing the application server to optimize when and how the EJBs are persisted, that is, written to the container database. Since the actual database writes and commits are often the most expensive part of and EJB-based transaction, this feature aids in making the persistence as efficient as possible.

Caching Data Across Transactions

The application server maintains a cache of entity beans for efficient transaction processing. This feature allows static or infrequently changed entity beans to remain in the cache across many transactions and can greatly improve overall transaction performance.

Web Services

Although not specifically part of the J2EE specification, Web services support is now a prominent feature set of the WAS V5 product. We found that the WAS help documentation (InfoCenter) provides a concise description of what Web services are:

- Web services are self-contained, modular applications that you can describe, publish, locate, and invoke over a network.
- Web services could be weather reports or stock quotes. Transaction Web services supporting business-to-business (B2B) or business-to-client (B2C) operations could be airline reservations or purchase orders.
- Web services reflect a new, service-oriented approach to programming, based on the idea of building applications by discovering and implementing network-available services or by invoking available applications to accomplish some task. This service-oriented approach is independent of specific programming languages or operating systems. Instead, Web services rely on pre-existing transport technologies such as HTTP and standard data encoding techniques, such as XML, for invoking the implementation.
- The key components of a Web service are as follows:
  - Simple Object Access Protocol (SOAP)
  - Web Services Description Language (WSDL)
  - Universal Discovery, Description, and Integration Protocol (UDDI)

Looking at this description, it is hard to see what is so new and revolutionary. After all, there are many mechanisms for invoking "services" over a network—for example, J2EE provided mechanisms such as servlets and EJBs or traditional distributed system protocols such as RPC and CORBA, and so on. What is new about Web services lies more in how they are defined and implemented: They can be implemented across many programming languages; they are designed to be self-describing; and they are evocable over the open Internet.

SOAP is the basic protocol for communication among Web services. It can be characterized as a remote procedure call (RPC) protocol over HTTP (although not restricted to HTTP alone). The fact that it's an RPC enabled over the open Internet is what's new.

WSDL defines how a Web service is cataloged and described so that it can be discovered and invoked by another program (Web service). UDDI serves as the directory for Web services, allowing the WSDL to be queried in an open fashion. What's new? It's the ability to discover and invoke Web services programmatically across the open Internet.

Underlying all of the protocols is XML, which is used as the data format language. A program that can parse and create XML can operate on any of the Web service protocols.

Beyond the basic set of Web service protocols, SOAP, WSDL, and UDDI, there are other protocols in various stages of definition and standards body ratification, which will extend the ways with which an application designer can make use of Web services. WAS V5 provides implementations of some of these newer protocols as well, such as Web Services Invocation Framework (WSIF) and WS-Security.

The Domino platform is not excluded from the Web service initiatives. It is easy to see various ways in which Domino functions can play in the Web services arena. Native Domino functions such as document handling, workflow, directory, and e-mail can be opened as Web services.

Application functions developed on a Domino base can be made available as Web services—we will see how it is possible to provide Web services even via LotusScript. All of this gives the application designer more tools with which to construct a Web service-based application.

WAS V5 Product Versions

As mentioned in the "Strengths and Weaknesses" section of this chapter, one of the weaknesses of WAS is its complexity. In our view, IBM has attempted to mitigate this weakness by offering WAS V5 product versions in increasing levels of function and complexity. From the basic Web application functions included with WAS Express, to the sophisticated functions of WAS Enterprise, the WAS V5 product set is structured to provide the appropriate set of functions (hence complexity) to meet a business's solution needs:

- **WAS Express** (servlets/JSPs, JavaScript, samples, embedded Apache, local OS authorization)
- **WAS** (WAS Express + EJBs, embedded MQ (JMX), SSO, LDAP, Tivoli Access Mgr, COM-to-EJB (ActiveX), external HTTP Server)
- **WAS Network Deployment** (WAS+load balancing, failure bypass)
- **WAS Enterprise** (WAS/XD+enterprise extensions)

Summary of What's New in WAS V5

WAS V5 represents a continuation of the evolution to a single, integrated, cost-effective, Web services-enabled, J2EE server foundation for foundation messaging flows and applications that offers customers

- One deployment model.
- One administration point.
- One programming model.
- One integrated application development environment.

With WAS V5, IBM enables customers to expand their business opportunity and productivity through a world-class infrastructure ready for On Demand Business. Version 4 made advances in usability, connectivity, and performance that allowed customers to more easily deploy Web applications as well as maximize the performance of those applications. In V5, IBM makes it possible for its customers to participate in the next chapter in e-business. The past five years of e-business were about Web-enabling content and applications. The next chapter is about On Demand Business: dynamically integrating those applications with existing data stores, other applications, and Web services that exist in heterogeneous environments, in multiple programming models, and on different sides of firewalls. WAS V5 provides a powerful foundation and integrated development environment that supports on demand business applications. It provides customers with the four essential characteristics of the on demand operating environment: open, integrated, virtualized, and autonomic. IBM recognizes that every deployment environment is unique, and for that reason, WebSphere Application Server V5 is a single application server code-base with multiple configuration options to support a wide range of scenarios, from simple administration of a single server to a clustered, highly available, high-volume environment with edge of network services.

Comprehensive Build-to-Integrate Platform

WebSphere provides a platform that allows the building of integration-ready applications that leverage existing software assets.

Enable Dynamic Application Interaction

WebSphere enables dynamic application interaction through native, high-performance JMS (Java Messaging Service), J2EE 1.3 Message Beans, and container managed messaging. JMS simplifies development by allowing loosely coupled, reliable, asynchronous interactions among J2EE components and legacy systems capable of messaging. Message beans save valuable programming time and skill in that they allow requests to be processed when they arrive, as opposed to code that checks for the arrival of messages. WAS V5 also supports container managed messaging. This feature further reduces required skill level and development time to create these asynchronous applications, allowing the EJB container to take care of core messaging aspects that would otherwise have to be coded in the message beans. This added support for JMS, Message Beans, and Container managed messaging simplifies development while not requiring in depth JMS skills.

Reuse and Integrate Disparate Systems and Applications

WAS V5 significantly reduces the complexity of interacting with back-end systems through advanced support for Java Connector Architecture (JCA). This feature simplifies the development necessary to connect different types of systems—for example, from single-phase resources, such as CICS, to two-phase resources like DB2.

Unleash Powerful Web Services

WebSphere Application Server V5 offers a complete Web services implementation allowing customers to create new business opportunities by exposing business and application services for integration to other divisions, business organizations, or platforms. WAS V5 allows implementation across many platforms, including iSeries and zSeries. The new version offers the ability to build, deploy, and securely externalize Web services for application consumption across the firewall.

Choreograph Application Interactions

WebSphere Application Server V5 also includes a Service Choreographer that allows businesses to build and deploy process flow quickly and efficiently. This new feature enables customers to expose J2EE-based applications, Web services, and legacy applications as reusable services and then mix and match these services to create a new flexible e-business application. This application can then, in turn, be exposed as a Web service. To facilitate further business efficiencies for customers, the new version supports long running flows with intermittent human interaction. Developers can build flows that can be interrupted prior to completion and then automatically restarted, in addition to flows that prompt users to perform a task or work list.

Support for Business Rule Beans and Compensation is added in the new version. Business Rule Beans allow a business analyst to dynamically update business rules without having to hand code application changes. This feature further reduces the cost of creating more efficient processes.

Finally, WAS V5 offers support for automated compensation to increase developer productivity by reducing the need to create complex logic to manage transactions. This feature gives the customer the ability to complete or negate a list of defined transactions that depend on each other to fully complete a unit of work. This ability lets customers visually define the process of the transactions as well as the appropriate actions to undo or commit the list of transactions in the event of a failure.

Agile Deployment and Administration

WebSphere Application Server V5 includes several significant improvements in the flexibility of administration, deployment, and management services.

Manage with Ease

A new XML-based administrator client that works over HTTP is one of many significant usability enhancements to WAS V5. With this client, the administrator can create and manage the WAS cluster while quickly and easily deploying new components, applications, and services. Databases are no longer necessary for administering WAS; however, for simplicity and cost savings, DB2 is included with WAS Network Deployment and Enterprise versions for use in session persistence. Additionally, management is made easier in WAS V5 with support for significant parts of JSR 115, the Java Authorization Service. It allows customers to plug in third-party authorization/roles engines to their WebSphere environment. Administrators can now easily manage a joint authorization engine and application serving environment.

Intelligent End-to-End Application Optimization

WAS V5 improvements allow customers to create competitive advantage and optimize price/performance while meeting the changing demands of dynamic e-businesses with industry leading reliability, availability, scalability, performance, and security.

Handle the Volume Dynamically

WAS V5 improvements allow administrators to work more efficiently and easily. Support for JMX (Java Management Extensions), which records and logs statistics on usage and resources, is included in the new version. JMX is a standard way of managing a J2EE environment and exposes the WebSphere administrative data to partners like Tivoli and others for management integration. This allows administrators to better manage performance through best of breed tooling that is tightly integrated with their application serving environment. The combination of JMX and Performance Monitor Interface (PMI) in WAS V5 allows customers to capture and manage more performance metrics than any other vendor. Customers can now capture not only performance metrics defined in the PMI but also WebSphere and application specific metrics. In addition to JMX, WebSphere Application Server V5 contains multiple components to improve the ability to handle volumes dynamically with high performance. For example, included in WAS V5 is the Tivoli Performance Viewer (previously named Resource Analyzer). This feature offers smart auto-tuning to simplify the administrator's job by automatically making recommendations to tune critical WebSphere parameters for maximized performance. Tivoli Performance Viewer saves time while enabling improved application performance.

Always On, Always Available

WAS V5 has new elements that decrease costly downtime. Additional load balancing is included to allow failover beyond the cluster level to the domain level, enabling high availability of critical enterprise Java applications. V5 includes improved, weighted workload distribution and failover of clustered application servers, isolation of application servers to avoid single points of failure, and a simpler and improved configuration infrastructure for much better coordination and integrity of related changes. The power of first failure data capture means that your application server will be able to report and analyze problems as they occur. In addition to these features, Application Profiling delivers powerful new capabilities that allow you to carefully optimize the performance of your applications without impacting any application source code. This capability offers a mechanism for instructing the same component to interact with the run-time infrastructure, such as a database, differently depending on the application that calls it. Therefore, you gain unprecedented control in defining strategies about concurrency, update intent, use of collections, and prepopulating component state and data buffers.

Instill Confidence with Security

WAS V5 offers outstanding security through its sophisticated infrastructure and through its extensive support of open, standards-based Java specifications. WebSphere WAS V5 will support the following:

- Java Authentication & Authorization Services (JAAS) to authenticate new principals and manage privilege information for a principal
- Java 2 Security Model to secure system resources

- JavaSecure Socket Extension (JSSE) to secure communication channels based on transport level security (TLS/SSL)
- Java Cryptographic Extension (JCE)
- Java Cryptographic Architecture (JCA) for Java cryptographic extensibility such as for PKI integration
- Common Secure Interoperability V2 (CSIv2) for interoperability between application servers

Together, these implementations enable you to implement standards-based security capabilities within your enterprise applications. Additionally, sophisticated enterprise topologies and infrastructure can be implemented by leveraging WebSphere pluggable security architecture. This includes the following:

- Pluggable user registries that enable you to exploit LDAP or custom registries
- Web single sign-on exclusively provided by WebSphere or through integration with front-end authentication end-points through Trust Association Interceptor (TAI) technology (Reverse Proxy Security Servers)
- End-to-end access to enterprise information systems using rich security features and exploiting the pluggable principal and credential mapping facility

The caching proxy provides plug-in support for Tivoli Access Manager that allows you to exploit the Access Manager infrastructure and an LDAP plug-in to exploit an LDAP-based repository for storing user authentication and authorization information.

Conclusion

IBM WebSphere Application Server is the foundation of the IBM WebSphere software platform for On Demand Business—a set of integrated, award-winning e-business solutions. WAS V5 provides a rich e-business application deployment environment with a set of application services including capabilities for transaction management, Web services, security, performance, availability, connectivity, and scalability. This new version is truly the next step dynamically integrating applications with existing data stores, other applications, and Web services that exist in heterogeneous environments, in multiple programming models, and on different sides of firewalls.

CHAPTER 4

Domino 6

> "Groupware handles three kinds of user interaction—the "three C's":
>
> 1. Communication—which generally means e-mail and related functions.
> 2. Collaboration—online discussion groups and common access to shared folders of shared data.
> 3. Coordination—the most complex part, which allows workers to jointly accomplish specific procedures and tasks.
>
> Lotus Notes combines all three aspects of groupware into one product, and has been elegantly designed to do so from day one. It includes e-mail, a database, a directory of users, security, tools to program custom applications, and support for remote access from portable computers."
>
> —*Fortune*, July, 1996

Domino 6 includes many new features to the Domino messaging and collaboration server, which-significantly improves the benefits of having a combined WebSphere and Domino system. The features discussed in this chapter reflect features through Domino 6.5, which was the latest version of Domino at the publication of this book. Many Lotus Collaboration tools are based on the Domino server—so even if you're going to use the WebSphere Portal as your launching point to Lotus Collaboration tools, a Domino server will usually need to be in the mix. See Appendix E on the WebSphere Portal for details on how the Domino server combines with the WebSphere Portal to provide a wide range of Lotus Collaboration tools. If you are a Domino developer, then you may want to skip this chapter because it is a discussion of the significant features of Domino 6 that will already be familiar to you.

What's So Great About This Version of Domino?

Since Lotus Notes first appeared back in 1989, there have been developers, administrators, and end users who have sworn by the product. The Lotus Notes server became known as the Domino server with release 4.5 in order to better differentiate the Lotus Notes client code from the server code.

New Features

This chapter describes the highlights of Domino 6 in general and the latest release of Domino 6 (version 6.5) when applicable. Additional details are given in Appendix B.

Server Installation and Setup Improvements

Domino 6 provides additional installation options for UNIX systems, support for multiple versions on a single UNIX system, and an improved Domino Server Setup Program.

New UNIX Installation Options

Domino 6 provides four new UNIX installation options:

- **Install template files.** Installs the Domino 6 system templates, overwriting any existing template files.
- **Create /opt/lotus soft link during installation.** Creates a soft link to the /opt/lotus location if you choose not to install in the /opt/lotus location; this option is available for single Domino server installations only.
- **Install service provider.** Sets up a service provider server after installation of a Domino Enterprise server.
- **Add data directories only.** Adds data directories to an existing Domino installation, creating a partitioned server, or adds additional directories to an existing partitioned server.

Multi-Version UNIX Support

Domino 6 supports installation of multiple instances of the Domino server, each with its own program directory—on a single UNIX machine. The instances can include the same release of Domino (for example, only instances of Domino 6) or instances of Domino 6 and one instance of an earlier Domino release. If you want all instances to be the same release, install a Domino partitioned server. Then all Domino partitions share one program directory, and in doing so, conserve system resources.

Domino Server Setup Program

In previous versions of Domino, server setup relied primarily on the databases setup.nsf and setupweb.nsf, which required either a Notes client or Web browser for setup. With Domino 6, a new Domino Server Setup Program written in Java lets you set up a server either remotely or locally. The setup program is available on all server platforms. To set up the server locally, start the server after installation to launch the setup program. To set up the server remotely, use another Domino server command line prompt, a Windows client system with or without the Domino Administrator client, or a UNIX workstation to run the program. The new Domino Server Setup Program allows you to customize server setup.

Domino Server Setup Profile

The Domino Server Setup Profile is a file that automates the server setup program. You can record a server setup from a server or Windows client and then run the setup profile on another server or client to setup a local or remote server with standard settings. The profile can be used to quickly configure a standard server, such as a mail server.

Server Scalability and Performance

Domino 6 includes a number of enhancements that improve server scalability and performance.

Server Startup and Server Performance

Domino 6 optimizes server start-up to speed up recovery and improve time. Improvements include

- View logging, a transaction logging of key views, which assures that you won't need to rebuild those views after a server crash.
- The ability to log information about the availability index, which reports back the health of the server. This index more consistently indicates the state of the clustered server and makes cluster load balancing easier to administer.
- Optimized process starts to improve server performance.
- Optimization of the agent manager start-up procedure to avoid opening all databases, which would slow server performance.
- Optimization of the schedule manager to reduce overhead on server startup.
- The inclusion of a persistent directory manager cache to enhance efficiencies.
- Enhanced cluster support, including automatic detection of software failures and automatic fault recovery, which is now available across all Domino platforms.

Replication

Streaming replication is a new feature that improves replication across all servers, and it especially improves mail server performance. Streaming replication involves a single server request, which then pulls in all the data (Notes documents and their attachments) into the database. This feature significantly reduces replication time and works in all Domino 6 client/server scenarios.

Streaming replication also means that when using the Notes client, you no longer have to wait until the replication is over before seeing replicated documents in folders. They appear individually as soon as they are pulled into the system, and you can begin to work on them before the database has finished replicating. In addition, documents are replicated in ascending size order.

Streaming replication consists of two components:

- **Object streaming.** When opening a note, object streaming retrieves all items and objects—including file attachments, images, and so on.
- **The stream mechanism.** When retrieving notes from the source database, the stream mechanism issues one high-level transaction to get all the notes and all their objects. It uses object streaming to retrieve all the objects.

Client/Server Interactions

Client/server interactions are also more efficient in Domino 6. For example, an advantage of the new streaming feature is that because attachments are also streamed, Open and Save operations are more efficient. Client/server interactions are also significantly improved by less frequent unread table exchanges, as there is a significant reduction in the number of bytes exchanged between a client and the server.

Another performance enhancement is the use of incremental view reading, used to update design information and in selected cases, user views. For example, when you move a document from one folder to another, the server is requested to provide just the incremental change it will take to fill the screen with the new view.

Network Compression

Domino 6 has introduced more new features to reduce network utilization. Network compression reduces the number of bytes sent during transactions by up to 50 percent. Connections across heavily loaded links such as WANs and XPCs will see the most benefit. To use network compression, you must enable it on both the client and the server. Additionally, Domino 6 incorporation of "network" streaming also reduces the overhead of large transfers.

Autodialer for Dialup ISP Connections

The autodialer coordinates dial-up Internet Service Provider (ISP) connections between servers. It links two Connection documents so that a server with a dialup Internet connection can connect automatically in case scheduled replication or mail transfer is initiated by another server.

Full-Text Search

The Domino 6 full-text search feature has been significantly improved. Most data is now updated in place. In addition, Domino uses the NSF buffer manager for memory services, which improves caching and balances memory between NSF and FT. Furthermore, a new search processor results in closer integration of text and field retrieval and significantly faster Boolean processing.

Formula Engine

The Domino 6 formula (compute) engine has had a major overhaul, resulting in computation performance up to two times faster than in previous Domino releases. This overhaul brings performance benefits in many areas including view refreshes, agents, and form rendering.

IMAP Server

In earlier releases of Domino, the IMAP server was based on a layered approach that relied on using additional Notes items and views to maintain the IMAP-specific data for messages. In Domino 6, the core database layer (NSF) has been enhanced to include native support for IMAP semantics, and the IMAP server has been redesigned to use these new capabilities. In addition, the IMAP server now has a new multi-threaded and data-streaming architecture for additional parallelism, providing much higher performance and scalability.

Automatic Fault Recovery

Available for the Windows Server and UNIX platforms, automatic fault recovery shuts down and restarts a server without administrator intervention after an exception occurs. Fault recovery uses operating system resources, such as message queues. When fault recovery restarts a server, it sends an automatic notification to whomever you specify when you set up fault recovery.

Event Generators and Event Handlers

Event generators replace probes and monitors. Event handlers were formerly referred to as event notification. You can use event generators to monitor server resources and network activity. Event generators gather information either by monitoring a task or statistic or by checking a server for access or connectivity. You can determine the criteria by which an event is created. When the criteria are met, an event is created and then passed to the Event Monitor task.

Event handlers determine which action to take when an event occurs. They can log an event or prevent an event from being logged, notify you when an event occurs, or forward an event for additional processing.

Improvements in Administration

Domino 6 includes several administration features that give you centralized control over Domino and help reduce your administrative tasks.

Policy-Based Management

Domino 6 introduces policy-based management, which significantly improves Domino 5 Setup Profiles and greatly simplifies administration, helps you maintain standard settings and configurations, and speeds up deployment of changes throughout an organization, business unit, or workgroup.

A **policy** is a collection of settings related to a class of end users that can be applied either when registering new users or retroactively to existing users. You can set and manage registration, setup, archiving, desktop, and security policies. You create new policies from the Configurations tab and apply them from the People & Groups tab of the Domino Administrator.

Policies are easy to set up and apply, and because they use a parent/child hierarchical model, they are easy to extend. The Policy by Hierarchy view shows the relationship among policies, subpolicies, and individual policy settings. A Policy Synopsis, which is available from both the People & Groups and Configuration tabs, shows you the effective policy settings for the specified people and groups (see Figure 4-1).

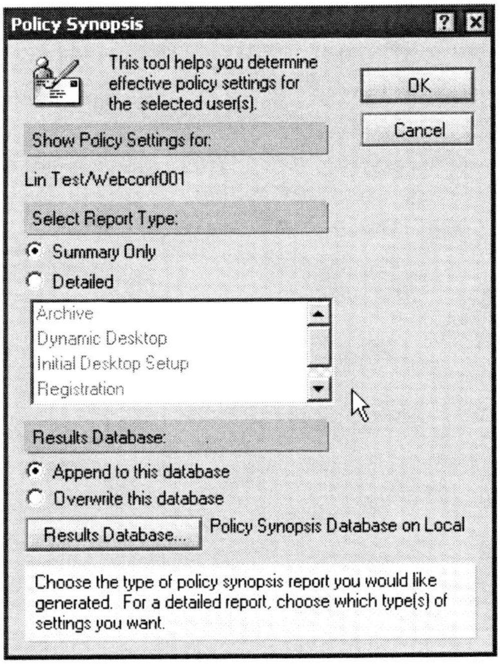

Figure 4-1 Domino 6 policy-based management.

Registration Policies

Policy-based registration options include such things as registration server, password options, mail server and template, Internet address information, ID and certifier information, and group assignments. When registering users, you simply choose the appropriate registration policy to apply all the policy's options to the users.

Setup and Desktop Policies

Setup and Desktop policies include the same settings, which include options for specification of a corporate Welcome page, dial-up connection information, server names, applet security, proxies, preferences, and more. The difference between them is how they are applied.

Setup policies are applied once to clients on their initial configuration; therefore, you use them for information that you do not want to keep updated on clients. Desktop policies are applied to all clients whenever a change in the policy occurs, so these policies include information that needs to be kept up-to-date.

Archiving Policies

Using policies, you can set and manage archiving settings for users, including allowing or disallowing users to create their own archive settings for their own databases. You can use more than one set of archive criteria and designate more than one archive destination.

Security Policies

Security policies contain password management and ECL fields. See the "New Security Features" section later in this chapter for more details.

Automatic Client Upgrades with Lotus Notes Smart Upgrade

One of the costs of deploying Domino in the past has been upgrading all the desktops in a system. A new feature of Domino 6, Lotus Notes Smart Upgrade, lets you install upgrades at the desktop level with the push of a button. Using a Smart Upgrade database on the server, the server Configuration document, and optionally, desktop policies, administrators control and maintain version deployment. When users connect to their home server, they are automatically prompted to upgrade to a new version of Notes. If they click OK, the new software is downloaded from the server, and the client is automatically shutdown, upgraded, and restarted.

xSP Administration

Domino 6 enhances administration features to support both Service Provider (xSP) administration of end users and organization administration of end users, securely. You also have the ability to generate bills and reports on a per organization basis. In addition, the new activity logging service provides consistent and complete reporting, which can easily be broken into organization reports. The HTTP log also can be easily broken into organization reports.

Deployment of Corporate Welcome Pages

You can create and automatically deploy corporate Welcome pages to your users. You create a corporate database to hold one or more Welcome pages and then link that database to the Desktop Policy document in the Domino Directory. You can control which Welcome page should be used and whether users can change their home pages.

Client Version Reporting and License Tracking

When a Notes 6 client connects to a Domino 6 server, it sends its Notes version, operating system platform, and machine name to the server. This information is added to the user's Person document, letting you see which version of Notes your users are running.

In addition, when a Notes client connects to a Domino 6 server through HTTP, IMAP, POP3, SMTP, or Lightweight Directory Access Protocol (LDAP), client information is collected and stored in a User Licenses database. An administration process request updates the License Tracking document in the Domino Directory with information from the User Licenses database. The License Tracking document is updated daily, so you can monitor the number of active Notes clients in your domain.

Console Innovations and Improvements

For Domino 6, the Web Administrator client has been revamped and improved to look and function like the (Win32) Administrator client, making administration from a Web browser more intuitive.

In addition, Domino 6 includes a separate, Java-based Server Controller that lets you control the Domino server from either the Administrator (remote) server console or a separate Java-based console called the Domino Console. This architecture allows you to control the Domino server when it is unreachable directly from the remote server console and facilitates issuing one command to multiple servers in one or more domains.

Both the remote server console and the new Domino Console include user interface improvements. New customization allows you to set console text, color, and highlight attributes for both the local and remote server consoles, making it easier to read and interpret the information that appears (see Figure 4-2).

In the remote server console in the Server\Status panel of the Domino Administrator, you can filter status messages for particular status levels and set stop triggers so that critical information pauses on the console screen. You can set new event notifications to help you monitor troublesome events more closely, and if you need more information on an error you receive, you can retrieve that information on-the-fly from the Domino Administrator. The server console can log to a text file so you can look at the output off-line rather than tie up the machine. You also can view OS platform statistics along with Domino server statistics. Finally, command-line help for most server tasks is now available.

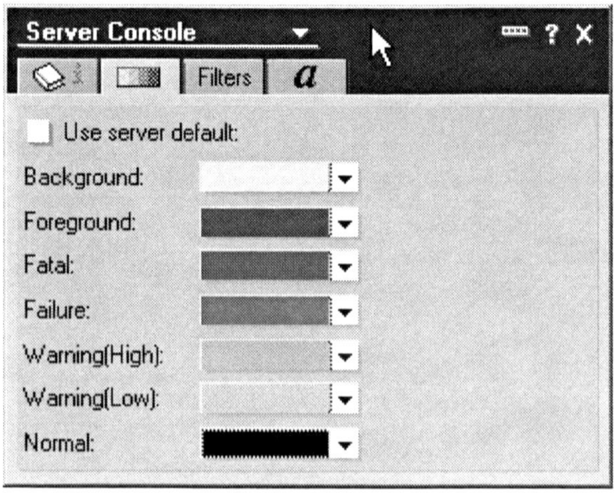

Figure 4-2 Domino 6 console customization feature.

Statistics Monitoring and Analysis

Statistics monitoring and analysis can help you plan and run individual systems, as well as your whole domain, more efficiently. In Domino 6, you can monitor performance statistic profiles using charts that display the statistics in realtime or historically. The Domino Server Monitor includes server profiles that monitor tasks and processes specific to a certain subset of servers.

IBM Tivoli Analyzer for Lotus Domino

It can be hard to know which information to monitor on a server and what indicates exactly when a server is "healthy" or in need of attention. The IBM Tivoli Analyzer for Lotus Domino (a separate product offering from Tivoli Systems) includes two integrated system-management tools: the Server Health Monitor, which offers real-time assessment and recommendations for server performance, and Activity Trends, which provides data collection, data exploration, and resource balancing. Using these tools, you can manage servers and databases, ensure better server performance, and plan for current and future needs.

> **NOTE**
> The IBM Tivoli Analyzer for Lotus Domino requires a separate license to use.

Server Health Monitoring is available from the Server Monitoring or Real-time Charting tabs of the Domino Administrator. Health ratings also appear as color-coded thermometers in a new column to the left of the server name in the By State view of the Server Monitoring tab (see Figure 4-3).

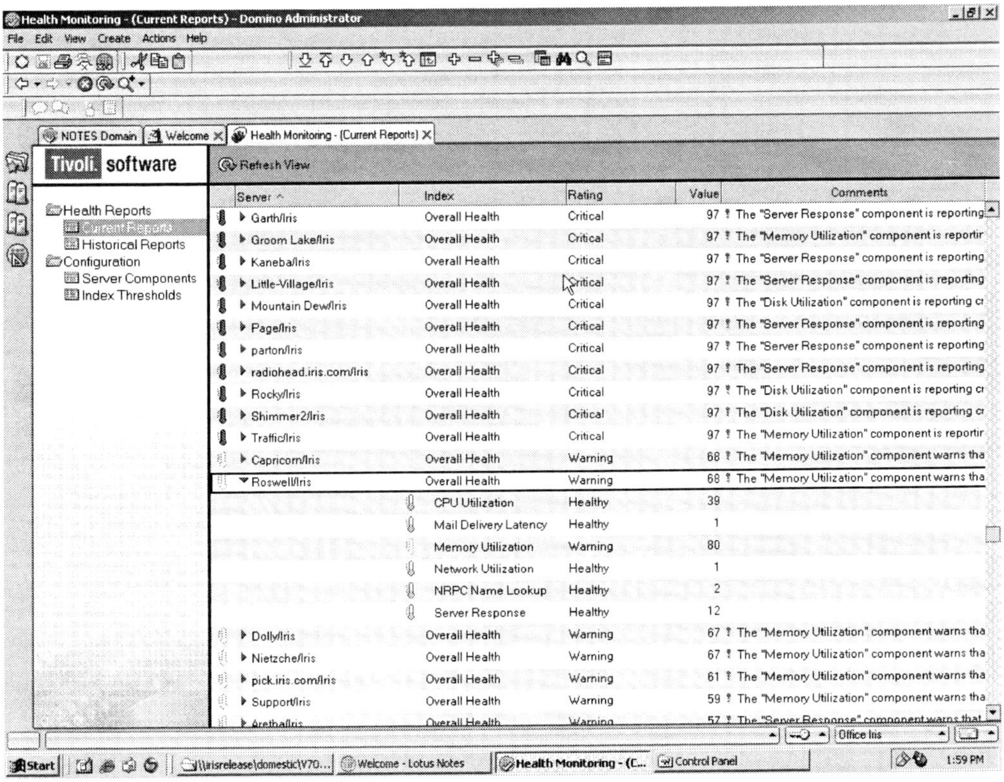

Figure 4-3 Domino 6 server health monitoring.

In addition, a Server Health Monitoring Report and the underlying metrics that contribute to a server's health rating are available, as well as Server Health Recommendation documents for servers with a Critical health rating (see Figure 4-4).

Integration of Server Health Monitoring with new historical charting features allows you to analyze past server health values, giving you insight into a server's health over a longer period of time. Additionally, you can view operating system level platform statistics such as CPU, memory, and disk I/O. The full set of network level statistics is also available.

Activity Trends presents server workloads by user, database, and protocol over time; makes load balancing recommendations; and can determine growth rate trends.

For Domino 6, supported platforms include Windows NT, Windows 2000, Solaris Sparc (version 2.8), and pSeries (AIX 4.3.3). Some iSeries (AS/400) platform statistics are also available.

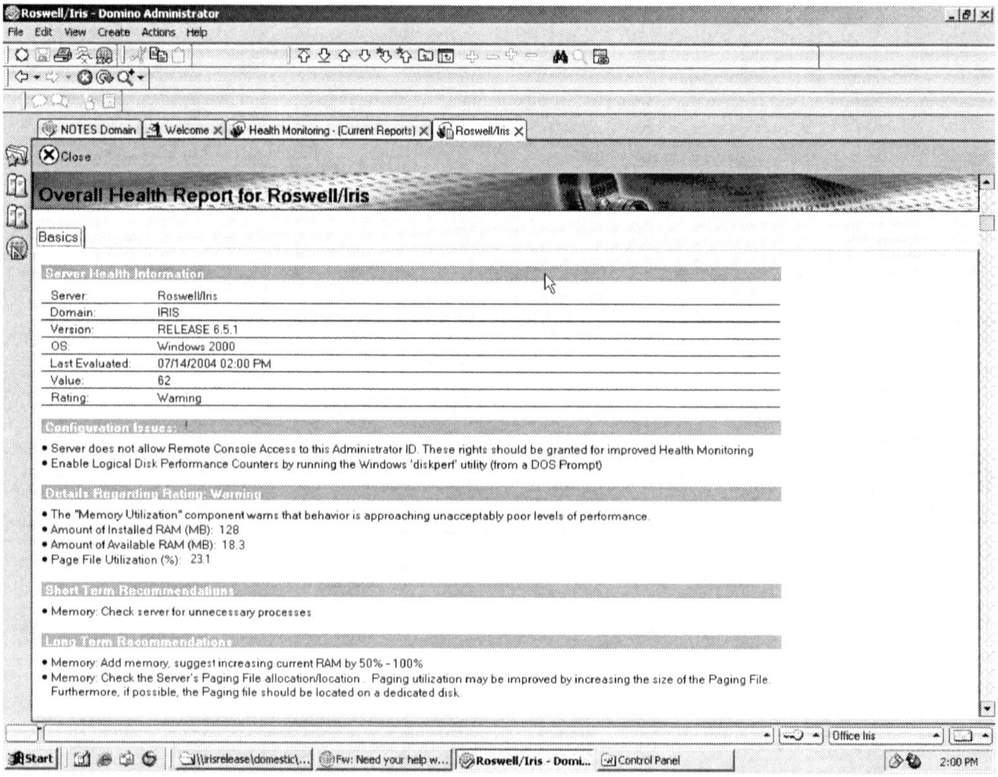

Figure 4-4 Domino 6 Health Monitoring Report.

Server Activity Logging

Domino Server Activity Logging has also been enhanced to include more detailed information about Notes sessions; databases; scheduled agent activity; and POP3, HTTP, SMTP, IMAP, and LDAP activity.

Enhanced Transaction Logging

Transaction logging—first introduced in R5—tracks all changes to a Domino database and writes the results to the transaction log. The logged transactions are then written to disk in a batch, either when resources are available or when they're scheduled. Domino 6 includes several key transaction logging enhancements. For example, the new View logging feature (Figure 4-2) lets you enable any view in any database for transaction logging. Note that this feature has a performance penalty, so you'll want to enable it only for complex views in critical applications.

Additionally, Domino 6 includes a new lock manager, which is designed to lock database elements required only for a particular function rather than locking all database elements in the database. This improves performance by allowing multiple operations (e.g., read or write) on the database simultaneously. Finally, an improved flush algorithm processes the transaction log and writes changes to the Domino database more often. This feature decreases user response time because the database is locked for write access for shorter intervals.

Enhanced Platform Statistics

Introduced in R5 for the Windows NT and Solaris platforms, platform statistics enable Domino to track operating system statistics. In Domino 6, this feature provides support for additional platforms, including pSeries, zSeries, iSeries, Sun Solaris, and Windows 2000. In addition, new system metrics are now available in Domino 6; you can access platform statistics via both the Domino Administrator client and the Web-based administrator client.

To view information about platform statistics, open the Monitoring Configuration database (events4.nsf). This database contains a complete list of platform statistics that you can track.

Other Administrative Features

Additional features that improve administration include the following:

- **Managing users, groups, and servers.** In Domino 6, you can find a user, group, or server in a domain using the new Find Name in Domain administration request. In addition, you can change a Web user name using the Domino 6 Administrator client.
- **Archiving.** Archiving in Domino 6 includes archiving to multiple archives, folder-based archiving, administrator control over archiving, and server-to-server archiving. See the previous "Archiving Policies" section for more information.
- **Extended administration server.** An extended administration server processes Domino Directory requests, allowing you to specify more than one administration server per Domino Directory.

Web Server Enhancements

The improvements to the Domino 6 Web server are geared toward enhancing performance and scalability and expanding the capabilities for Web application development and deployment.

The rewritten HTTP server provides HTTP 1.1 persistent connections, improved session handling, better denial of service attack handling, and more administrative control over things like URL length and number of path segments.

The redesigned Web server task supports an Internet Sites view with Internet Site documents in the Domino Directory. The Internet Site documents contain most of the information from the Domino 5 Server document that affects the HTTP stack as well as some new settings. All servers that share the same Domino Directory (that is, are in the same domain) share the same Internet Site documents in the Internet Sites view (see Figure 4-5).

Figure 4-5 Domino 6 Internet site document.

Also in Domino 6 there is a new Web Site Rule document: HTTP response header. Web Site Rules documents appear as responses to the Web Site documents in the Domino Directory. Web Site Rule documents let you relocate or reorganize your sites without breaking links or Web browser bookmarks. The HTTP response header rule type adds an Expires header or custom header to HTTP responses that match specified URL patterns and response codes.

Additionally the HTML generation engine is now more standards compliant and includes the ability to generate pages in XHTML.

WebDAV Support

Domino 6 furthers your ability for collaborative application development by supporting Web Distributed Authoring and Version (WebDAV). WebDAV provides a development environment for controlled and safe simultaneous development efforts. Teams of developers using tools that support WebDAV can open, edit, and save file design elements to and from a Domino database without risk of overwriting code. This means that the NSF can serve as the common repository for elements developed in third-party tools, as well as in Designer 6.

> **NOTE**
>
> Macromedia's Dreamweaver 4.01 is necessary for WebDAV support; this upgrade can be downloaded from the Macromedia Web site. Microsoft FrontPage 2000 is not a supported WebDAV client.

Domino Custom Tag Library

Domino 6 provides a custom tag library for those developers developing J2EE applications in third-party tools. They allow quick access to Domino databases and Domino objects such as views, forms, and fields, so you have access to Domino data and services without having to write low-level Java code. The tags are based on the JSP 1.1 standard and are usable with Web application servers that support that standard.

Web Preferences

Web preferences include settings for time zone, date/time format, and number format, which are stored in cookies in a user's Web browser. You enable Web preferences in the Internet Site documents. Users set preferences via a URL similar to `http://servername/$preferences.nsf`.

Single Sign-On

Single sign-on (SSO) for Web browsers allows you to sign on to a Domino or WebSphere server once, and then have access to any SSO-enabled Domino or WebSphere server in your domain without signing on again. In addition, you can have multiple Web SSO Configuration documents in a Domino Directory or domain.

Support for WebSphere Third-Party Web Server Plug-Ins

Domino 6 supports the WebSphere plug-ins that allow you to use a third-party Web server as a front-end to a Domino server. The initial release of Domino 6 supports the plug-ins for Microsoft IIS and the IBM HTTP Server. This feature replaces the "Domino for IIS" architecture that was provided in Release 5.

Server Cluster Enhancements

Many enhancements have been made to clusters, including the following:

- Making the Cluster Administrator a server thread so that it automatically starts the Cluster Replicator and Cluster Database Directory Manager.
- Ensuring the server availability index gives a more accurate indication of the availability of each server in a cluster. (You no longer need to use the Notes.ini setting Server_Transinfo_Normalize to improve accuracy.)
- Adding new settings to control the number of active Cluster Replicators.
- Using the Domino 6 Server Monitor to monitor all servers in a cluster.
- Allowing cluster replication to ignore database size quotas.
- Making activities such as user registration and database replication and deletion cluster-aware.
- Adding new Cluster Replicator commands for better control over cluster replication and information gathering.

Changes to Directories

A major goal of Domino 6 was to make Domino easy to integrate in a multi-directory environment. Large enterprises are beginning to see the advantages of a centralized directory configuration. It gives them more control and less overhead and, in the end, is easier to manage.

With Domino 6, you have the option of moving from a distributed directory architecture and making Domino the central directory. If you do this, you only need to store the complete Domino Directory, with all of its person and group information, on one central server. (You actually will want to reserve at least two servers to be used as central servers in the event that one server becomes unavailable. Automatic failover of directory servers is now completely built into the product.) You can then store the smaller configuration directories with Domino-specific data on the other servers in your domain. The centralized directory information is available to all users, but you save on disk space, because you no longer need to store the whole directory on each server. You also save on time, as you are no longer required to replicate your directory across all the servers in your domain.

Lotus has also enhanced the implementation of LDAP capabilities and improved the performance of LDAP directory access. For example, a new Domino LDAP Schema database helps you maintain and extend the schema. There is an automatic schema maintenance process and true object class inheritance, and directory schema can be imported via LDIF files. The new LDAP upgrade service lets you migrate person and group entries directly from an LDAP directory server into the Domino Directory. Support of arbitrary distinguished names, new LDAP configuration settings, activity logging for the LDAP service, and multiple values in the Hostname field in the LDAP Directory Assistance document for LDAP server failover are among the many LDAP improvements in Domino 6 we've made. In addition, you can now choose cluster failover as your mechanism of choice for directory assistance failover. Using this mechanism gives you the added capability of load balancing.

You can authenticate Internet clients for IMAP, POP3, LDAP, and NNTP clients as well as HTTP clients using a secondary directory (Domino or third-party LDAP). You also can use a secondary Domino Directory to maintain user names and groups that you don't want to store in names.nsf. Secondary Domino Directories also can be queried when resolving groups specified in database Access Control Lists (ACLs).

You also can create a multiple organization Domino Directory, using extended ACLs (xACLs) to ensure users have access to only their organization's information. Also, the new xACL controls allow enterprises the ability to delegate administration to regional administrators without giving them manager access. You can configure these regional administrators to allow them to administer only directory objects within their own organizational units.

Domino 6 provides greater efficiencies with a new directory lookup cache that is significantly better than previous caches. It's effectiveness for mail sending and routing lookups, for example, are over 95 percent. Other directory changes to note are as follows:

- Domino 6 can use LDAP, NameLookup, or both to serve up directories.
- Additional LDAP server configuration settings are available on the Configuration Settings document.
- A directory indexer task created by the database indexer task updates views in the Domino Directory.

Working with Active Directory

If you are using Windows 2000, administering users and groups can be synchronized between the Domino Directory and Active Directory. ADSync lets you register, synchronize properties and passwords, and rename and delete users and groups in the Domino Directory when you perform such actions in Active Directory.

You can migrate users and groups from an Active Directory to a Domino Directory using the Active Directory Domino Upgrade Service (ADDUS). This migration tool uses a search base, attribute field mapping, or custom LDAP filters to migrate users and groups.

Domino Hosting Features

The Domino 6 server includes new hosting features that allow multiple organizations to be transparently hosted by a single logical Domino server. Clients from different hosted companies access their data from the same physical server securely, using standard Internet protocols (IPs). This new server feature simplifies server administration and application support and satisfies the needs of the xSP market. The major Domino components have been modified to support the hosted organization environment.

> **NOTE**
> If you enable the xSP configuration, the entire domain runs in xSP mode to ensure the proper security environment. You may want to set up a separate test domain to try out Domino's hosting features.

Addressing Models

xSPs can choose between two different IP address configuration models. For each server and protocol on a server, a hosted organization can have its own unique IP address, or a single IP address can be shared across multiple organizations.

Multiple Organization Domino Directory

The multiple organization Domino Directory feature can significantly reduce the complexity of server administration. The administrator works with only one server, yet each organization on that server can function as if it is hosted by its own unique server. For example, each organization has its own HTTP application and file locations. The server also has organization-specific authentication controls.

The Domino Directory template has been modified to allow granular configuration control for each hosted organization. A new feature in Domino Administrator allows an xSP to register a new organization—creating the hosted configuration, producing a new certificate, creating a subdirectory, and implementing the security mechanisms (database ACLs, .ACL files, and xACLs) automatically.

Security

Because in a hosted environment the Domino Directory is a database that is shared between multiple organizations, security is a critical element. Each document in the Domino Directory is controlled by xACLs (extended ACLs) to allow or disallow access. The existing database ACLs and the new .ACL file feature ensure that organization-private databases remain secure. In addition, file protection documents for the Domino Web server provide additional access control for files accessed via HTTP. Multiple organizations hosted by the logical server also can access shared databases.

Protocol Support

For Domino 6, it is possible for xSPs to provide the following services to their customers: IMAP, POP3, LDAP, SMTP, HTTP, SSL, and IIOP.

DOLS

xSPs also support Domino Off-Line Services (DOLS).

Mail Routing

The Domino router has been modified to support multiple organizations simultaneously on the same physical/logical server.

Activity Logging for Billing

Data about the Domino server is collected in the log.nsf file with a new Server Activity Logging feature. The xSP can access log.nsf via an API. Data is collected on a per server basis and can be configured per protocol. Each record contains the organization name, and the xSP can determine the appropriate billing model for its customers.

Database Server Utility Programs

To provide more granular control, the database server utility programs such as compact, fixup, updall, and design now allow a directory to be specified. This means, for example, that an xSP administrator can configure program documents in the Domino Directory to have compact run on Company One's databases at 2 a.m. and Company Two's databases at 3 p.m.

Scalability

Scalability for a hosted environment include the following features:

- Support for a configuration-only directory to improve server performance
- Qualified name lookups per organization in the Domino Directory to provide improved name lookup performance for any size directory
- Support for the use of a network sprayer to provide load balancing or failover capabilities

New Security Features

In the 1990s, Lotus was one of the first companies to use public key encryption, and security continues to be one of the greatest strengths of Lotus products. Domino 6 recognizes that today's computing environments are heterogeneous, using different clients (for browsing and messaging), different servers, different security protocols, and even different security vendors, each providing a different security component, such as certificate authorities, single sign-on servers, and firewalls. For example, a company might run both Notes and Outlook clients for secure messaging, with Domino and Exchange backends, respectively. Those clients might, in turn, be issued certificates from VeriSign. To maximize this kind of environment, Domino 6 provides support for new security standards. For example:

- **Support for the PKCS#11 standard for Smartcards.** Smartcard support provides additional protection for your user credentials because the user's ID is locked such that both a Smartcard (with the ID's password) and a Smartcard PIN are required to unlock it. You can enable and disable Smartcard support in the User Security dialog box.
- **Support for the S/MIME v3 feature, S/MIME capabilities.** The Notes client can receive messages that have specified algorithms and key lengths and will use that information when sending encrypted mail.

New Certificate Authority

The Domino certificate authority for Domino 6 includes an optional certificate authorization (CA) process that provides both a unified mechanism for issuing Notes and Internet certificates and an integrated registration of Notes and Internet keys. The certificate authority process is a "locked box" task that runs on the server. Administrators enable Notes and Internet certifiers to use the CA process to take advantage of the following features:

- **Ease of granting certificate authority.** Administrators select a few parameters, such as certificate duration, and designate administrators who are authorized to use that particular certifier. They then load the certificate authority server task to enable the new certifier to use the CA process.
- **Separation of certificate authority and registration authority roles.** Administrators can delegate certificate approval/denial process to lower-echelon administrators (registration authorities), who can perform certification tasks without access to the certifier ID or password.
- **Creation of Certificate Revocation Lists (CRLs).** CRLs maintain information about expired certificates, and their use helps administrators maintain the integrity of their organization. Certificate revocation status can be checked before someone chooses to trust a certificate. A CA administrator can easily revoke a certificate if the subject of the certificate leaves the organization or if the certificate has been compromised. CRLs are published on a regular schedule and are posted in the CA's Certifier document in the Domino Directory.
- **Simplified certificate request process.** There is a new Web interface for certificate request management.

Additionally, the Notes client has been extended to allow users to manage their certificates through the User Security dialog box.

A Site document (created for each Internet protocol so that a server can obtain the SSL security settings for that protocol) contains settings controlling the use of CRLs. These settings are not part of the Server document but are passed into trust policy code during certificate chain evaluation.

Delegated Server Administration

Domino now includes several levels of server administrators ,so that you can securely delegate different levels of administration to different administrators. For example, some administrators may have full access when using a remote console while others may not.

Notes and Internet Password Management

Domino's password management features provide the administrative functions you need to protect your Notes and Internet environment. You can automatically synchronize Internet passwords with Notes ID passwords by setting this in a security policy. In addition, you can more easily manage Notes and Internet password quality and length, as well as control expiration period, change intervals, and in the case of Notes passwords, password history.

Admin ECLs

You can now push Admin ECLs to clients dynamically on an as-needed basis, making it easier to deliver timely updates and to update clients who get the default ECL during setup because they were disconnected from the directory. In addition, the key string `<ECLOwner>` is added to the Admin ECL to enable the current user to modify the ECL during the client ECL refresh.

Messaging

New products, including iNotes Web Access and Domino Everyplace Servers, extend access to Domino's messaging infrastructure, from desktop to laptop, to the Web, to cell phones and Personal Digital Assistants based on the Palm, Windows CE, and EPOC operating systems. In addition, Domino 6 includes new features for managing and controlling your messaging infrastructure. For example, in Domino 6:

- Using the server's Configuration document, you can set up system mail rules that let you specify message criteria and actions to be taken for all messages processed on the server. This allows you to deny and quarantine messages. You might filter mail contents to stop the flow of SPAM through your routers, for example.
- Mail journaling works with the system mail rules to save copies of messages to a Notes journal database. You can save all messages that pass through MAIL.BOX, or you can specify criteria for selecting certain messages to save. For instance, you can specify people, groups, or domains from which or to which you want to save messages.
- New mail file quota management options provide you additional tools to control the size of users' mail files so that you can better manage disk storage. You can specify that mail be held for users whose mail file exceeds a specified quota limit, with quota warning and error notification to the mail file owner.
- The IMAP server now supports the Namespace extension, so an IMAP client can now view folders in another user's mail file or the public folders in a shared database. This means that users can have delegated access to another user's mail via IMAP, in addition to access via the Notes client.
- You can track SMTP servers that permit third-party relay using DNS Blacklists (DNSBLs). You can specify which DNS Blacklists to check and which action to take when mail is received from a host listed in the DNSBLs. Mail from servers found on a DNS Blacklist can be tagged or rejected, giving you more control over mail from hosts that may be sending or relaying potential SPAM.
- A set of Extension Manager hooks are available for the SMTP protocol that can be used to process incoming SMTP messages. While primarily intended for third party tool vendors, you can use the API to develop custom tools and applications that process incoming SMTP mail prior to it being processed by the Domino Messaging server.

- Anti-relay restrictions prevent unauthorized relays on your Domino SMTP servers. The inbound relay controls define the hosts from which and to which an SMTP server can relay messages.
- Shared mail now supports configuration of multiple shared mail databases and directories and can associate a user mail file with more than one shared mail database. When you install Domino 6, it automatically creates a shared mail database.

Domino Off-Line Services

There are many new enhancements for Domino Off-Line Services (DOLS) to help users work more efficiently off-line, including the following:

- Taking the directory catalog off-line
- Encrypting a subscription
- Loading customized subscriptions onto the DOLS Client PreInstaller CD
- Sharing files among subscriptions
- Preventing compaction and full-text indexing of subscriptions
- Pushing settings off-line to users, so users don't need to reinstall a subscription to update their settings
- Preventing users from modifying synchronization schedule settings
- Using a passthru server to connect to the DOLS server
- Providing an optional or secondary TCP/IP address configuration to access the DOLS server
- Overwriting the user's off-line ID each time the user installs a subscription to keep IDs updated and consistent
- Supporting Domino xSP server configuration
- Supporting single sign-on authentication to the DOLS server
- Supporting the server platform
- Providing cleaner integration with iNotes templates
- Improving external authentication coverage
- Improving proxy server configuration coverage, such as reverse proxy, DMZ, passthru server, and dual IP addresses
- Enabling accessibility features

In addition, DOLS supports the Notes multi-user configuration, and Web Control and iNotes Sync Manager provide keyboard shortcuts for most options.

New Designer Features

Layers, Notes Elements to/from HTML

Domino 6 Designer offers two new features to make it easier to create sophisticated Web pages and form designs. The layer feature allows page or form elements to be grouped together and then combined to form a single page design. Layers can be thought of as special types of page or form elements, which can be stacked to create a single page or form design.

It is now possible to easily create pure HTML source from Designer 6 pages or forms and work with the HTML source in external HTML editors or with an HTML editor pane built into Designer itself.

Shared Resources

With Domino 6 Designer, two new types of design elements are able to be defined as shared resources. Shared resources are those elements which are created outside of the Domino database itself such as images, files, and applets and that can be used by several design elements within the database. Cascading Style Sheets (CSS) and Data Connection Resources (DCR) definitions can be created as shared resources with Domino 6.

CSS resources can be imbedded into page and form (subform) design elements to control their appearance within a web browser. Support for CSS resources allow Web pages created in Domino Designer to utilize CSS files created for nonDomino Web applications. Most CSS properties are supported by Domino Designer.

DCR definitions allow relational database (RDBMS) tables, views, or procedures to be accessed directly from an external source. A DCR can be defined via a dialog in Domino Designer in which the designer specifies the system, database, userid, password, table name, and so on to be accessed. Then that DCR can be associated with a field on a form such that the data for the field is obtained from the external database.

JavaScript Enhancements

The support for JavaScript has been enhanced in both the Domino 6 Notes client and Designer. Many more JS features are now supported by the Notes client itself, making it possible to code Web pages with JS that work for both browsers and Notes clients. New events have been added to the Notes Document Object Model (DOM) such as onLoad, onUnload, and onSubmit, which make it more consistent with the JS DOM.

In the Designer client, it is now possible to store JS as a shared resource library and reference it across design elements (pages, forms) within a Notes database and even across servers (as long as the servers are in the same Notes domain).

XML Enhancements

Domino 6 Designer is a significant step toward full, native XML support. A Document Type Definition (DTD) is provided for Domino design elements. This DTD, referred to as the Domino DTD, defines the XML for all Domino design elements. Using the Domino DTD, it is possible to have Designer generate an XML document for any design document. The XML produced according to the Domino DTD is called DXL (Domino XML) and can be used to express Domino design elements is various way—as HTML, text, or other XML formats.

The Designer client offers three new utilities to work with DXL: an exporter, viewer, and transformer. The exporter utility generates DXL files; the viewer opens DXL in an XML-capable browser (such as IE6); and the transformer applies specified XSL transforms to DXL.

Chapter 5

Using WebSphere and Lotus Collaboration Together

"Success depends upon previous preparation, and without such preparation there is sure to be failure."

—Confucius, ancient Chinese philosopher, 551 BC–479 BC

WebSphere and Lotus Collaboration can be used together in several ways. WebSphere can be used together with Domino; WebSphere can be used with the new Lotus Workplace components; and WebSphere can be used with the WebSphere Portal Server and the Lotus Collaboration Central components. Of course, WebSphere also can be used with any combination of these options. This chapter explores the benefits of these different ways of using WebSphere and Lotus Collaboration together.

A common installation setup for WebSphere and Domino together is to have DB2 and WebSphere on one computer and Domino on another computer. This configuration allows separation of the components so that, for example, WebSphere and DB2 could be installed behind a firewall, with the Domino HTTP server in a DMZ. A DMZ (Demilitarized Zone) involves the formation of multiple firewalls to add additional layers of protection between a company's database and the Internet. A typical DMZ configuration includes: An outer firewall between the public Internet and the Web server or servers processing the requests originating on the company Web site; An inner firewall between the Web server and the application servers to which it is forwarding requests. Company data would reside behind the inner firewall. In other words, the business logic and data so important to the company are protected behind a double layer of firewalls.

This chapter will first discuss the basic hardware and software platforms that are used for a combined WebSphere and Domino system. Then the use of the WebSphere portal with Lotus Collaboration is described. Finally, some recommendations on moving Domino data and applications to the WebSphere Portal are given. The detailed steps for using WebSphere Application Server 5 and Domino 6 together are given in Chapter 6, "Building a Combined WebSphere and Domino System," and Appendix C.

Using WebSphere and Domino Together

Platforms

Because both WebSphere and Domino can be run on several platforms, we will discuss the most popular platforms for these products—Windows, UNIX, and Linux.

Windows

The minimum hardware required for the WebSphere and Domino computers for the Windows platform consists of

- Pentium II or higher with a minimum of 256MB RAM (at least 512MB recommended).
- At least 600MB free space on the hard drives used to install the products. Here are the hard drive space requirements after installation:
  - DB2 475MB
  - Domino 300MB
  - IBM HTTP Server 20MB
  - WebSphere 220MB

The DB2 space requirements can be reduced by 12MB if you choose to install the DB2 Administration client on a separate machine. The space requirements for Domino also can be reduced to a certain extent by choosing to install fewer components. For example, it is possible not to install the help files. These files alone require 50MB. Also the space for WebSphere specified here includes documentation files of about 60MB, which would not be needed in a production installation.

The Windows software required can be Microsoft Windows NT, Windows 2000, Windows 2003, or Windows XP. Either workstation or server code can be used—although usually server code would be used. In addition, the normal assumption for Web servers applies in that the configuration would be for TCP/IP networking with a fixed IP address for each computer.

UNIX (AIX)

The minimum hardware required for the WebSphere and Domino computers for the UNIX (AIX) platform is the following:

- IBM RS/6000® or RS/6000 SP™ running IBM AIX, Version 4.3.3 with the 4330-10 recommended maintenance package; IBM AIX, Version 5.1 with the 5100-02 recommended maintenance package or 5100-03 and APAR IY36884; or IBM AIX, Version 5.2
- IBM RS/6000 604e workstation at 375MHz or faster
- CD-ROM drive
- Minimum of 512MB available disk space for installation (including IBM Software Developer Kit)
- Minimum of 256MB memory, 512MB recommended
- Support for an appropriate network interface

IBM AIX, Versions 5.1 and 5.2

- 192MB memory (256MB or more recommended)
- 1GB hard disk space (1.5GB or more recommended)

Linux

Domino is supported by IBM on several Linux distributions for Intel processors. The supported distributions for versions prior to Domino 6.02 are Red Hat Linux, Version 7.2 or SuSE Linux, Version 8.0. In addition to these, Domino versions 6.02 and later are supported on Red Hat Advanced Server, Version 2.1 (uniprocessor only) and United Linux, Version 1.0. (It is possible to install and run Domino on other Linux distributions that provide the same level of Linux kernel and key library software as the supported distributions, but only the distributions mentioned here are supported by IBM. Other distributions could be used for experimentation, development, or testing work.)

For Domino on Linux, IBM recommends at least 192MB of system memory and 1.5GB of disk space for acceptable performance. Based on our experience, to run any substantial application code on Domino requires more system memory—the more the better. We'd recommend at least 512MB of system memory. An installation of Domino code requires approximately 500MB of disk space on the Linux filesystem.

Software Levels

Red Hat Linux Operating Environment on Intel

- Red Hat Linux Advanced Server 2.1, SuSE 7.3 or SLES 7.0, based on kernel 2.4; Red Hat 8.0; UnitedLinux 1.0; Connectiva Linux Enterprise Edition, powered by UnitedLinux 1.0; SCO Linux Server 4.0, powered by UnitedLinux 1.0; or SuSE SLES 8.0, powered by UnitedLinux 1.0
- Intel x86 processor at 500MHz or faster
- CD-ROM drive
- Legacy application support RPMs installed (provided on SuSE and Red Hat distribution CD-ROMs)
- Minimum of 300MB available disk space (including IBM Software Developer Kit)
- Minimum of 256MB memory, 512MB recommended
- Support for TCP/IP and an appropriate communications adapter

Red Hat Linux on IBM zSeries Operating Environment

Red Hat Advanced Server, Version 2.1 (using uniprocessors only) and UnitedLinux, Version 1.0SP2 on 390

- 128MB memory (192MB or more recommended)
- 1GB hard disk space (1.5GB recommended)
- IBM zSeries® server running Linux distribution SuSE SLES 7.0 or Red Hat Linux 7.2, based on kernel 2.4
- G5, G6, or higher processor
- Legacy application support RPMs installed (provided on SuSE and Red Hat distribution CD-ROMs)
- CD-ROM drive
- Minimum of 512MB available disk space for installation
- Minimum of 256MB memory, 512MB recommended

Domino

Software requirement

Lotus Domino Versions 6.0 to 6.5.1

AIX

Software requirement

IBM AIX®, Versions 5.1 and 5.2

Linux

- Red Hat Advanced Server, Version 2.1 (using uniprocessors only)
- All "United Linux/Powered by United Linux 1.0 SP2 for Linux on Intel distributions." That includes SuSE Linux Enterprise Server (SLES) 8.0, Turbo Linux Enterprise Server, Connectiva Linux Enterprise Server.
- United Linux/Powered by United Linux 1.0 SP2 for S/390

Administration

WAS Administration

The WebSphere Administrative Server (admin server) for WAS V4 runs in a JVM on each node in a WebSphere Administrative Domain (see Appendix A for WAS V5 admin details). The admin server is responsible for providing run-time support for JNDI service, security, transaction logging, Location Service Daemon, and EJB workload management. The admin server also provides system management support initiated from administrative interfaces, including the Administrative Console, XMLConfig, and WSCP. Communications with the admin server take place using RMI/IIOP.

The second major function of the admin server is to provide a single administration point for all of the objects within the WebSphere domain. The admin server allows for administration of information stored in the WebSphere repository.

Types of System Administration Clients

Administrative Console (admin console)

The admin console is the graphical user interface for systems administration. It uses the bootstrap server and the name server in the admin server to perform JNDI (Java Naming and Directory Interface) operations. User operations on the admin console are relayed to the EJBs on the admin server and possibly indirectly to another admin server if the request affects another node. If security is enabled, the security runtime in the admin server uses the security server in the admin server to authenticate the user.

If the admin server fails during admin console initialization, the console will fail to bootstrap to the admin server. This prevents the console from starting up altogether. Failover is achieved by restarting the admin console to bootstrap to a different admin server. The command-line options are as follows:

- adminclient -host host [-port port]
- where the default port number is 900

After the admin console successfully starts up, workload management of the admin server takes effect. If an operation on the admin console fails, retrying the console allows the workload management runtime to reroute the request to a different admin server to complete the operation. Note that requests involving the applications on the node where the admin server failed, such as querying the state of the servers on that node, will fail until the admin server is restarted. But requests affecting other nodes will succeed.

XMLConfig

XMLConfig allows the system administrator to perform application management functions, such as application installation. If the admin server fails, XMLConfig operations will fail as well. If the operations of XMLConfig affect the node where the admin server is running, the admin server needs to be restarted before those operations can succeed. However, if the operations do not affect the applications on the same node as the failed admin server, XMLConfig may be rerun to bootstrap to a different admin server with the following parameter: -adminNodeName host.

WebSphere Control Program

WebSphere Control Program (WSCP) is used by the system administrator to perform administration operations, such as stopping and starting servers. If the operation affects the node where the admin server fails, the admin server needs to be restored before the operation can complete. If the operation does not affect the node of the failed admin server, WSCP can be rerun to bootstrap to a different admin server by setting the following properties either from the command line, or via the configuration property file: wscp.hostName=host.

Domino 6 Administration

Domino 6 allows you to apply and manage standard corporate policies by providing a new technology called policy-based system administration. This feature gives you a way to centrally define, organize, and manage user settings throughout your organization. All information required for policy-based system administration is stored in your Domino Directory, giving you a single place to control administration activity, from user setup to mail archiving.

Policy-Based System Administration

In broad functional terms, policy-based system administration is exactly what its name implies—a way for you as a Domino administrator to manage and apply corporate policies for your employees. These corporate policies define the way Domino enforces user settings, which ensures that your established corporate standards and practices control how your technology works (and not the other way around).This important Domino enhancement on policies was also touched on in Chapter 4 and is discussed in detail in Appendix B.

In Domino 6, you enforce corporate policies through individual Policy documents and Settings documents in the Domino Directory. To create a policy, you use a Policy document to specify which Settings documents to include.

A Policy document defines a set of corporate information you want to apply to your users. It does this by specifying which Settings documents to include in the policy. Each Settings document contains detailed information applicable to a specific area of Domino administration, for example archiving or security. This information defines user settings for that area. You can consider policy-based system administration as a way to create and apply rules within your user community. The Settings documents contain the rules; the Policy document organizes these rules. So when you create a Policy document, you have a single place to list these rules. You can then apply the rules to establish and enforce administrative standards by distributing them throughout your organization. To change an existing policy, all you need to do is edit the Policy document and/or one or more Settings documents. No more running around to each user's computer, making the same change over and over—you can now do this all centrally.

You can use policy-based system administration to manage five major areas of Domino administration: archiving, desktop, setup, registration, and security. Again, please see Appendix B for details on Domino 6 Policy Administration.

Management

WAS Management

Managing Administrative Agents

If your system uses administrative or repository services, you can specify settings for the services. Steps for this task are to

- Use the settings page for an administrative service to configure administrative services.
- Use the settings page for a repository service to configure administrative repositories.

In order to understand this task, it is important to go through the details of **cell** and **node**.

Cells

Cells are arbitrary, logical groupings of one or more nodes in a WebSphere Application Server distributed network. A cell is a configuration concept, a way for administrators to logically associate nodes with one another. Administrators define the nodes that make up a cell, according to the specific criteria that make sense in their organizational environments. Administrative configuration data is stored in XML files. A cell retains master configuration files for each server in every node in the cell. The nodes and servers also have their own local configuration files. Changes to a local node or to a server configuration file are temporary—if the server belongs to the cell. While in effect, local changes override cell configurations. Changes to server and node configuration files at the cell level are permanent. Synchronization occurs at designated events, such as when a server starts.

Nodes

A node is a logical grouping of managed servers. A node usually corresponds to a physical computer system with a distinct IP host address. Node names usually are identical to the host name for the computer. A node agent manages all WebSphere Application Server servers on a node. The node agent represents the node in the management cell.

Managing HTTP Sessions

IBM WebSphere Application Server provides a service for managing HTTP sessions: Session Manager. The key activities for session management are summarized below. Before you begin these steps, make sure you are familiar with the programming model for accessing HTTP session support in the applications following the Servlet 2.3 API (see, for example, `http://www.onjava.com/pub/a/onjava/2001/03/22/servlets23.html`, for good background information on this API).

Steps for this task include the following:

- Plan your approach to session management, which could include session tracking and session recovery.
- Create or modify your own applications to use session support to maintain sessions on behalf of Web applications.
- Assemble your application.
- Deploy your application.
- Ensure the administrator appropriately configures session management in the administrative domain.
- Adjust configuration settings and perform other tuning activities for optimal use of sessions in your environment.

Managing Using Command-Line Tools

There are several command-line tools that you can use to start, stop, and monitor WebSphere server processes and nodes. These tools only work on local servers and nodes and cannot operate on a remote server or node. To administer a remote server, you can use the wsadmin scripting program connected to the deployment manager for the cell in which the target server or node is configured.

Managing Object Request Brokers

An object request broker (ORB) is a middleware technology that manages communication and data exchange between objects. ORBs promote interoperability of distributed object systems because they enable users to build systems by piecing together objects—from different vendors—that communicate with each other via the ORB. The implementation details of the ORB are generally not important to developers building distributed systems. The developers are only concerned with the object interface details. This form of "information hiding" enhances system maintainability since the object communication details are hidden from the developers and isolated in the ORB.

For WAS, default property values are set when the product is started and the Java Object Request Broker (ORB) service is initialized. These properties control the run-time behavior of the ORB and can also affect the behavior of product components that are tightly integrated with the ORB, such as security. It might be necessary to modify some ORB settings under certain conditions. In every request/response exchange, there is a client-side ORB and a server-side ORB. It is important that the ORB properties be set for both sides as necessary. After an ORB instance has been established in a process, changes to ORB properties do not affect the behavior of the running ORB instance. The process must be stopped and restarted in order for the modified properties to take effect.

Domino Management

Since Lotus Domino is a combination of groupware and e-business applications that together give you e-mail, shared freeform databases, and Web site support, Domino management must cover a wide range of capabilities. In addition to the "built-in" management tools for Lotus Notes and Domino, there are several very comprehensive third-party products available such as CA Unicenter Management for Lotus Notes/Domino, and BMC PATROL.

Using WebSphere Portal with Lotus Collaboration Features

Over the past several years there have been many new design features announced making it easier for Notes/Domino and WebSphere to be used together. When IBM released WebSphere 3.5, single sign-on was a new feature that stirred more than a passing interest among users. Before long, and without a lot of fanfare, the Domino 5.0.5 release introduced single sign-on as well. From that moment on, the platform integration between WebSphere and Domino was a fact.

During 2003, the release of the Lotus Workplace concept confirmed the crucial role of Lotus components within the WebSphere portfolio. Also Lotus announced in 2003 that WebSphere Portal (WP) would be the User Interface and Access "engine" for Lotus products.

These are many signs that the convergence between Lotus and WebSphere is proceeding rapidly. It's also becoming too important for Lotus and WebSphere users to continue to view it from the parochial comfort of one product set or the other. If you're coming from strictly a Lotus perspective, you've probably noticed IBM's movement of collaboration products and components to the WP environment. What does this mean to Lotus users? How do Lotus's collaborative features increase the value of WP, and what new possibilities does this movement create? What do WP and J2EE mean, why a portal, and what does this mean for the future of Lotus products?

If you're a WebSphere fan, it may be the case that your portals are missing an important collaborative element that could save your company significant amounts of money, make company employees more productive, and make your customers happier with your products or services. This is a collaborative element already inherent in the Lotus side of the picture. Lotus is all about collaboration and human interaction software and the process of helping users collaborate and learn wherever they work—across organizational boundaries and independent of time and place.

This section offers some additional perspective on the convergence of Notes/Domino and WebSphere, gives Lotus users a high-level introduction to portals and the components available for inclusion in them, discusses some specifics of the Lotus strategy as it relates to Web portals, and shows some of the kinds of functionality that can take your applications and company to the next competitive level.

Before we start to take a closer look at the possibilities of Lotus products within the portal, let's take a step back, take a look at some portal basics, and see how WP is evolving into the vehicle that can fulfill the promise of Lotus's transition to J2EE.

Portals and Integration

Basically, portals provide a single point of access to a variety of applications. For example, on a single page, users can access Domino calendaring, a WebSphere order-processing application, and a knowledge-based customer support application. This data is available to users in one place without requiring them to click through multiple screens and remember each screen in order to handle customer issues. In addition, using single sign-on within the portal framework lets users avoid logging on multiple times to access all the applications available via the portal. Figure 5-1 shows how diverse applications can be brought together within a portal environment.

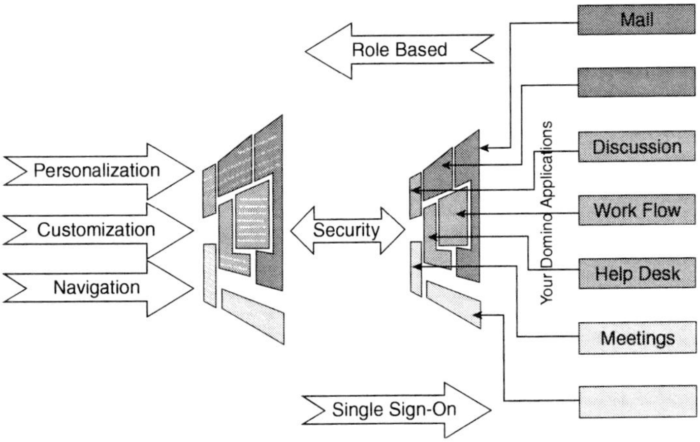

Figure 5-1 Lotus Collaboration influences portals and vice versa.

Portals, however, also provide additional valuable functions, such as security, search, collaboration, and workflow. Complete implementation of these added capabilities lets portals deliver integrated content and applications in a collaborative workplace. Combined with WP tools and user interfaces that help users access information and applications and manage the selection and personalization of content, portals are evolving to become the next-generation desktops, or communities, that can deliver e-business applications to all kinds of devices via the Web.

Portals can reduce costs in various ways. Users can leverage existing programs and can more easily find and share information securely. And this easy integration is crucial. Simplifying relationships between programs and users can help employees become more productive if they can find or link information more easily. Another way of stating this is—Collaboration, Communication, and Coordination, the three C's of Lotus. But now they're applied and available in the WebSphere open-standards environment.

The business need to improve business process coordination is driving demand for collaborative solutions in the form of "pluggable" components that extend current CRM, ERP, and other back-end applications and make that information more widely available. As adoption of Web services accelerates, this places even more emphasis on the value of plug 'n' play portlets that share information across enterprises.

For those of you who are new to Lotus, it's important to understand that Lotus products focus on the human side of business and bring to the WebSphere side of the table solutions that help people organize information and facilitate secure communications between fellow employees and other partners. This means that a convergence of Lotus and WebSphere will bring ease of use and communication benefits that WebSphere alone doesn't offer today.

WP seems destined to become the most important vehicle within the WebSphere portfolio, providing a controlled transition from Lotus proprietary components into J2EE pluggable collaborative components, resulting in a new platform for collaboration and human interaction: the Lotus Workplace as shown in Figure 5-2.

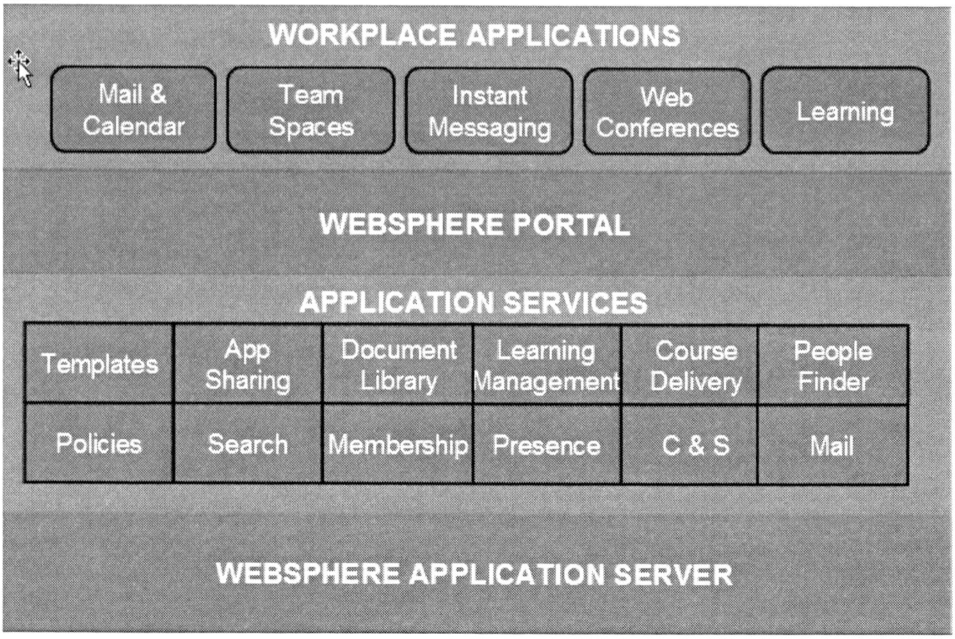

Figure 5-2 Lotus Workplace blends Lotus Contextual Collaboration and WebSphere features.

WebSphere Portal Offerings

In order to understand the concepts behind the Lotus Workplace strategy, let's first focus on the portal offerings available at the end of 2003, followed by the current integration capabilities of Lotus products into WP.

In simple terms, WP runs as an application on top of the WebSphere Application Server, inheriting all the scalability, security, and performance provided by this foundation layer.

The WP product actually consists of three major offerings, each designed to provide the infrastructure you need to build and deploy highly scalable portals. All three offerings share a common framework (the portal server) plus additional products and services. The Enable offering is a foundation for the portal solutions. Extend adds even more collaborative functionality from Lotus.

By the end of 2002, IBM launched Portal Express, an offering tailored for the small-to-medium business (SMB) market. The most important segmentation is between Portal Express and Portal Express Plus because the Plus version adds collaborative capabilities.

WebSphere Portal Extend integrates tightly with Lotus's collaboration and knowledge management products (e.g., Lotus Team Workplace (QuickPlace), Lotus Instant Messaging (Sametime), and Lotus Discovery Server) by providing portlets and services that access these products seamlessly from the portal.

Lotus Collaboration Features in WebSphere Portal

WebSphere Portal V4.2x and V5 versions include several functions and enhancements that support collaboration. Besides the fact that the Domino server can function as both the HTTP server and as the LDAP server, several portlets are available to connect to, as follows:

- Domino Mail and Calendaring
- Lotus Domino Web Access (iNotes)
- Domino databases
- Instant Messaging (Sametime) functionality (ranging from simple portlets such as a buddy list or who is where on this page to Sametime Awareness integration into other portlets)
- Team Workplace (Quickplace)
- Lotus Domino Document Manager (Domino.doc)
- Extended Search
- Discovery Server

All these portlets take advantage of the single sign-on concept to integrate all these products and WebSphere, so users need to log on only once when entering the portal. Other features such as personalization and tailored navigation features are also part of the portal concept available in WP.

In addition, WP now includes the Collaboration Center. This is a set of ready-to-use collaborative portlets for finding, connecting, and working with people:

- **People Finder portlet:** An online company directory and organizational navigator. People Finder lets you find any employee by name and see the employee's contact information, background, areas of expertise, and context within the company's organizational chart (manager and peers).
- **My Lotus Team Workspaces (QuickPlace) portlet:** Lists your workplaces, which are provided by Lotus Team Workplace (QuickPlace) right on the portal page. You can search across all of the team workspaces you belong to, or you can quickly see what's new in a workplace, join an existing workplace, or create a new workplace.
- **Web Conferencing (Sametime) portlet:** Provides integrated tools for managing on-line meetings. From within the portlet, people can join existing on-line conferences, see active meetings they need to join, or schedule new meetings.

As the Collaboration Center example shows, the WP environment is already leveraging a growing number of Lotus product features. The goal of the Next Generation strategy is to facilitate integration more easily between the collaborative capabilities of the Lotus products and custom or third-party business applications. As WP is all about integration "on the glass and behind the scene," the portlet principle sets the stage for user-oriented process integration.

One area in which this concept is emerging is that of contextual collaboration, in which employees collaborate with each other "in context" of business processes, starting with B2E (Business to Enterprise) but just as easily leveraged to B2C (Business to Consumer) or B2B (Business to Business) in virtual places. The Lotus Workplace strategy is to take this contextual collaboration to the next level. IBM is implementing this vision, and it's clear that the Lotus products' move to modular collaborative services is extremely important for IBM. Instead of providing the currently available pure product portlets, over time more components will become available and offered as services to other applications.

The signs of this trend is apparent in the Collaboration Center for WebSphere Portal 4.2x and 5.0, which provide integrated tools for managing online meetings and instant team workplaces, complete with an online directory, a people finder, online awareness, and instant messaging. See Appendix E for details on setting up the WebSphere Portal for collaboration.

Lotus Workplace

In the Lotus Workplace platform, IBM is delivering components within a standards-based J2EE platform. Here we discuss the main elements of the Lotus Workplace strategy. By taking from different products the best-of-breed components (see Figure 5-3), partners, administrators, and users can assemble these componentized capabilities into horizontal functions called Collaborative Centers (e.g., a team-room or a task or process-specific messaging place). All these components can also be aggregated using the WP technology and made immediately accessible to a wide spectrum of devices such as browsers, rich clients, and pervasive devices.

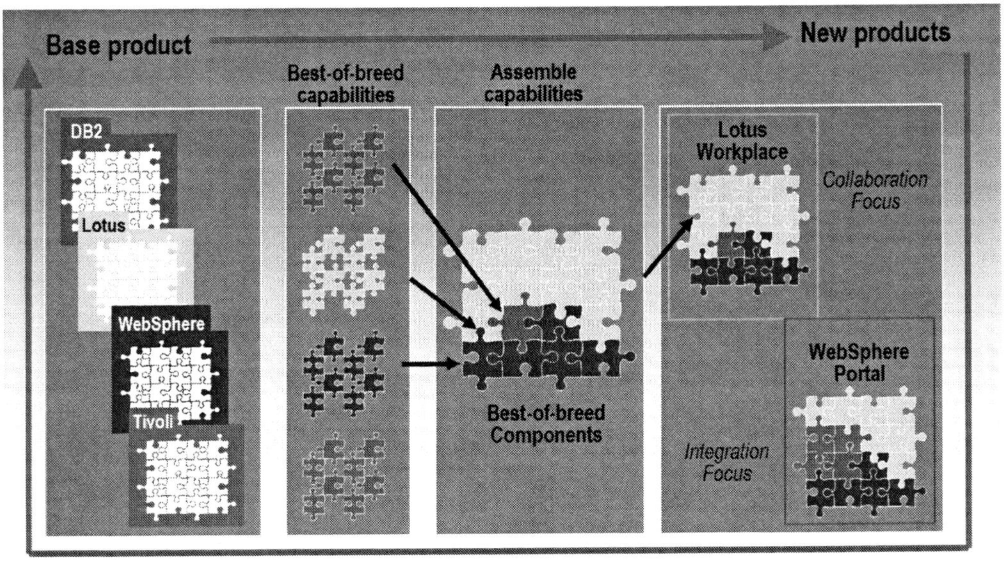

Figure 5-3 IBM/Lotus Workplace Strategy uses "best-of-breed" components.

To illustrate this process, let's take the example of the calendar functionality currently available in the Lotus Product portfolio. This calendar service facilitates creation by a developer of a calendar picker that can be offered as a collaborative service. By aggregating more of these collaborative services together using portal technology, developers can build horizontal Collaborative Centers, or, if differently aggregated, task-specific Collaborative Places such as messaging, learning, or custom-designed spaces. In the future, IBM-provided application templates will facilitate such service aggregations.

These Collaborative Places are the environments where people dynamically interact with integrated business processes, peers, partners, suppliers, and customers. They simplify access to people, processes, and information and provide end users a perspective they need to accomplish their tasks based on the context of their specific role. In other words, a "place" is nothing more than an internal model that represents a user's work. It holds projects and tasks the user works on, tools the user works with, and means of connecting easily to other people. So for an end user from the HR department, for example, a place could contain various collaboration services, such as Web conferencing, a calendar picker, personalized access to Domino.doc, Instant Messaging (Sametime) buddy list, some Domino database information, SAP fetched records, and a messaging component. The Workplace Platform, by using the portal framework in this way, provides a consistent user experience independent of whether it is a Notes 6.5 client, a Web browser, a rich J2EE client, or a mobile device. See Appendix F for details on setting up the Lotus Workplace.

Continuing WebSphere Portal Enhancements for Collaboration

Announced in November, 2003, Lotus has a series of already-defined Lotus Workplace solutions, such as Lotus Workplace Messaging, Lotus Workplace Content Management, Lotus Workplace for Retail Operations, and Lotus Workplace for Business Controls and Reporting. IBM is renaming Notes/Domino 6.5.1 and the upcoming Notes/Domino 7 to Lotus Workplace 6.5.1 and Lotus Workplace 7, respectively, to show that the company's commitment to the Lotus product line is stronger than ever.

A good example is Lotus Workplace Messaging. This Workplace delivers a messaging solution that lets organizations extend enterprise e-mail services to users currently without a dedicated workspace or e-mail access, as shown in Figure 5-4.

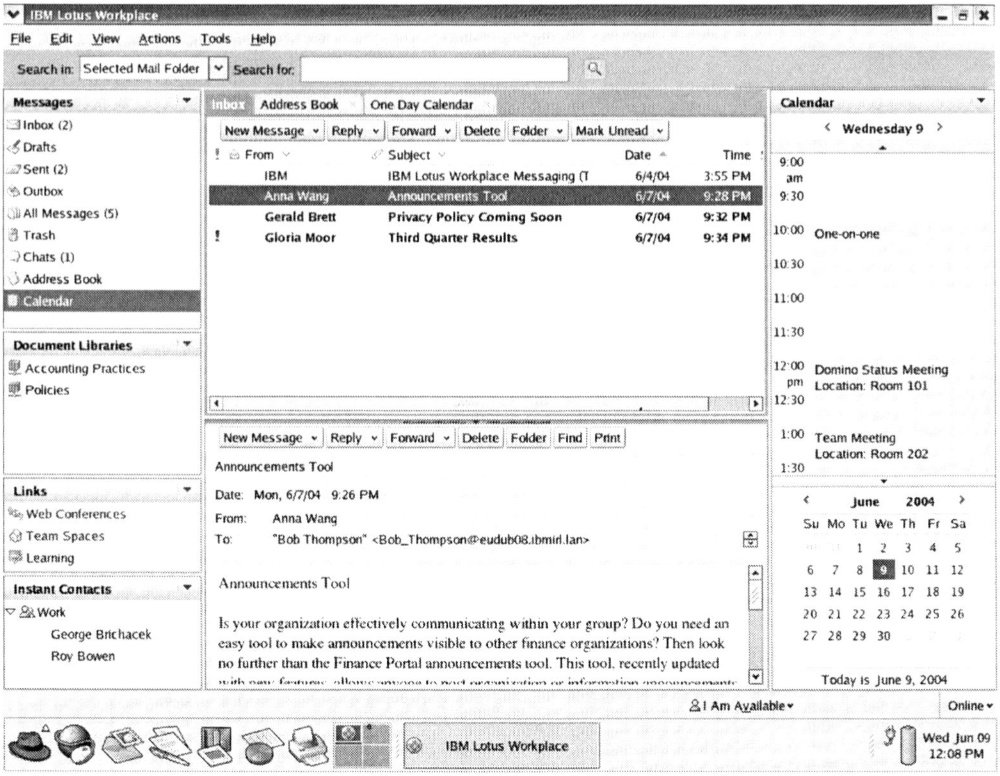

Figure 5-4 Lotus Workplace Message can provide workers with a tailored mix of features.

The ongoing integration of Lotus products and components using the WP framework and movement of Lotus toward WebSphere's J2EE environment isn't going to stop. As the Lotus and WebSphere worlds become more blended over time, the need to start blending skill sets is growing. This means WebSphere professionals getting better acquainted with Lotus technology and Lotus professionals learning about the J2EE world of WebSphere. This is necessary not only to enable integration of various components in infrastructures, but also from a development perspective, because Domino's rapid application development tools are becoming more visible in the development environment of WebSphere Studio.

The convergence of WP solutions with the Lotus brand pillar demonstrates IBM's focus on the importance of Lotus's products within not only WebSphere but also the entire IBM Software Group Portfolio. The recent and upcoming releases of Lotus Workplaces confirms this commitment to make the three C's of Lotus available in the J2EE portfolio of WebSphere products.

Moving Domino Data and Applications to the WebSphere Portal

One key to creating a successful, highly utilized portal is to provide end users with custom views into the applications and data they need to do their jobs effectively. However, creating these custom views, or portlets, can be time-consuming, requiring developers to write Java code that connects the various data sources, business processes, and existing systems with the portal vendor's specialized interfaces.

This problem is especially pronounced when it comes to leveraging Domino applications within a portal framework. Companies can have hundreds of custom Domino applications containing highly valuable data and logic but not have the Java or J2EE expertise to write the code needed to present these applications in a portal.

For example, suppose a company wants to utilize an existing Domino expense report management application from within their portal and enable portal end users to view, create, edit, and delete expense documents. Suppose this company also wants to be able to reuse, rather than rewrite, the existing agents within the Domino application.

For even the most seasoned J2EE developer, creating a portlet to meet these requirements would be time-consuming and code-intensive. The developer would first need to learn the Domino and Portal APIs and then write the following code (at a bare minimum):

- Java class to handle events
- Bean(s) to handle passing of data from the portlet to the JSP for rendering
- Java code for retrieving Domino Views and Documents, updating the Domino Documents, and remotely invoking the agents
- Java code for obtaining, releasing, and pooling of Domino Session objects
- Java code to handle authentication
- JSP for the portlet's user interface
- Deployment descriptors (web.xml and portlet.xml)

This doesn't include enabling portal administrators or end users to customize or personalize this portlet. That would require the J2EE developer to write additional JSPs, create encoded URIs, and develop the code that listens for and handles the various portlet modes (e.g., view, edit, or configure).

To reduce the effort and time needed to create and customize Domino portlets, developers can use different third-party graphical tools that shield them from needing to learn or implement the required APIs. Some of these tools also automate the creation of the Java code by abstracting development with wizard-like UIs. An example tool is the Bowstreet Portlet Factory for WebSphere. This type of tool can simplify the portlet development process, making it easy for even novice developers to rapidly create, customize, deploy, and maintain Domino portlets. By leveraging a tool such as the Portlet Factory, Domino developers can rapidly create standard, J2EE-compliant portlets that that display Domino Views and let end users view, create, edit, and delete Domino Documents. And because tools such as Portlet Factory offers a graphical environment, Domino developers can quickly learn to build robust J2EE portlets within a very short amount of time. The next section provides an overview of the Portlet Factory tool and describes how to use this product to create and customize the expenses portlet described here. The idea is to describe how tools such as Portlet Factory can be used to simplify the building of WebSphere portlets based on your existing Domino data and Domino applications.

Architecture and Concepts

Bowstreet Portlet Factory for WebSphere is a tool and framework for rapidly creating custom portlets for the WebSphere Portal (WP) environment. With Portlet Factory, developers build portlets by snapping together a sequence of highly adaptive software components called Builders. Builders generate the code (including Java, JSP, and XML) for specific application functionality. Portlet Factory includes a set of Builders that automate the creation of code for page controls, such as buttons and links. Portlet Factory also includes Builders for orchestrating page and process flows and for integrating with common back-end systems, such as Domino, SAP, or any JDBC-compliant database.

To create portlets using Bowstreet Portlet Factory, developers specify inputs to the wizard-like Builder interfaces, rather than entering code. Once a Builder is populated and saved, the Portlet Factory regeneration engine dynamically generates the appropriate J2EE code, which the application server can then execute. In this way, developers can capture and automate the process of building dynamic portlets instead of explicitly coding each portlet. Also developers can easily iterate during development by simply going back into a Builder and changing the input values.

The Bowstreet Portlet Factory snaps into the WebSphere Environment:

- **IDE.** The Portlet Factory developer tool, called the Designer, plugs right into the WebSphere Studio family of products as well as the open-source Eclipse platform.
- **Application Server.** The Portlet Factory regeneration engine is installed as an enterprise application (EAR file) on WebSphere Application Server.

- **Portal Server.** Applications created using the Bowstreet Designer integrate with WebSphere Portal via the Bowstreet Portlet Adapter Web Archive File (WAR). This is a standard WAR that includes the Portlet Factory classes (JAR files).

Creating a Portlet That Enables Viewing and Editing

To build the expenses portlet described in the preceding section , this section describes the use of the Domino View & Form Builder. To create the Domino portlet, follow these steps:

1. Create a new model from the Designer (a model is the Portlet Factory name for the container that holds all the Builders), by selecting File>New>Bowstreet Model in the Bowstreet Perspective within WebSphere Studio.
2. Select the Domino View & Form Builder from the Builder Palette.
3. Supply the Builder with the proper inputs, such as Domino Host Server, Database name (*.nsf file), View name, Form name, and so forth, and save it.

It's a relatively straightforward task to populate the Builders' inputs because they have numerous pickers and check boxes that help simplify development. For example:

- Once a developer enters a Domino host server, the Builder retains this value in the drop-down.
- Developers can select from pickers the appropriate Domino database and elements, including Views, Agents, and Forms.
- Using the run-time credentials select list, developers can specify the mechanism used to capture and utilize end user credentials (i.e., user name and password). As such, developers can quickly enable single sign-on by leveraging LTPA tokens or the WebSphere Credential Vault. Alternatively, developers can have the Builder generate the code that prompts end users for their login information and then authenticates the users to Domino.
- Developers can add paging to the portlet by checking the "Paged Data Display" box and specifying the rows to display per page. This feature is useful for Domino views that contain a large number of documents. Thus, rather than displaying all documents on the first page of the portlet, the portlet will display only the specified number of rows per page. The portlet will also include paging buttons that help end users traverse through the documents.
- To create a portlet that enables users to filter their views by category, developers simply need to check the "Show category selector" box. This automatically adds a select list to the portlet that contains the categories associated with the selected Domino View.

- Developers can build portlets that enable end users to view, create, edit, and delete documents by simply checking the appropriate box in the Builder (e.g., "Enable document create").
- Developers can also browse for HTML files and stylesheets.

After populating the desired inputs, you can save the Builder and examine the code that it created. The Builder generates a schema (view_schema) that represents the structure of the selected Domino view. It also generates a schema to describe the Document (view_documentSchema), as well as the JSP and supporting code for the retrieving and formatting of the View and Document data.

This Builder creates all of the code needed to support full-text search, paging through views, single sign-on, handling of Domino Session objects, caching of the View data results, invoking of agents, and creating, editing, and deleting of Domino documents. Furthermore, much of this code is also wrapped within helper methods (e.g., getViewData, sendDocument, updateDocument, createDocument, runAgent, and setComputeWithFormEnabled). Developers can call (pick) these methods from any Builder.

Developers can then run and test the model right from the Designer window. Also without requiring any additional modifications, developers can run the model as a portlet within WebSphere Portal. Developers simply need to make a copy of the Bowstreet Portlet Adapter and point it at the appropriate model. As mentioned earlier, the Bowstreet Portlet Adapter is nothing more than a WAR file that gets uploaded to the WebSphere Portal server once during the installation of the Portlet Factory.

Changing Portlet Functionality

It's important to note that Builders aren't simply "one-time use" wizards. At any time, developers can go back into any Builder and make a change and the code automatically updates. For example, suppose that after filling out and saving the Domino View & Form Builder, the developer wants to use a different view and also add category searching. The developer would simply need to open up the Builder, pick a different view, check off the "show category selector" box and click OK. The Builder then automatically regenerates the proper code.

Furthermore, if the developer wants to add additional functionality to the portlet, all that's needed is to add new Builders to the model. For example, using a single Builder can add sorting to any column. Adding another Builder can modify the naming and order of columns. In fact, any aspect of the portlet behavior can be modified or enhanced by adding additional Builders to the model, including portlet-to-portlet communication, look and feel, and application flow. In short, you don't have to settle for choices someone else has made for you.

Serving Multiple Constituencies

There is usually quite a bit of extra work required to create customized portlets (i.e., portlets that can be configured by portal administrators or personalized by end users). By using a tool such as Portlet Factory, developers can repurpose a single code base to serve multiple constituencies without requiring additional code changes or redeployment of files. In other words, developers don't need to write any additional code to handle new variations; they get the variations for free.

Bowstreet's profiling functionality makes this possible. Profiling makes it easy to pass different inputs into the Builders to generate new versions of a portlet. Essentially, a profile is a level of abstraction above Builders. The inputs to the Builders in a model can be captured within profiles. When a developer applies a profile to a model, Bowstreet dynamically generates a new version of the portlet application. This helps developers generate different behaviors from a single portlet—for example to enable application customization, user entitlement, or personalization—by simply creating a new profile for a portlet application.

We could create a Manager profile for our expenses model that would run the "approve" agent when a manager edits and saves an expense report. We would simply need to profile the "Agent to run on Save" Builder input, passing the name of the approve agent when the profile is Manager and leaving the input null when the profile is default. Similarly, we could profile the delete functionality, enabling delete when the profile is Manager and disabling delete when the profile is default.

Developers can expose profiles to a portal in several different ways. For example, applications can automatically select profiles based on portal user login information. So if someone belonging to the Manager group logs in, the application can use the Manager profile to apply to the expenses portlet. Profiles can also be exposed in a portlet's Edit or Configure modes.

With tools such as the Portlet Factory, the process of reusing and repurposing the same model for different constituencies without having to write new code can be significantly simplified. As a result, developers no longer have to deal with the maintenance of keeping up multiple versions of a portlet or the problem of maintaining complex patterns of conditional logic.

CHAPTER 6

Building a Combined WebSphere and Domino System

"WebSphere Application Server is available for AIX, Solaris, HP-UX, Windows NT, and Linux (Red Hat and SuSE)."

—IBM U.S. Customer Announcement Letter, March 2002

"Lotus Notes is truly cross platform."

—Forrester, January 1997

In this chapter, we describe various ways to build and combine the WebSphere Application Server (WAS) and the Lotus Domino Collaboration Server (Domino) products to present, from the application user's viewpoint, a single system image. We first discuss options for setting up WAS and Domino server systems. In particular, we present common ways WAS and Domino server systems are placed in a multiple tiered network configuration behind the Internet facing firewall. We then consider installation planning, including hardware and operating system requirements. Finally, we discuss product installation and migration issues.

Throughout the chapter, we refer to WAS and Domino features in the current product versions as of this writing, namely, WAS V5 and Domino R6.5. As a reference aid, we present the details for the installation steps for WAS V5 and Domino R6.5 on different computers in Appendix C, "Detailed Steps for Building a Combined WebSphere and Domino System."

Options for a Combined Server Configuration

There are various options for how to build the combined WAS and Domino system. These options differ mainly in what functions the WAS and Domino servers take on and where they are placed in the usual tiered network configuration. A common network configuration for providing a Web application to Internet users consists of three tiers (or layers or zones), each separated by firewall systems. The Internet facing tier is often referred to as the "demilitarized zone," or DMZ, and usually consists of load balancing or authentication systems. Web and application servers sit in a zone behind the DMZ, and data servers in a "data layer" behind that. The multi-tiered configuration provides increasing security protection from the DMZ through the data layer. The server systems are placed within the tiers according to security and inter-system performance considerations.

Depending on the roles they play, WAS and Domino usually sit in either of the two network tiers behind the DMZ. As a Web application server (for servlets, JSPs, or Web services), WAS is often located in the application zone. When WAS serves primarily as an EJB container, it is placed in the data layer, so as to be close to (or co-resident with) the database server. Domino used as an HTTP server can be placed in the application zone (or even DMZ if necessary). However, it is most common to find Domino in the data layer being used as a data or document repository or collaboration server.

In a production application environment, it is most often the case that WAS and Domino are installed on separate computers. This is done simply to allow the WAS and Domino systems to be placed in separate network tiers. It is also done for scalability and performance reasons and to make system management easier. Although with more sophisticated hardware platforms, it is possible to have all system components installed on one physical system, for example using hardware partitioning to create separate server images on a single system.

In some cases, such as for development or testing, it is convenient to install Domino and WAS on a single computer. The steps for installing Domino and WebSphere on the same computer are basically the same as for separate computers, although there are some simplifications in the install procedure when both applications share the same hardware platform. Those simplifications are described in the configuration and setup details given later in this chapter and in Appendix C.

WAS and Domino Installation Planning

Hardware/Operating System Requirements

Both Domino and WebSphere run on a variety of hardware and software platforms. In particular, Domino runs on Windows 2000/2003, AIX, Solaris, HP-UX, Linux, Netware, OS/2, OS/400 (IBM iSeries), and z/OS (IBM zSeries), while WebSphere runs on Windows 2000/2003, AIX, Solaris, HP-UX, and Linux for Intel and IBM iSeries, pSeries, and zSeries processors.

Linux is becoming an operating system to seriously consider for both WAS and Domino. Linux on Intel or AMD servers is popular for low to mid-range servers (up to eight processors) because of generally lower cost. Moreover, Linux spans the widest range of server types, from blade server to enterprise mainframe so that common system administration resources can be applied. It is important to note that as a complete operating system, Linux is provided in various combinations of specific component (especially kernel and library) versions, referred to as "distributions." For both WAS and Domino, IBM provides support for only certain Linux distributions, namely RedHat, SuSE, and distributions conforming to the United Linux 1.0 standard. It is likely, however, that WAS and Domino will run without problem on Linux distributions similar to these, such as Mandrake Linux, which is similar to RedHat.

The basic hardware requirements are what you would expect for large, sophisticated system applications. System memory size is the most important hardware requirement. Both WAS and Domino specify a recommended amount of 512MB, but for best performance you should provide the most memory you can afford. Processor speed is recommended to be at least equivalent to a 500MHz Intel Pentium® processor but again should be as high as affordable.

Disk space requirements vary by the specific operating system platform but are in the range of 600-800MB for WAS and 500MB for Domino. The disk drives and interfaces themselves should be high-performance, as usually found in most server systems. Careful consideration should be given to the use of external storage systems such as storage area networks or network file systems as repositories for WAS or Domino data because they may degrade application performance.

A set of hardware requirements often overlooked in the design of WAS and Domino systems is that related to the network hardware. The speed and bandwidth capacity of the network routers and firewalls should be sufficient so as not to introduce delays in the data paths to or between the WAS and Domino system components.

WAS and Domino Product Coexistence

Domino and WAS have long had product features that allow them to work together, but in order for the features to work properly, the correct versions of Domino and WAS are required. Sometimes the versions must be at a specific service update level. For example, Single Sign-On (SSO) is one very important feature for applications built on Domino and WAS. (The SSO feature lets Domino and WebSphere recognize when either server has already authenticated a user, as is described in detail in Chapter 12, "Security and Single Sign-On.") The first version for which SSO could be enabled for Domino was R5.0.5. The first version of WAS (Advanced Edition) for which SSO could be enabled was version 3.5 with service update (fix pack) number 1, or version 3.5.1. The SSO feature of iSeries was first available with Domino for AS/400 5.0.6a and WAS 3.5. Servlet access to Domino databases was first available on iSeries with Domino for AS/400 5.0.4 and WAS 3.0.2. The ability to use Domino HTTP Server to access WAS resources was first available on iSeries with Domino for AS/400 5.0.5 and WebSphere 3.5.1.

In order to have Domino and WAS work together, the requirements are as follows:

- A minimum Domino level of R5.0.5
- A minimum WebSphere level of version 3.5.1

But this is somewhat ancient history now. Significant new releases of both Domino and WAS were made available at the end of 2002 with the Domino R6 and WAS V5 products and continue on almost a yearly basis.

HTTP Server Considerations

Another area of interoperability between WAS and Domino is the HTTP server plug-in, which enables J2EE servlet and JSP requests to be forwarded to WAS for processing. WAS provides various plug-in modules for different HTTP server products such as Apache, IBM HTTP Server, Microsoft IIS, and Domino's integrated HTTP server. In the case of the Domino HTTP server, the WAS plug-in is provided as part of the Domino product package. Various plug-in modules are provided corresponding to the different WAS versions and operating systems. Table 6-1 lists the set of plug-in modules provided by WAS V5.

Table 6-1 Plug-in Modules Provided by WAS V5

System	Web Server	Plug-in Executable File
Windows	IBM HTTP Server	mod_ibm_app_server_http.dll
	Lotus Domino	libdomino5_http.dll
	Apache	mod_app_server_http.dll
	iPlanet (Netscape)	libns41_http.dll
	Microsoft IIS	iisWASPlugin_http.dll
IBM AIX	IBM HTTP Server	mod_ibm_app_server_http.so
	Lotus Domino	libdomino5_http.a
	Apache	mod_app_server_http.so
	iPlanet (Netscape)	libns41_http.so
	HP-UX IBM HTTP	Server mod_ibm_app_server_http.sl
	Lotus Domino	libdomino5_ http.sl
	Apache	mod_app_server_http.sl
	iPlanet (Netscape)	libns41_http.sl
Solaris	IBM HTTP Server	mod_ibm_app_server_http.so

System	Web Server	Plug-in Executable File
	Lotus Domino	libdomino5_http.so libdomino6_http.so
	Apache	mod_app_server_http.so
	iPlanet (Netscape)	libns41_http.so
Linux	IBM HTTP Server	mod_ibm_app_server_http.so
	Apache	mod_app_server_http.so

Note that there is no WAS plug-in provided for the Domino HTTP server on the Linux platform. Most of these modules are also shipped with the Domino product (except for the HP-UX modules).

With the recent versions of WAS, the communication mechanism between the WAS HTTP server plug-in and WAS itself has been changed from a proprietary mechanism to HTTP. This change has greatly simplified the configuration and setup of WAS and the plug-in. The use of HTTP makes it easier to configure the network layer, especially firewall systems, to support the plug-in connection to WAS. The implication that WAS now provides its own HTTP server is true. Although the WAS HTTP server is far from a full-fledged HTTP server, it is sufficient to provide a quick and easy means for accessing WAS without having to set up a separate HTTP server. As we will see later, the built-in WAS HTTP server is also used to provide a Web administration interface instead of a separate administration process and corresponding client.

Given that the Domino HTTP server is the only HTTP server that can provide Web access to Domino databases (like a client's mail or address book database), the situation arises where you need to provide both a "standard" HTTP server and the Domino HTTP server. In this case, you can simply configure the Domino HTTP at a different port or (virtual) host from the other HTTP server. Then the application can supply URLs for Domino resources specifying this alternate port or hostname. Another approach available on Windows systems is to use Microsoft IIS as the HTTP server and configure it to use a plug-in that redirects Domino requests. This IIS Domino plug-in can be used in addition to the IIS WAS plug-in.

For certain IBM platforms, namely the iSeries (OS/400) and zSeries (z/OS), it is difficult to run multiple HTTP servers because the HTTP server is essentially built into the operating system. In these cases, there are also HTTP plug-in modules available for Domino on the iSeries platform (refer to the IBM Redbook, "Domino and WebSphere Integration on iSeries," for details) and the zSeries, similar to the Domino plug-in for IIS. These Domino plug-ins can be run in combination with the WAS plug-ins on these platforms.

See the section on configuring the Domino HTTP server plug-in later in this chapter for details on how the HTTP plug-in can be configured to work with WAS.

Networking Considerations

Prior to installing WAS and Domino, it is a good practice to establish your network naming conventions. For most applications, you will want to use fully qualified host names for the WAS and Domino servers, that is, names consisting of a TCP/IP host name and domain name. If you don't have a domain name system (DNS) server set up, it is possible to configure each server operating system with the host/domain name to IP address mappings (e.g., via the /etc/hosts file on both Unix and Windows systems). Certain features in WAS and Domino require fully-qualified host names to work, especially SSO.

If Domino servers don't already exist in your network, you'll want to establish a Domino naming convention for the servers. Here you'll need to pick a common Domino domain name and server names for the planned Domino servers. For those new to Domino, the Domino domain name is altogether different than a DNS domain name, although conceptually they are quite similar. The Domino servers will need to be given fully qualified TCP/IP host names in addition to their Domino server names.

If you plan to use SSO across your WAS and Domino servers, you will need to put all of the WAS and Domino servers in a single DNS domain and all of the Domino servers in a single Domino domain.

One area of network planning often overlooked until deployment is that of firewall rule configuration. As mentioned previously, WAS and Domino servers are usually deployed to a tiered network where the tiers are separated by firewall systems. As a result, a Web application can easily span every tier, and the communication between the components running on WAS and Domino will need to be permitted to pass through the firewalls. Since the firewall systems control TCP/IP sessions by their source and destination host addresses, and ports and TCP/IP protocols are usually associated with specific ports or port ranges, you should understand which protocol types must be allowed to pass to, from, or between WAS and Domino. Besides HTTP, WAS and Domino servers may make use of several other Internet (and proprietary) protocols, such as LDAP, SMTP, DIIOP, SSL, etc. TCP/IP protocols used by other components such as database servers or chat servers must be taken into account as well.

Domino Server Configuration and Set-Up Considerations

The Lotus Domino R6 Server family consists of the Domino Mail Server, the Domino Application Server, and the Domino Enterprise Server. Most likely you will want to do more with Domino than just use its HTTP server to integrate with WebSphere, so you will want to install either the Application or the Enterprise server.

The Domino server supports various features, especially Internet protocols, which may be useful for your planned applications. These features and protocols include the following:

- **Access to the Domino directory via the Lightweight Directory Access Protocol (LDAP).** LDAP access may be used with SSO. (We discuss SSO in detail in Chapter 12.)
- **Simple Mail Transfer Protocol (SMTP).** Web applications could make use of SMTP, especially via the JavaMail API, for sending e-mail.

- **Post Office Protocol (POP).** An e-mail client-to-server protocol.
- **Internet Mail Access Protocol (IMAP).** Another e-mail client-to-server protocol.
- **Domino Internet Inter-Object Protocol (DIIOP).** DIIOP is a communication mechanism for invoking Domino functions in a client-server model. It is required if J2EE components running on a WAS server make use of the Domino Java API to access remote Domino servers.

Depending on your application requirements, you may want to enable or disable a specific set of Domino features or protocols.

WAS Configuration and Set-Up Considerations

The deployment options for a WAS installation are many. With just the WAS Advanced Server edition, it is possible to set up multiple application server instances on a single system. This can be done even among different WAS versions. Or it can be done where the instances are at the same level and share a common runtime. With the Network Deployment edition, WAS can be installed and managed across multiple systems. A network deployment consists of "cells" containing "nodes" containing application server instances. The WAS deployment options are flexible enough to accommodate almost any installation requirements. You can plan a deployment to suit your needs—be they performance, capacity, or compatibility with previous deployments. We do not cover the deployment options in detail in this book and restrict our discussion to a single-server deployment.

The WAS administrative interface allows an administrator to manage WAS nodes, cells, and servers, as well as the applications running in them. It provides the means to monitor application server activities, manage security, and adjust the configuration. With WAS V5, the administrative interface is now a Web interface, the administration application is a Web application running on the application server, and the configuration data is maintained as a set of XML files on the server's filesystem. It is accessed directly on the server by specifying the server's host name and the port for the administrative application. Note that with the use of XML files to store configuration data, WAS is no longer dependent on a database system such as IBM's DB2. WAS requires a database system only if applications require it, such as for EJB entity beans.

For SSO support, there are a number of considerations involving the set up of the user registry used for user authentication. Please refer to Chapter 12 for the details.

General Installation Procedure for WAS and Domino

This section describes in general steps the installation of WAS and Domino on separate systems. The steps presented here will give you an overall idea of how to build each server. For a detailed description of the installation for WAS 5.0 and Domino 6.5, see Appendix C. The steps are as follows:

1. Check for necessary hardware and software prerequisites.
2. Verify the network configuration.

Working with the WAS system:

3. Log on as a user with administration privileges (e.g., as "root" for Unix systems).
4. If a database is required, install the database software.
5. Install WebSphere via the install utility (Java GUI).
6. Install the latest service update (fixpack) and any relevant e-fixes.
7. Start WAS, and verify that WAS resources are accessible. The "snoop" servlet sample can be used for this.

Working with the Domino computer:

8. If not available on any system, install a Notes Administration client.
9. Install the Domino server.
10. Using the Notes Administration client, configure the Domino server, and populate the directory.
11. If the Domino HTTP server will be used with WAS, ensure the plug-in module and configuration file are available to the server, and configure Domino to use the plug-in via the Administration client. Modify the plug-in configuration file if necessary.
12. Verify Domino resources can be accessed. If the plug-in was installed, verify that WAS resources can be accessed (e.g., the "snoop" servlet).

There is nothing in the installation procedure that prevents WAS and Domino from being installed on the same computer.

Hardware and Software Prerequisite Details

In this section, we list the hardware and software requirements for the computer you want to install on, as well as the different product software levels required.

Platform Hardware:

- Pentium III or higher, with an absolute minimum of 256MB, but 512MB is recommended for WebSphere 3.5. We used 512MB for the computer on which we installed WebSphere and DB2 and 385MB for the computer on which we installed Domino for our testing, with satisfactory results.
- At least 600MB free on the drive or drives used to install the products. The disk space requirements of the products after installation are DB—475MB, Domino—300MB, IBM HTTP Server—20MB, and WebSphere—220MB.

The DB2 space requirements can be reduced by 125MB if you choose to install the DB2 Administration client on a separate machine. The space requirement for Domino also can be reduced to a certain extent by choosing to install fewer components. For example, it is possible to not install the help files. These files alone require 50 MB. Similarly, the space for WebSphere specified above includes documentation files of about 60MB, which would not be needed in a production installation.

Platform Software:

- Any of the operating system software supported by WAS and Domino. For Microsoft Windows, either workstation or server code can be used. TCP/IP networking with a fixed IP address for each machine.

Product Software Levels

The product software levels we used were as follows:

- DB2 Universal Database Version 8.1 Enterprise Edition plus Fix Pack 1
- Domino 6.5
- WebSphere Application Server V5

Higher versions of the products should also work. The highest currently available product level should be used except where specifically stated otherwise. Care should be taken, however, to check WebSphere for recent e-fixes.

Creating a User with Administration Rights

WAS, Domino, and DB2 must run under the permissions of a user or as system services. For testing purposes, it is more flexible to use a user ID with rights to run as an extension of the operating system than load the products as system services (however, WAS and DB2 must run as system services). For a production system, these products should run as services so that they will automatically load when the system is started without operator sign-on.

Installing WebSphere Application Server V5

You can either install WebSphere from a product CD or one large installation file. If you use a product CD, the installation program should start automatically when you insert the CD. If not, you have to run the setup.exe program on the CD. If you have one big file, simply start the installation by running the file. It will automatically unpack itself and start the installation program. Note that the installation process itself requires 70MB or more free in the system temporary directory, even if installation is to another drive or volume. Step by step details are given in Appendix C.

Installing and Configuring Domino 6

We now turn our attention to our Domino computer and the installation of Lotus Domino.

The Lotus Domino R6 Server family consists of Domino Mail Server, Domino Application Server and Domino Enterprise Server. If you want to do more with Domino than just use its HTTP stack to integrate with WebSphere, you should install the application or the enterprise server.

To use the Domino HTTP stack and enable Single Sign-On, WebSphere V5 requires Domino Server R5.0.5 or higher. Step by step details for the installation of Domino Enterprise Server R6.5 are given in Appendix C.

You need to install the Domino administration client on your server or another workstation (Lotus recommends using a separate workstation for administration). This will allow you to change the server's settings easily, especially those in the Domino directory; although much, but not all of the testing we describe can be managed by direct access to the server's text console.

We will not describe the installation of Domino Administrator in detail. Be sure that the **Administration Client** is selected for installation; you can accept all other default values during the installation. One of the first things you can do after installing Domino Administrator is allow your users to run Java programs on the Domino server.

Configuring Domino to Use the WebSphere Plug-In

The plug-in itself is provided as a shared library (DLL for Windows, .a for AIX, .so for Solaris, and as of this writing no plug-in exists for Domino on Linux) and a default configuration file: plugin-cfg.xml. The plug-in files are packaged with both the Domino and WAS product. The plug-in files can be found under the Domino install path: ./Domino/Data/domino/plugins/was5.

There are two essential steps to installing the plug-in: 1) specifying the plug-in library file to the Domino HTTP server and 2) specifying the location of the plug-in configuration file to the plug-in itself. The first step is straightforward and simply requires a change to the Domino server document in the Domino Directory database (names.nsf). You specify the plug-in library file as a DSAPI filter file on the Internet Protocols/HTTP page of the Server document.

The second step is not as straightforward and is operating system-dependent. The plug-in library itself looks for the plug-in configuration file via an operating system environment variable (registry key on Windows). Unfortunately, these variable names are not well advertised in the WebSphere or Domino documentation. For IBM iSeries systems, the configuration file location is specified in the Domino notes.ini file. See Chapter 10 and Appendix C for the details on how to specify the plug-in file location and the plug-in configuration settings themselves.

Once these settings have been made, you can restart the Domino HTTP server task to have it load the WebSphere plug-in as a DSAPI filter. If the settings are correct, you will see the following messages at the Domino console:

```
HTTP Server: Java Virtual Machine loaded
WebSphere HTTP DSAPI filter loaded
HTTP Server: DSAPI WebSphere HTTP DSAPI Filter Loaded successfully
HTTP Server: Started
```

Migrating to Domino and WebSphere from Earlier Versions

In this section, we discuss the issues around migrating from earlier versions of Domino and WebSphere products. Since both products are in their fifth or greater major release, many customers will be moving to the latest versions from earlier installations.

Migrating to Domino 6

Before you integrate Domino with your WebSphere server, it would first be a good idea to update your Domino server to the latest version (that would be Domino 6.5 when this section was written). This section gives some useful information on how to go about upgrading your version of Domino.

Version 6 of Lotus Notes and Domino from IBM was a significant upgrade and should be the basis for your Domino platform. If you have been using an earlier version of Domino, this section describes the steps for migrating Domino R5 servers to Domino 6. Here are the benefits and interoperability issues of upgrading to Domino 6.

Interoperability Issues

When planning the type of migration to undertake, you must decide whether to go for the "big-bang" approach in which you migrate all servers in parallel (typically over a weekend or holiday) or whether to stick to a more practical (and more often chosen) alternative of phasing in the migration. The latter approach involves running and maintaining a mixed environment (R5 and Domino 6 servers) for a period of time.

The best practice is to first migrate the servers to Domino 6 and then migrate the clients over time. Using this approach lets the servers stabilize before you migrate the clients (this method is low risk, proven, and recommended in most scenarios for practical reasons).

The Domino 6 Administration client can administer a mixed release of servers; however, some limitations exist. The Domino 6 release notes and Administration Help database (`http://doc.notes.net/domino_notes/Rnext/help6_admin.nsf`) define these restrictions in detail.

Migration Preparation Steps

The following steps are essential in the preparation process. Skipping them will result in a failed migration or will significantly increase the risk of a migration that appears successful but has hidden issues that users will eventually discover.

1. **Understand Domino 6 and its enhancements.** You'll probably need to undergo this exercise to define the business case for migration, but it's important that everyone in the administration and migration teams are aware of the benefits migration provides.

It's recommended that you install the Domino 6 Administration and Notes clients on your administrator machines long before you undertake an actual upgrade. Be careful not to conduct any design changes in the R4 or R5 databases, using the Domino 6 Designer client until a complete migration has been successful (otherwise, you may encounter potential problems and even corruption of the design components).

2. **Carefully plan the schedule of the upgrade.** The migration of servers will almost certainly occur during evening hours or over a weekend. Make sure you have sufficient time to complete all tasks and that you can put precautions in place should issues arise. Thoroughly understanding the upgrade process and remembering to not attempt too much at once are essential for a successful plan.

3. **Conduct a full system and application audit.** This audit should cover both the servers and the applications that run on the servers. Typically, you should document the results in a table or Notes database that can be shared by or published to all stakeholders (e.g., administrators, designers, and application owners) for sign-off.

   For each server, you should capture the following information: server name, IP address, network ports enabled, Domino version, OS level and service packs, purpose (e.g., mail, app), user types (e.g., browser, Notes client), physical location, type (e.g., hub, spoke), and the servers that act as the administration server and directory server.

   It's also important to work with key people in the administration team and user community and document any existing issues with the servers (that way, problems cannot be blamed on or attributed to the upgrade.

   For each application, you should capture the following information: application name, owner, purpose, user community, method of access (e.g., browser, Notes client), template used, level of changes (if any) to the standard templates, code duplicated from templates, use of undocumented or unsupported coding features (these could change with Domino 6), whether the app is a mail-in database, all creative workarounds, all accessible Notes databases and external databases or files, all Notes API or OLE functionality, and a list of servers this database replicates to (and the use of any selective replications).

4. **Conduct a test server migration.** Not only is setting up a test environment a good way to learn about and play with the Domino 6 functionality, but it also ensures that you conduct a trial upgrade of the server and that you test applications thoroughly before migration. After all, you want to know the product is stable in your environment and with your applications, data, policies, and procedures.

   By setting up a test environment that closely reflects that of your production environment, you'll know your upgrade procedures work, the results are as expected, and above all, you've practiced the migration before touching production systems. If test hardware is an issue, consider using the virtual server product VMWare (`http://www.vmware.com`) to create multiple virtual machines on a single server. This product is ideal for testing a multi-server migration.

5. **Test application migration.** Once you're happy with the upgrade process, you should thoroughly test key business applications (including mail) to ensure that functionality hasn't changed. It's essential that you copy the application and production data so you can test this data in isolation. By doing so, you'll know this data works in the test environment before you upgrade. Ad hoc testing by the migration team is often insufficient, so you should work with the application's business owners to draw up a structured test plan and sit with a knowledgeable user of that system as he conducts the test (with you capturing any unexpected results or variations)

   Next, you need to consider the results of the application audit conducted in Step 3. You'll have to make changes if the standard templates have been modified or if undocumented features or creative workarounds have been made in the coding. Working with the application's developers before testing will help you prepare the application for migration.

Migration Steps

The following steps define the best practices and considerations for conducting the actual migration. Note that Lotus will support a mixed-client environment with a Domino 6 server infrastructure because organizations will often upgrade the servers first (as according to best practices) and then upgrade clients over time.

1. **Upgrade the Domino Administrator client.** Be sure you upgrade the Domino Administrator clients used to administer the Domino network to the latest Domino 6 release long before you conduct the subsequent migration steps. Doing so will give you time to familiarize yourself with the Domino 6 Administration client variations. It's important you consider any interoperability issues that may exist in your environment; you can establish such issues by carefully reading the Domino 6 Administration Help database.

2. **Upgrade the Administration server and Domino directory.** Before shutting down the Administration server, purge the Administrations Request (AdminP) database by issuing the tell AdminP process all command. Once all processes have been completed and you've emptied the AdminP database, you can shut down the server and install the Domino 6 server software.

   To properly administer a Domino 6 server, you should also upgrade the Domino directory (but be sure you always migrate the Administration server first). To upgrade the Domino directory, start the server; it will immediately recognize that the Domino directory isn't running on the Domino 6 design and will prompt you to upgrade the design. Answering "Y" (yes) to this prompt lets the server upgrade the design and start as normal.

Once the server is running, the final task in upgrading the design is to upgrade the database format (known as On Disk Structure or ODS) to the Domino 6 format and rebuild all views in the Domino directory (Names.NSF). To do this, type the following two commands in the Domino 6 server console:

```
load compact
```

and once compact has been completed, type:

```
load updall -r
```

Keep in mind that this can be a time-consuming process, as its duration depends on the number of databases and the size and complexity of the views. Because it's hard to estimate the duration of an upgrade, you should leave plenty of time to ensure the conversion of all databases and the rebuilding of all views. During this process you should test client access to the server to make sure no errors occurred on the logs.

Once you've compacted all the databases and rebuilt all the views, two more steps remain: upgrading the Administration Requests database and the Web Administration client. To upgrade the Administration Requests database, open Admin4.NSF, and replace its design with the template titled "Administration Requests (Release 6)." If Admin4.NSF is replicated to other servers in the domain, then force a replication to update their design, too.

The Web Administration database (WebAdmin.NSF) upgrades automatically when you install Domino 6 because the upgrade process deletes the existing database and creates a new one based on the Domino 6 template (WebAdmin.NTF). It's important that you check the Access Control List (ACL) to ensure it has your specific entries.

3. **Replicate the Domino directory.** Once you've upgraded the Administration server, replicate the newly updated Domino directory to all servers. Next, rebuild the views of that database on each server (using `load updall -r`).
4. **Upgrade hub servers.** The next set of servers to upgrade are the hub servers. Because the Domino directory is already upgraded to Domino 6 on these servers, you won't be prompted to upgrade the Directory when these servers are started for the first time. To upgrade your hub servers, perform the following steps:

   Shut down the servers in order by entering the following commands on the console:

   ```
   tell router quit
   tell SMTP quit
   tell HTTP quit
   quit
   ```

   Complete a full backup and a test restore of each server. Be sure to take soft backups of key files—Desktop.DSK, all ID files, Log.NSF, Names.NSF, Mail.BOX, all secondary address books, .DIR files, Notes.INI, and all templates—by copying them to a temporary directory. Now install the Domino 6 software. Before restarting the server, complete the following tasks, where appropriate:

If the server runs Network News Transport Protocol (NNTP), remove this task from the Notes.INI (NNTP is no longer supported on the server but is on the client). If you didn't use the certification log in R5, then you must create one manually, as this is compulsory in Domino 6. Create a new database called Certlog.NSF from the certification log template.

Restart the Server and Check for Errors

Upgrade databases to the new ODS format by typing `load compact` on the server console. Once the compact has completed, rebuild all the database views and full-text indexes by typing `load updall -r` on the server console.

1. **Upgrade spoke servers.** The final servers to upgrade are the spoke servers (e.g., mail, applications, and Web). The process to upgrade these servers is identical to that of upgrading the hub servers.
2. **Upgrade clients.** Before users can take advantage of Domino 6 mail and application functionality, you must upgrade the client software to Domino 6. Before you install Domino 6 on the client, it's important to ensure that no antivirus software is running, that users have both read/write access to the Notes program and Data directory, and that users are logged in to the computer with administrative rights. Once you've checked all these items, you must choose how you want to upgrade the client, as Domino 6 presents several new options:

   - **Upgrade by mail.** This feature sends an e-mail to nominated users to upgrade their Notes client. Users conduct an upgrade by clicking the Install Notes button, which launches the Domino 6 client setup from a shared installation directory.
   - **IBM Lotus Notes Smart Upgrade.** This process notifies users to update their clients to the latest release and uses Policy and Settings documents to help manage the updates. When users log in to their home server (running Domino 6), Lotus Notes Smart Upgrade compares the release number of the users' client against the release number specified in the source version field in the Smart Upgrade database. Users are then prompted to upgrade their client, or they can delay the upgrade for a specified grace period. Please note that Smart Upgrade is a Domino 6 feature. You must first be using Notes and Domino 6, so only subsequent upgrades can take advantage of it.
   - **Administrative upgrade.** This process uses the Microsoft Windows Installer and copies the installation code to a shared file server that users access through the network. Users launch the installation program and then install Domino 6 locally to their machines.

Once you've upgraded the client, you can use the new Policies feature in Domino 6 (which replaces setup profiles) to configure Notes client settings for users. These settings include customized welcome pages, bookmarks, and Execution Control List (ECL) settings. To successfully implement policies, be sure you disable any previous setup profiles from R5.

3. Upgrade mail templates. You're now ready to upgrade the clients' mail templates to Domino 6 so they can leverage the full functionality of the Domino 6 mail client. At this stage, you can also upgrade the mail templates to support iNotes, if required.

   The easiest way to upgrade mail templates is to use the mail conversion utility. To run this process, you must first shut down the router on the mail server by issuing `tell router quit` on the console, and then establish which Domino 6 mail template you require (or you can select a customized one). The default is Mail6.NTF for Domino 6 mail and iNotes60.NTF for Domino 6 mail with iNotes.

   Next, Domino 6 must find all mail templates of a specific type (the source) and convert them to the destination. To do this, type `load convert mail\*.nsf stdr46mail mail6.ntf` at the server console, where stdr46mail is the source template and mail6.ntf is the Domino 6 template you've chosen. You'll find variations on the use of this command in the Domino 6 Admin Help database. Once Domino 6 has converted the mail templates, you can rebuild the databases' indexes using the `updall -r` console command.

   Finally, to re-enable the router, type `load router` at the console to ensure that mail can be routed. Now, users can start using their new mail templates. It's a good idea to test the full features of mail and calendar/scheduling before signing-off the server as working.

   The actual migration of your R5 server environment to Domino 6 isn't a complicated process—it just takes time and careful planning. The only big areas of risk are the amount of testing you need to complete and the level of customization you've made to the standard Lotus templates. You should consider these issues when preparing a schedule and be sure to allow enough time and resources for contingency and a worst-case scenario that would require a complete rollback. Once you've migrated the servers and upgraded your clients, developers can start leveraging the full functionality of the Domino 6 environment.

Migrating to WebSphere V5

Before you integrate Domino 6 with your WebSphere server, it would first be a good idea to update your WebSphere server to the latest version (that would be WAS V5.1 when this section was written). This section gives some useful information on how to go about upgrading your version of WAS.

Significant changes have occurred between WebSphere Application Server (WAS) versions 3.5 or 4.0 and 5.0. Even if we go back to WAS 3.5, it does contain a number of elements from the

J2EE specification. However, it's not a J2EE-compliant application server. WAS V5, on the other hand, implements the J2EE 1.3 specification. In this section, we'll focus on the migration of V3.5 to V5 since the migration of V4 to V5 is much simpler.

The changes brought in with J2EE introduce a number of direct and indirect issues. Many applications require some changes to code, and the application development tools have changed dramatically. With the introduction of J2EE, the way that applications are packaged is also different. Changes to packaging have an impact on the way that builds are done as well as the process and skills involved with deployment. Code and packaging, however, are only a small part of the overall migration process.

Migration Overview

Application developers tend to think of migration only in terms of the application code; administrators think only of the production runtime. It is important to think of migration in comprehensive terms. This means considering (at least):

- The development environment
- Developer skills
- Administrator skills
- Runtime environments
- Required and recommended changes to the application code

A comprehensive assessment is a key element of a successful migration. With careful assessment, you can very accurately plan your migration and put the necessary resources into place. Assessment typically takes between five and 10 days and should involve somebody familiar with WAS 5.0 and WebSphere Studio Application Developer (WSAD).

Bring concerned parties together. Communication between different groups often breaks down over time. The migration assessment is a good opportunity to bring these groups together. For example, developers can often gain a greater appreciation for the issues faced by the administration staff, and vice versa.

High-Level Architecture Review

With the gathering of concerned parties comes an excellent opportunity to describe the existing applications. The architecture review sets the stage for other parts of the assessment by bringing the "big picture" into clearer view. A high-level architecture review should not take more than a couple of hours.

Application Code Review

An obvious part of the assessment process is a review of the application code. This doesn't need to be a full-blown code review; in the context of a migration assessment, you're generally concerned only with discovering the use of non-J2EE-compliant code.

In general, code quality has no direct impact on migration complexity. Complexity is a factor only when significant redesign of an application is required or when the complexity makes it difficult to understand the code.

An application code review typically takes one to two days for each application.

Review the Development Environment

With the change to WAS V5 comes a completely new development environment in the form of WSAD. (VisualAge for Java (VAJ) is not recommended for V5 development.) As part of the assessment, you should review the current development environment to determine how you can adapt the existing practices, tools, and skills to build a new development environment around WSAD.

Assess Development Skills

Software developers familiar with WAS V3.5 may need some skills upgrades to familiarize themselves with J2EE 1.3 and WSAD.

Review the Runtime Environments

Most organizations support a number of runtime environments, each with a different purpose. The configuration of these environments (potentially including one or more development test, system test, preproduction, and production environments) should be reviewed to ensure that the existing topology, configurations, and hardware are adequate for WAS V5. A review of security requirements and implementation is also an important part of this stage of the assessment. Review of the runtime environments typically takes at least half of a day.

Review the Build and Deployment Processes

Review the existing processes to identify nuances that you may need to consider when the current approach is adapted to use J2EE 1.3 packaging and associated WAS V5 deployment. With the changes introduced with J2EE, it is very likely that many deployment-time tasks (e.g., servlet configuration) will be moved into the build process.

Review and Understand Schedule Issues

Scheduling is always an issue with migration. Most organizations have applications that are constantly changing in response to external business drivers. Working a migration plan around other deliverables can be a daunting task that requires special attention.

Development Environment

If you currently use VAJ, the development environment will likely be one of the biggest changes you make as part of your migration. VAJ 3.5 (or 4.0) is generally not recommended for development of applications targeting WAS V5; for this purpose, IBM recommends one of the WebSphere Studio family of products, such as WSAD. IBM no longer provides technical support for the VAJ products.

The WebSphere Studio family includes several products:

- WebSphere Studio Site Developer (WSSD)
- WebSphere Studio Application Developer (WSAD)
- WebSphere Studio Application Developer Integration Edition (WSAD-IE)
- WebSphere Studio Enterprise Developer (WSED)

WebSphere Studio development tools are very different from VAJ. VAJ is a relatively closed environment intended primarily for Java development. But WebSphere Studio development tools are intended for all kinds of development, including components to build and edit Java, J2EE, and Web site content (HTML, images, and other resources).

WebSphere Studio development tools gain their flexibility through a large collection of "plug-ins" that provide all functionality. The base product contains a very comprehensive set of plug-ins that provide everything from Java editing and debugging to J2EE component construction and testing (via a WebSphere Test Environment plug-in) to XML editing and manipulation (via XSLT).

WebSphere Studio is built using open-source Eclipse technology, so it can be extended with Eclipse-compatible plug-ins (WebSphere Studio 5.0 is based on Eclipse 2.0). A large open-source community has already constructed hundreds of plug-ins that provide a variety of different kinds of functionality. Chances are, if you need WebSphere Studio to do something that it doesn't do already, somebody in the open-source community is already working on a plug-in.

All the WebSphere Studio products except WSAD provide J2EE application development support with increasing levels of extended support. WSAD provides everything you need to build, test, and deploy J2EE applications. WSAD-IE extends the J2EE support with support for integrating J2EE with legacy applications using JCA connectors, business process choreography, and other features. If you're accessing host systems using Enterprise Access Builders, then WSAD-IE may be the best choice for you. WSAD adds support for additional legacy enterprise development and includes additional rapid application development tools.

With the impending retirement of VAJ, its integrated team environment Visual Team Server (also known as "Envy") is also retired. WebSphere Studio doesn't support Team Server as a source code repository, so as part of the migration effort, you must select a new source code management (SCM) tool.

WebSphere Studio development tools support several third-party SCM solutions, including Rational ClearCase, CVS, and others using vendor-supplied plug-ins. With this change, all application artifacts (Java code, JSPs, XML, and property files) can now be stored in the same SCM, which makes it much easier to manage these resources and keep them synchronized. Unfortunately, there is no built-in mechanism for extracting an existing version history from Team Server into an alternate SCM tool, so this process is largely manual. (It is possible to write a VAJ tool that can automate some part of this process.)

If you're already using a third-party SCM with VAJ, you may need to modify your existing project folder and artifact hierarchy as part of the migration. Version 5 of the WebSphere Studio development tools requires that application code be structured along J2EE packaging lines. For example, Web components must be in a Web Project, and Enterprise JavaBean (EJB) components must be in an EJB Project. Each of these projects has its own folder hierarchy. Modifying existing project hierarchies in an SCM (while maintaining the history of the various components) can be difficult and time-consuming.

Your development team may use other tools as well. You'll need to evaluate these tools to determine their appropriateness for WAS V5 development. Replacements may be required. For example, WebSphere Studio products provide tools for XML authoring, which may be appropriate replacements for third-party XML tools.

Developer and Administrator Skills

Because WebSphere Studio development tools are very different from VAJ, expect your developers to require some time to fully develop the skills required to use the product effectively. Typically, five days of classroom training are sufficient for developers who are familiar with enterprise application development using VAJ. Every migration plan should include mentoring for the development team to guide them through their first few weeks with J2EE and WebSphere Studio tools.

Due to fundamental changes to the administration model, scripting, and tools, administrators will also require training. With the adoption of J2EE packaging, administrators have less to do when deploying an application, but configuration and maintenance of the application server instances are still their responsibility. Much of the terminology has changed with 5.0, and XML-based configuration files have replaced the administration database. Managing a collection of WebSphere nodes is still a task that requires dedicated administrators with specialized skills.

Even skilled administrators will benefit from formal training. Five days of classroom training are typically enough to arm seasoned WAS administrators with the information they need to install, configure, and maintain WAS V5.

Runtime Environments

When you think of runtime, the production runtime environment springs immediately to mind. Most organizations, however, have more than one runtime environment. It's common to have development test, system test, performance, and preproduction environments as well. Other environments with other purposes may also exist.

Rarely is the migration of a runtime environment a simple matter of shutting everything down, installing new hardware, and starting everything back up. This may be possible for development test and system test environments, but other runtime environments have restrictions that make such practices impossible. In almost every case, the production runtime environment must stay in service while it is migrated. Very often, very limited downtime is acceptable for even the preproduction environment.

Migration of a production runtime is even more complex if you have multiple applications with different development teams and different delivery schedules. In this sort of environment, it may be necessary to run both versions of WAS concurrently for a time. Unfortunately, WAS V5 cannot run applications packaged for V3.5 without modification. (Version 5, however, can run applications packaged for V4 without modification.)

Depending on your runtime configuration, it may be possible to do a "rolling migration." In this scenario, application servers are upgraded individually or in small groups using the following steps:

1. Select one or more (possibly an entire domain of) application server nodes.
2. Remove the selected nodes from the load balancer's routing tables.
3. Remove the nodes from service (shut down).
4. Remove WAS V3.5 or make it dormant on the selected nodes.
5. Install, configure, and test WAS V5 on the selected nodes.
6. Install, configure, and test the migrated application code.
7. Activate WAS.

The process is repeated for the remaining nodes.

This is a very simple overview of a rolling migration and may not necessarily be appropriate for your configuration. (This scenario assumes that you have at least three application servers providing at least 50 percent more capacity than you require.) The first node to be migrated would become the first server in a new WAS V5 cell. As additional nodes are migrated, they are added to the cell. Each WAS V3.5 domain you have in your current configuration will likely become a cell in your new configuration.

Any migration plan should include a back-out strategy. In the above steps, we suggest that you make V3.5 "dormant." Rather than completely removing the installation of 3.5, you can instead simply shut it down (and configure your system not to start it) and remove its plug-in from the HTTP server. If later testing reveals that the migration didn't work, it is then a relatively easy matter to reinstate the 3.5 instance.

If you have multiple applications, the rolling migration may take weeks or even months (with the unmigrated applications remaining on the V3.5 nodes). If you have a single application, or if all your applications are migrated before the runtime migration, the rolling migration may be accomplished in a matter of hours or days. In either case, you must consider that you will need to support (at least for some period of time) a runtime environment that contains a mix of nodes running WAS V3.5 and V5.

The mix of versions has some interesting implications. To start, if two different versions of an application are running on different versions of WAS, you must make sure that the application appears consistent to the user. That is, the application should look and act the same on both application server versions.

For example, if your applications make use of `HttpSession` state, you may consider introducing (if you have not already done so) server "stickiness" to ensure that all requests from a particular user are directed to the same server instance. Unfortunately, `HttpSession` state cannot be shared by different versions of WAS. This also has a potential impact on failover.

If your application is split physically across tiers, you may need to consider interoperability options. It is possible to connect from servlets running in WAS V3.5 to EJBs running in V5, but several fix packs are required to make this work.

As part of the migration, you may need to update your hardware, operating systems, and other components (such as the HTTP server). WAS V5 prerequisites are well documented.

Try your runtime migration plans in a test environment before starting the migration of your production runtime.

From a system administration perspective, migration is more than just installing new products. Some of your existing practices and process may need to be considered as well. The scripting languages of WAS V3.5 (WSCP or XMLConfig) are no longer supported with V5. They're replaced by the evolving JMX standards. Any scripts you currently use to administer your servers will need to be revisited and likely rewritten (there are currently no tools available to automate this).

Planning Is Key

Because every organization and application is different, every migration experience will be different. We've outlined many of the issues you must consider. More specific information about migration is detailed in the IBM Redbook, ''Migrating to WebSphere V5.0: An End-to-End Migration Guide," at `http://www.redbooks.ibm.com`.

Migration involves more than just the code or any single environment. Even if no change is required to your application code, migration will still take time when you consider all the other aspects.

Application Code

Application code is the first thing that most people think of when considering migration. Organizationally, the structure of the code often requires significant change; J2EE requires that the various application artifacts be structured in a particular way. However, the amount of actual change required to the code itself may be very small. Most of your Java code should migrate without change. Even some of the J2EE artifacts such as servlets, Java ServerPages (JSPs), and Enterprise JavaBeans (EJBs) may work without modification on the new version.

As part of the migration of application code, you must consider the following:

- Code organization
- Changes to the specifications
- Use of proprietary technologies
- Third-party libraries

J2EE defines a very specific packaging structure that separates the various artifacts into different archives. Web artifacts (including servlets, JSPs, and Web content such as GIF and JPG files) are packaged in a Web archive (WAR) file. EJB artifacts, typically consisting of the EJB types, are packaged in an EJB file (also known as a Java archive (or JAR) file. An enterprise archive (EAR) file is used to collect all the application components (WAR and JAR files) into a single file.

Other libraries (JAR files) used by the application can be located in the EAR file, as well. The archive can also contain application client archives, which are another J2EE module type. Note that an XML deployment descriptor is packaged with each module. The deployment descriptor describes the organization of the components in the module and how they're configured at runtime.

Specifications

The various specifications that make up J2EE have evolved somewhat since WebSphere Application Server V3.5. Some of the specifications (e.g., JDBC) have changed very little. Others (e.g., EJBs) have changed dramatically. The extent to which the changes will affect you depends, of course, on which parts of the specifications your code makes use of.

Some minor changes have been made to the servlet specification. Methods such as `HttpSession.getValue(String)` have been deprecated in favor of renamed equivalents such as `HttpSession.getAttribute(String)`. Perhaps the most significant change to the servlet specification has to do with session scope. The more recent versions of the servlet specification limit the `HttpSession` scope to the Web module. That is, it's not possible (according to the specification) for multiple Web modules to share `HttpSession` data.

In previous versions of the servlet specification (including those supported by WebSphere Application Server V3.5), a single `HttpSession` scope is shared by all applications. If your application code depends on this semantic, it must be changed. To make migration easier, WAS V5 provides an option that permits `HttpSession` state to be shared by multiple Web modules packaged within a single .ear file. This deviation from the specification is provided as a convenience, and its use should be limited.

Version 1.2 of the JSP specification, as supported by WebSphere Application Server V5, is largely compatible with version 1.1 (supported by WAS V3.5), and JSPs coded to that level of the specification should migrate with little difficultly. However, WAS V3.5 supports three versions of the JSP specification. Any JSPs written to the 0.91 specification may require significant modification if bean tags are used (scriptlet and expression tags should migrate relatively easily).

As part of the migration of servlets and JSPs, you must build a web.xml deployment descriptor. At very least, all the servlets in your Web module should be named and have initialization parameters provided (WebSphere Studio tools provide support for doing this). We strongly recommend that you name your servlets in the deployment descriptor; IBM discourages continued use of the WAS class name server, and it is disabled in WAS V5 by default.

EJBs have changed considerably between WAS V3.5 and V5; the most significant changes involve entity beans. With EJB 1.1, XML-based deployment descriptors replaced the serialized deployment descriptors supported by previous versions. WebSphere Application Server V3.5 actually supports "EJB 1.0+" by including some parts of the EJB 1.1 specification, which may mitigate some of the impact of the migration.

There are two ways of getting existing EJBs into a WebSphere Studio project. The existing EJBs can be packaged into an EJB 1.0 .jar file (with source code), and that .jar file can then be imported into an existing WebSphere Studio EJB Project. When importing the .jar file, WebSphere Studio tools recognize the serialized deployment descriptor and will create appropriate entries in the deployment descriptor for the contained beans.

This process does a reasonably good job of importing basic EJB information, such as the types involved, JNDI names, and transaction attributes. It does not consider Container-Managed Persistence (CMP) entity bean-persistent field mapping information, however. If your application makes use of CMP entity beans, you should consider using the EJB 1.1 Export Wizard included with VisualAge for Java 4.0. This Wizard generates an EJB 1.1-compliant .jar file (with source) that can be imported into WebSphere Studio projects. The exported .jar file contains all the deployment information required to make CMP entity beans work, preserving the existing persistent field mappings and finder queries. Of course, the Wizard generates EJB 1.1-compliant beans that must be upgraded to 2.0. The J2EE Migration Wizard available in WebSphere Studio products (except WSSD) can help in this regard.

Once the code is imported into a WebSphere Studio project, additional changes may be required. Neither of these migration processes actually modifies any code, so you may need to address other changes to the specification (`java.rmi.RemoteException` should not be thrown by business methods in EJB 2.0, for example). Many of these changes are only deprecations that you can address at some later time. Others, however, must be addressed before the beans can be deployed and tested.

Servlets, JSPs, and EJBs will likely take up most of your migration effort, but they probably won't be the only things you must consider.

Proprietary Technologies

In this context, "proprietary technology" refers to features not governed by standards. The Connection Manager provided with WebSphere Application Server V3.0 is an example of an IBM solution that was provided to fill a gap in the J2EE standards at the time. The Connection Manager was deprecated in WebSphere Application Server V3.5.x in favor of standards-based data sources, but it's still supported in WAS V5. If your application makes use of this technology, it will eventually need to be changed.

Perhaps the most commonly used example of proprietary technology is the various code builder technologies included with VisualAge for Java. For most of these builders, the generated code works with little or no modification on WAS V5, but there is no support (other than directly editing the generated Java code) in WebSphere Studio development tools.

Third-Party Libraries

All versions of WAS treat third-party libraries the same as your application code. Libraries composed of "just Java" code should continue to work without modification on WebSphere Application Server V5. However, libraries that have dependencies on particular versions of WAS may need updating. To complicate matters, most library vendors don't provide source code or permission to change their code, which can potentially leave you at their mercy to provide support.

As part of the migration effort, you should contact third-party vendors and ask them to provide compatibility statements for WAS V5. Then, you should test their compatibility with V5 extensively. This should be done relatively early in the migration cycle so you can find alternatives, if necessary.

Reference Cycles

One of the most common general issues encountered when migrating code from VisualAge for Java to WebSphere Studio development tools is reference cycles. A reference cycle occurs when a class in one project refers to a class in a different project that refers back to a class (possibly the same class) in the first project.

Reference cycles between classes in different packages or projects are generally considered to be bad programming because they introduce tighter coupling between components. WebSphere Studio tools' build process stops when a reference cycle between projects is detected. VisualAge for Java has no problem with reference cycles and does nothing to discourage their use.

Unfortunately, though, this is one area where WebSphere Studio tools don't provide much help. Third-party tools such as SmallWorlds (`http://www.thesmallworlds.com`) can help to identify reference cycles and help you find potential solutions for them.

Other WebSphere Migration Aids

Because every organization and application is different, every migration experience will be different. This section has outlined many of the issues you must consider. More specific information about migration is detailed in the IBM Redbook, "Migrating to WebSphere V5.0: An End-to-End Migration Guide," at `http://www.redbooks.ibm.com` (see the Bibliography for more details on this Redbook).

CHAPTER 7

What to Put on WAS vs. Domino

> **"'Domino and WebSphere: So Happy Together.' Both Domino and WebSphere are thriving because they solve different customer problems."**
>
> —*E-Pro Magazine*, **June 2001**

When looking to combine features of both Domino and WAS (or J2EE) in a single application, designers and developers often find that some application elements could be supported on either WAS or Domino. This chapter explores those areas where WAS and Domino offer comparable function and tries to provide some insight as to which platform may be more appropriate for a particular set of requirements. Although there are certain cases where there is a clear choice of where best to place a function, it is our experience that there are also cases where there is no clear choice in terms of function alone, and other factors such as development skills or familiarity may be the basis for the final choice. In fact, these other factors, especially development skills, may override any purely functional benefits.

At a high level, the type of application is an important factor in deciding on which platform to build. As discussed in earlier chapters, the strengths of Domino lie in collaborative applications. Domino is good at applications involving user interaction, providing a scalable document data store, rich content handling, and a robust security model. WebSphere Application Server (WAS) is ideal for transactional applications, highly leveraging a relational data store, and providing enterprise-level J2EE application hosting. WebSphere Portal excels at application and data aggregation, providing a rich framework for Web-based applications.

Assuming that we have an application that has both collaborative and transactional elements, let's review some of the key features of each platform to help understand which parts may be supported by which platform.

Lotus Domino Server

Lotus Domino is a comprehensive application platform for collaboration that handles both connected and disconnected requirements for data and applications. Many customers initially purchase Lotus Domino for the built-in enterprise e-mail, calendar, and scheduling applications, making those types the most-widely deployed collaborative applications. The majority of customers, however, exploit the document-sharing and form capabilities that support core business processes, which enable employees to work together efficiently and securely. Domino is comprehensive; it provides the complete infrastructure needed to create, test, deploy and manage distributed, multilingual applications including directory, database, application server, administration, security, connectivity, Web server, e-mail server, calendaring engine, and so on—all in one system.

WebSphere Application Server

The WAS is IBM's core product for J2EE application support. IBM has differentiated WAS from other J2EE application servers in its scalability. From its inception, WAS was geared toward large-scale J2EE applications. It provides features supporting highly distributed applications and their management and support for the latest J2EE and Web services standards. It also addresses scalability and performance with support for deployment across multiple servers, load-balancing services, and automated performance tuning. Today, WAS is available in several configurations across a wide range of platforms, allowing any size of J2EE application to be deployed in the most cost-effective way possible.

In contrast to Domino, WAS does not contain its own data manager. Instead, it relies on a relational database being available in the environment for application data storage or for J2EE components themselves, such as entity EJBs. One way that WAS provides access to those data sources is by leveraging the J2EE Connector Architecture (JCA). JCA defines a set of service contracts that a connector developer can expect will be available to the adapter at application runtime. The three basic services defined in JCA V1.0 include connection management, transaction management, and security management.

Also unlike Domino is the fact that WAS relies on external sources for user directory (or what it terms "user registry") functions. For WAS (and J2EE), the user registry is required for user authentication relative to providing secure access to application components. WAS can be configured to use two categories of user registries—local; for example, the Windows Active Directory or Unix user account file (/etc/passwd), or LDAP; that is, directories supporting the Lightweight Directory Access Protocol. The fact that Domino can provide LDAP access to its directory enables it to be considered for the WAS user registry. For single sign-on (SSO) support between Domino and WAS, and LDAP user registry must be used. We discuss the options for directory configuration in detail in Chapter 12, "Security and Single Sign-On."

Domino Features Enabling Applications to Integrate with J2EE

Domino ships with features that facilitate the integration of Domino data with J2EE applications. You can reap many benefits by combining Domino and J2EE applications including, most importantly, the addition of human interaction into scalable and transactional applications built on J2EE. When integrating Domino with a J2EE-based application, there are two key ways to access Domino functions:

- Domino Objects for Java
- Domino Custom JSP tags

The Domino Objects for Java are essentially the Domino Java API. These objects allow for Java object-based access into the Domino data store for the key database functions of create, read update, and delete—or CRUD actions. The Domino Objects for Java API allows access not only to Domino data but also to key Domino services, such as registering users, running agents, sending mail, and others. The Domino Objects for Java can be used by experienced Domino developers for adding Domino specific function to J2EE applications. For example, a servlet or JSP can be enhanced with Domino function by using the Domino Java Objects within the servlet's or JSP's Java code.

The Domino Custom JSP tags are for use from within a JSP. These tags are XML tags embedded in a JSP and provide access to Domino data, data validation, and flow control. The Domino Custom JSP tags are built on the Domino Objects for Java and are intended for use by experienced JSP and JavaScript developers. These developers can use the Domino tags to build a J2EE application that uses Domino data and services, without having to become familiar with the pure Java Domino APIs.

Lotus Domino Toolkit for WebSphere Studio

Starting with Domino 6.0.2, the Lotus Domino Toolkit for WebSphere Studio provides developers with an intuitive tool to implement the Domino custom JSP tags into their WebSphere applications. The toolkit is a set of plug-ins that enhances the WebSphere Studio Application Developer (WSAD) user interface. It enables J2EE developers to integrate Domino functions into Web applications without the complexity of the Domino Java API or even a detailed understanding of the Domino Custom JSP tags.

The toolkit adds a new view to WSAD, a "Domino view," which allows you to work easily with Domino database objects such as forms, fields, views, and agents. Using this new Domino view within WSAD, you can access a Domino database (locally or remotely) and display the database's forms, fields, views, and agents. These objects can be dragged and dropped onto JSPs being developed, and the equivalent Domino custom JSP tag code is inserted. Using the Lotus Domino Toolkit enables developers without an in-depth understanding of Domino function to easily incorporate that function into their applications.

XML

Domino has extensive XML capabilities that can be used from a J2EE application. A J2EE application can use the Domino Java API to retrieve Domino documents in XML format without knowing specifics about the data stored in Domino. Then, the J2EE application can process the XML and transform it to the specific format needed by the application. If the application needs to update the Domino data, the same interface can be used to update the database by passing back XML.

Web Services

Emerging Web services standards simplify and broaden application integration by defining a standardized means to make service requests and receive response data over common Internet protocols. The lingua franca of Web services is the XML based access protocol known as Simple Object Access Protocol (SOAP). Designers express the application or service interface using XML. Domino can host Web services that expose Domino data and function using Java or LotusScript. Using a combination of a J2EE server such as WAS, appropriate SOAP classes and the Domino Java Objects, developers can expose desired portions of their current Domino applications as Web services. To do so requires some Java development skills and knowledge of the Domino object model, which we discuss in detail in following chapters. For pure J2EE applications, WAS ships with the needed SOAP classes, and the WebSphere Studio development environment has wizards for creating, consuming, managing, and deploying Web services.

Domino Data in a Relational Database

Domino has a capability, via the Lotus Enterprise Integrator (LEI) feature, to store all document data in a relational database, such as DB2, Sybase, or Oracle, rather than in a Domino NSF database file. The benefit is that Domino maintains its security controls and the database design, but since the document data is stored in the relational database, document access is improved due to the performance efficiencies offered by the relational database. This access method is referred to as Advanced RealTime connectivity. Advanced RealTime connectivity allows significantly better scalability and performance, since the relational technology is used for data storage versus the Domino NSF file format.

The Advanced RealTime connectivity feature can be used as a means to integrate Domino data with J2EE applications. As mentioned previously, J2EE applications running on WAS can access relational data stores directly using JDBC or EJB interfaces. Instead of using Domino provided mechanisms to access Domino data stored in NSF files, the application can utilitize relational data access methods. A designer can achieve high performance by employing standard data access techniques such as connection pooling, caching, and minimizing JNDI lookups.

J2EE allows for the abstraction of databases and their connections from application code by accessing a business object layer rather than the database directly. This layer, in turn, accesses the database itself. The same business objects can be accessed by different applications. The database design is decoupled from the application itself, creating easier maintenance going forward.

WAS (J2EE) Functions Compared to Domino

In this section, we discuss the main areas where WAS and Domino overlap in function and present the conventional wisdom (and our view) about using one over the other.

Servlets vs. Web Agents

At a high level, J2EE servlets and Domino Web agents are pretty much equivalent—they provide a means to run server-side programs triggered by client-side (browser) events. The Domino "Web agent" function is new with Domino 6 and basically extends the traditional Domino agent function to include triggering from Web events. The traditional Domino agent was restricted to server-side programs triggered by time or by Domino database events. Both servlets and Web agents can return HTML to the client.

Domino Web agents can be coded in either Java or the Domino-specific languages of LotusScript, formulas, or simple actions. In fact, Web agents are the only context in which LotusScript may be used for Web applications. Thus, Web agents may be easier to implement for Domino developers. The Web agent, since it runs as part of the Domino server, has direct access to Domino data elements. The biggest drawback to using Web agents for server-side functions is that the agent must be loaded each time it is invoked. This is done by the Domino agent manager, which does not provide for agent caching. For Web agents written in Java, an entire JVM must be started to run the agent. If the design calls for frequent use of the agent, loading on each invocation usually leads to unacceptable performance.

Servlets, on the other hand, are architected to be run efficiently by the J2EE application server. Servlets are loaded either on server start or on first invocation of the servlet and are simply passed subsequent requests by the application server. If the servlet function needs to access Domino resources, this can be accomplished via the Domino Objects for Java. The servlet designer has a choice of how Domino is accessed, which can be either locally (where Domino is running on the same server as WAS) or remotely. Naturally, there is overhead in accessing Domino remotely, but this can be tailored so that the access is not needed for each servlet invocation. We discuss strategies for accessing Domino via Java and J2EE in more detail in Chapter 10, "Accessing Domino from J2EE."

Which to choose then? We usually adhere to the common guideline of implementing server-side functions as servlets, especially when the functions will be invoked frequently. If the functions are infrequent and require a fair amount of or sophisticated Domino processing, then Web agents are appropriate.

Java Server Pages (JSPs) vs. Domino Forms

Another area of overlapping function between WAS and Domino is the ability to generate and process HTML content, especially forms. A designer can include forms in a Web application by using either JSPs managed by WAS or Domino database forms served by the Domino HTTP engine. A third possibility is available now with the support for Domino Custom JSP tags, where any Domino database form can be converted into a JSP and handled by WAS.

The choice between WAS and Domino here is not clear-cut. There are several factors that could affect the decision. Probably the most influential ones are the following:

- **Overall application look and feel.** If the application is predominantly JSP- or Domino-based, then it makes sense to stay with the same rendering. Although if there is any JSP content, it may be easier to convert Domino Web content to JSPs rather than the other way around.
- **Development skills.** If the development experience lies with one product, then it will likely cost less to develop using that product.
- **Unique feature rendering.** There are certain cases where attempting to convert a Domino form to a JSP will not be of perfect fidelity. This can occur when using the Domino Custom JSP tags. (In Chapter 10, we discuss ways to generate JSPs from Domino forms with greater fidelity.)

WAS vs. Domino Servlet Engine

Before the advent of WAS, the Domino server product provided a means for plugging in a servlet engine to handle servlets in addition to the standard Domino resources. So, there is the possibility of using a servlet engine under Domino versus WAS for servlet processing. The choice here is clear-cut. WAS should handle all J2EE resources. It's difficult to concieve of any advantage to having Domino handle servlets.

Keeping an Eye on the Future of Domino and WAS

For IBM, the future of both WAS and Domino is clearly all about how best to integrate the product families in terms of both function and developer communities. Clearly with recent product versions, this process has started. How may this integration affect application design decisions that need to be made today? We consider this question in the following section.

Toward Open Standards

A key issue in the Web application server and services marketplace is the choice between the use of open standards vs. proprietary programming interfaces. Or, to be blunt, J2EE versus .NET. As is well documented, an open standards foundation is a major part of IBM's strategy in the marketplace. WAS, being an implementation of J2EE and now Web services standards, will continue down the open standards road. Domino, having a longer product history, will continue to incorporate open standard interfaces.

Although proprietary technology is an important differentiator in the market, even Lotus has long supported key Internet standards. Domino delivers enterprise messaging while supporting an open standards-based application development architecture that includes Java, JavaScript, XML, HTML, LDAP, SMTP, POP3, IMAP, and SSL. Domino has embraced Java as both an external interface, providing Java classes for all its key interfaces, and as an internal language that can be used in place of Lotus' BASIC-like programming language, LotusScript. It is natural to expect that the integration of the WAS and Domino products will occur along these open standards lines.

Browser-Based Clients

Another sign of the impact that the Web application model is having on Domino are the iNotes and, more recently, the Lotus Workplace products. Traditionally, Lotus Notes was a client/server architecture based on the feature rich Notes client and the Domino server. With Domino R5 and iNotes, browsers became a prominent client. Lotus iNotes for Web Access, despite its browser dependencies, was an important step along the path to provide Notes client function via the Web browser. The Lotus Workplace product is the latest step along this path.

Browser-based client function is important for applications that integrate several, often disparate types of function on a single browser screen—the so-called Web portal. For example, using the WebSphere Portal Server product, you can integrate transaction processing under WebSphere with e-mail, calendar, and other Domino functions. Thus, you can now construct a Web application that looks and feels like an extension of the iNotes application, resulting in a single, browser-based desktop from which users can access all the applications that they use on a daily basis.

It is likely that IBM will continue to provide client function via the browser and enhance the WAS and Domino functions that can be presented in a portal context.

Convergence of Function

The integration between WAS and Domino also will occur with functions that span both products. Security is the prime example. IBM has implemented single sign-on (SSO) support between Domino and WebSphere. SSO lets users authenticate once when accessing both Domino data (databases, documents, fields) and WAS resources, such as HTML, servlets, JSPs, and EJBs. This is accomplished via sharing an encryption mechanism used for passing user credentials and sharing a common user registry using LDAP.

Another area of common function is data storage. We can see that Domino's data storage technology (the NSF database) has already benefitted from synergy with other IBM database products, especially DB2. Domino R5 provided faster recovery and improved reliability for its data store using transaction logging. We can expect this sort of synergy to continue, perhaps resulting in convergence of Domino's document data store with the relational model.

IBM and Lotus will continue to meld their product families into one. Lotus K-station (Domino's portal server) has already been merged into WebSphere's portal server as a single portal solution with the benefits of both products. IBM's recent move to allow WebSphere workloads on its iSeries Dedicated Servers for Domino (DSD) line of servers is a good indication that IBM understands how important this integration is.

Another notable item is the Domino Collaboration Objects for Java. IBM and Lotus say their intent is not to extend Domino but to provide a more intuitive interface to Domino objects with fewer Java methods and classes. The current array of objects includes the following:

- CalendarEntry (to search calendars and create new entries)
- Mail (to compose and send e-mail)
- Login and workflow initiation services

As for Domino's traditional programming environment, don't assume IBM will abandon LotusScript. Lotus Notes and Domino have an enormous customer base of users and developers using LotusScript-based applications. Continued support for this language is unquestionable. As new classes and properties are developed for LotusScript, equivalents for Java will follow. Obviously, the intent is that J2EE (Java) programmers can be trained to work with Domino without having to learn a new language.

Evolution of WebSphere

So, if the future of Domino is WebSphere, what's the future of WebSphere? Since WAS is IBM's implementation of the J2EE standard, the evolution of J2EE will largely determine what WebSphere becomes.

IBM is committed to building the most complete and best-performing application server based on the J2EE standard. And because many areas aren't covered by the standard, IBM can be creative and add value by supplying add-on features that make WebSphere more enterprise ready. IBM is also adding application products and proprietary extensions to the base WebSphere product to meet needs expressed in the marketplace.

IBM updates WAS every six months or so with a major release. Each version of WebSphere gets a little easier to install and manage, but the real activity is in the extensions and applications. Consider the following:

- With WebSphere MQ, IBM implements the Java Messaging Services (JMS) standards. Numerous connectors to CICS and other legacy systems and middleware are also planned or shipping.
- IBM has rewritten Net.Commerce into Java to run under WebSphere as the WebSphere Commerce Suite—a high-powered storefront system.
- IBM has grafted speech recognition technology (as well as its vaunted national language support) onto WebSphere.
- IBM has implemented portals, wireless support, and B2B capabilities as Java applications that run in the WebSphere environment.
- IBM is moving its legacy connectivity products to the WebSphere environment with offerings such as WebFacing on the iSeries and Host Publisher.
- IBM has leveraged Lotus' expertise with its ERP Connectors and introduced several WebSphere adapters that are based on Lotus LEI connector technologies.

WebSphere is the logical place for IBM to create new applications because they can be leveraged over the entire eServer line. Expect a steady stream of WebSphere-based products from IBM over the next few years. Watch areas like content management, system management (Tivoli is already starting to play in these areas with its Policy Manager), collaboration, and application accelerators.

Tooling for developing WebSphere applications is also advancing rapidly with the new WebSphere Studio products. Based on the open-source Eclipse project, IBM has replaced VisualAge for Java and the old Studio product and added new support for XML and Web services. For some of the IBM hardware platforms, such as the iSeries, IBM is adding platform-specific extensions to aid in application development and integration.

A big question is the Lotus Domino Designer client. It's hard to see how IBM will merge this with the WebSphere tooling. You may see, however, improved Domino deployment options for objects developed in the WebSphere Studio tools.

IBM has already repositioned several product lines and placed them under the WebSphere umbrella. Is Domino next? Will we see a WebSphere Mail and Collaboration Server? Assuming that there will be a new integrated WebSphere/Domino Server in the future, it will be worthwhile to keep an eye on WebSphere and J2EE evolution.

A Web Conferencing Example Using Both Domino and WAS

In this section, we discuss the design of a project that provides an "on demand" Web conferencing service and that incorporates both Domino and WAS product functions. It illustrates various ways in which Domino and WAS-based products can be combined to build applications providing complex function without a lot of custom application development.

The basic function of the service is to allow anyone with Internet access to register, schedule, and join browser-based conferences. The conferences include chat, multi-media, and whiteboard sharing functions. The Domino-based Lotus Instant Messaging (née Sametime) Room Server product is used for hosting the Web conferences. To register the Web conferences, the WAS based WebSphere Everyplace Subscription Manager (WESM) product is used. The Lotus Enterprise Meeting Server (EMS) product, which is a J2EE application, thus WAS-based, is used to manage the meetings and perform load balancing for the Sametime room servers. The EMS keeps track of logging into and exiting meetings for registered users. The Sametime room servers run on Domino servers, as they don't require transaction capability. The user registry is provided by the IBM Directory Server product and is shared across the room, subscription, and meeting servers. There was also a requirement for an outbound mail function, and this was provided by a stand-alone SMTP server. The project was designed for 24×7 availabilty, so many of the server nodes were built as clusters using an IBM clustering product, HACMP.

Figure 7-1 shows the overall architecture of the service and where the different types of servers are placed in the network tier layout. We point out that due to product requirements, not all the WAS servers are at the same WAS version. WESM runs on WAS V5, but the EMS product supports only up to WAS V4. For this reason, the WESM and EMS servers were built on separate server systems.

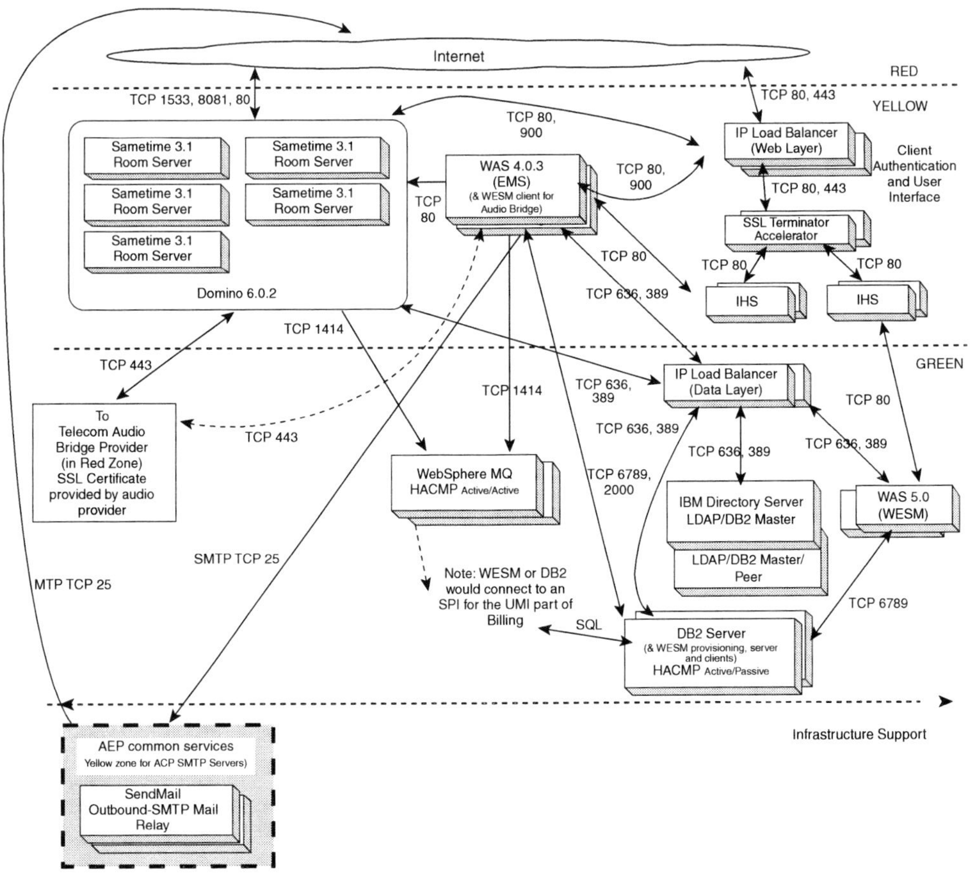

Figure 7-1 Web Conferencing logical server architecture.

Directory Master-Peer Design for High Availability

The directory function of the Web conferencing service had two key requirements—that it be shared among the functional servers (e.g., for single sign-on) and that it be highly available. The approach taken was to place the directory server behind a shared load balancer and to make use of a master-peer replication topology provided by the IBM Directory Server product. Figure 7-2 depicts the directory server configuration.

In the master-peer approach, the directory client will read/write to the master. If the master server fails, the client requests are directed to the peer server until the master is made active again.

The replication process keeps the data in the master and peer directories synchronized. Whenever a write (or update) is to be done against the directory, it goes to the master. After the write/update is complete, it replicates it to peer server immediately. When the master server is down, the write/update goes to peer server, and it will be queued for replication once the master is restarted.

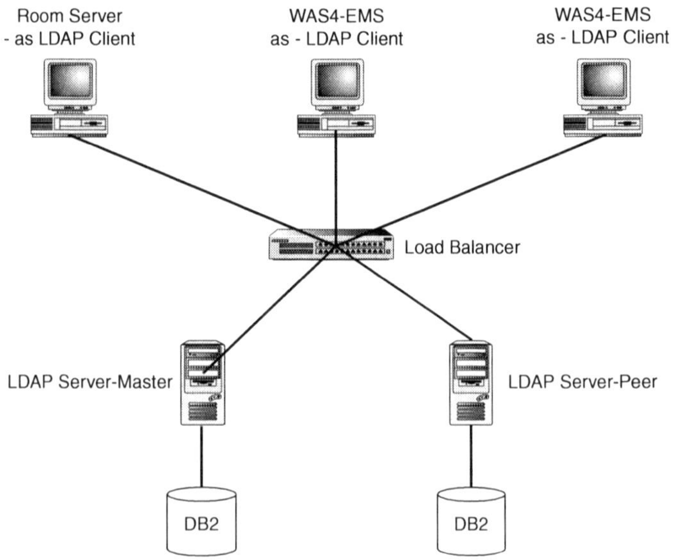

Figure 7-2 Design for the shared directory.

For directory sharing, the Domino, WESM, and EMS servers are configured to the virtual IP address of the directory server. As part of their user management functions, the EMS and WESM applications perform write/update operations to the directory server.

WebSphere MQ Application Details

EMS communicates with the Sametime servers using the Java Message Service (JMS). This JMS communication is implemented using IBM WebSphere MQ message queuing. The JMS communication supports load balancing, server independence, and failover necessary to provide central meeting management for all of the clustered Sametime servers. It also allows centralized logging, statistics, and configuration change notification.

The Sametime Room servers must be able to communicate to EMS/WAS using TCP/IP port 900 (WebSphere 4.x JNDI port). The EMS and Sametime Room servers must be able to communicate to WebSphere MQ using TCP/IP port 1414.

WebSphere MQ Cluster Design

Because the WebSphere MQ (WMQ) server acts as the blood supply of the service, availability is a key criterion for these servers. It is also important that the design allow for the easy addition of MQ servers. IBM's High Availability Cluster Multi-Processing (HACMP) product (for AIX) is a control application that can link servers into highly available clusters. Clustering servers enables parallel access to data, which can help provide the redundancy and fault resilience required for business-critical applications.

The WMQ clusters reduce administration and provide load balancing of messages across instances of cluster queues. They also offer higher availability than a single queue manager because following a failure of a queue manager, messaging applications can still access surviving instances of a cluster queue. However, WMQ clusters alone will not provide automatic detection of queue manager failure and automatic triggering of queue manager restart or failover. HACMP clusters provide these features. The two types of cluster can used together to good effect.

By using WMQ and HACMP together, it is possible to further enhance the availability of the WMQ queue managers. With a suitably configured HACMP cluster, it is possible for failures of power supplies, nodes, disks, disk controllers, networks, network adapters, or queue manager processes to be detected and automatically trigger recovery procedures to bring an disabled queue manager back on-line as quickly as possible.

An HACMP cluster is a collection of nodes and resources (such as disks and networks), which cooperate to provide high availability of services running within the cluster. Hopefully, we've been making a clear distinction between such an HACMP cluster and a WMQ cluster, which refers to a collection of queue managers that can allow access to their queues by other queue managers in the cluster.

A "mutual takeover" configuration is one in which all nodes are performing highly available (movable) work. This type of cluster configuration is also referred to as "Active/Active" to indicate that all nodes are actively processing critical workload. An Active/Active cluster configuration is used for the Web conferencing service.

The load balancing is based on client connections and not the message distribution. Each Sametime room server has a connection to either of the cluster nodes through connection channel. If the queue exists on the queue manager node to which it is connected, it will always go there for the duration of the connection. Figure 7-3 depicts the overall cluster design.

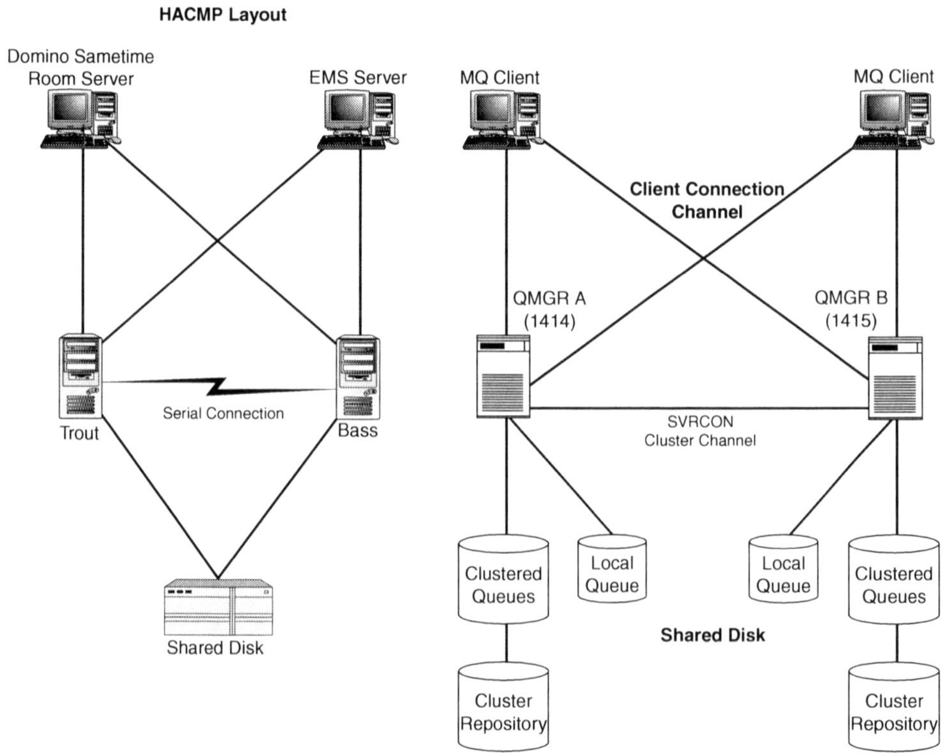

Figure 7-3 WebSphere MQ HACMP cluster (active-active) design.

CHAPTER 8

Development Tools

> **"Knowing is not enough; we must apply. Willing is not enough; we must do."**
>
> **—Johann Wolfgang Von Goethe, great 18th-century German poet**

In this chapter, we present an overview of the development tools used to build WebSphere and Domino applications and discuss how they can be used to build integrated Web applications involving both products. We focus on the features of each tool set that can be used to build the application's integrated parts. Some of the recent toolkits provided by IBM for the WebSphere Studio Application Developer product address WebSphere and Domino integration, and we discuss these in detail.

When discussing development tools in the context of WebSphere and Domino, there are several issues that arise because of the long history (at least in Web years) of each product as an application platform on its own. Obvious to those with a background in the Lotus Domino products, but perhaps not to those with a Java or J2EE background, Domino does predate J2EE and has a mature development tool in Domino Designer and a legion of developers with hard-won skills in LotusScript and Formula languages. On the other hand, with the advent of J2EE, there are now many thousands of developers with pure Java skills combined with Web server, application server, database, XML, and other skills.

Perhaps foremost on the minds of Domino developers is the question of which development tool IBM will champion in the world of WebSphere and Domino application development. In the case of J2EE developers, perhaps there is the concern over how they can leverage their Java skills to work with Domino functions without having to learn yet another tool.

We feel there is no simple answer to these concerns at this point in the evolution of WebSphere and Domino; however, the situation is far from bleak from either a J2EE or Domino developer viewpoint. In this chapter, as we present the current state of development tools for the products, we'll hopefully alleviate some of the development tool concerns.

WebSphere Studio Application Developer (WSAD)

The WSAD product has emerged in the last two years as IBM's flagship integrated development environment (IDE) platform for J2EE application development. From a simple stand-alone Java program to a multi-tiered, distributed, team-oriented, enterprise class application, WSAD provides the tools necessary to build, test, and deploy almost any conceivable J2EE application. In what follows, we will refer to the most recent WSAD version at the time of this edition, WSAD V5.0.

WSAD and Eclipse

IBM took an interesting and, in our view, bold step during the early development of the WSAD product line. The core of the WSAD IDE platform was released as an open-source software project, called Eclipse (`www.eclipse.org`). The Eclipse software (largely written in Java itself) provides the framework to create an IDE for almost any type of application development. This is the heart of the Eclipse design. As stated on the Web site: "The Eclipse Platform is an IDE for anything, and for nothing in particular."

The Eclipse software package is not just an inert framework. The download includes the basic IDE framework and components (editor, builder) for Java development, so it is a capable Java IDE to start with. But it is its ability to accept various add-in components ("plug-ins") that give Eclipse its power. Any one or any organization or company can develop a tool for Java, WebSphere, or Domino development and offer it as an Eclipse plug-in developers can use.

IBM's motives behind the Eclipse project are an interesting topic for discussion on their own. We will resist this urge except to cite one that we believe is important in the context of WebSphere and Domino development. It is the Eclipse architecture of tools and plug-ins that IBM wanted to expose to developers and get them writing software for. A tool written for Eclipse is a tool that will work with WSAD. We will discuss below how IBM itself has utilized the Eclipse tool plug-in architecture to provide Domino development functions within WSAD.

Eclipse is written in Java, and the plug-ins are Java entities delivered in a .jar (Java Archive) file. The actual connection between a plug-in and Eclipse is described in an XML file. This file carries information, such as which other plug-ins this plug-in depends on, which extension point the plug-in connects to (where it hooks into Eclipse), or which classes within the plug-in are associated with each extension point. In short, the XML file defines how the plug-in will be bound to the Eclipse environment. Each plug-in is stored in its own folder within a special plug-ins directory. When Eclipse starts up, it traverses this directory, reading all the XML files and building an internal registry of all available plug-ins. The plug-ins are not loaded from the start, but are instead loaded only when needed. Once a plug-in is activated, it is never removed.

With this plug-in format, Eclipse 2.1 can deduce how plug-ins are hooked in without having to load any Java code. Consequently, Eclipse launches more quickly than it otherwise would. And because an Eclipse plug-in has no life cycle, you avoid complex life-cycle management issues. Because multiple plug-ins can attach to a given extension point, however, it also means that the first time an action requires one or more plug-ins to be activated, response can be slow as Eclipse activates the series of "attached" plug-ins. In addition, a running Eclipse session will consume more and more memory as plug-ins are loaded over time; however, the Java VM (virtual machine) garbage collector will mitigate the forever-upward memory usage.

Eclipse Plugin Central, a portal where developers can find Eclipse-based software tools, is located at `http://www.eclipseplugincentral.com`.

The Lotus Domino Toolkit for WebSphere Studio

The Lotus Domino Toolkit for WebSphere Studio (LDTWS) is an Eclipse plug-in that provides the WebSphere Studio developer with views and tools to integrate Domino database design elements such as forms, views, and agents into a J2EE application. Specifically, the LDTWS provides a set of JSP tags that use the Domino Java classes to access NTF databases on the same server as the JSP or via a separate Domino server system. The Domino JSP tags are packaged as two tag libraries that can be included in a Web project or individual JSP file within WSAD.

At the time of this writing, there are two versions of the LDTWS. LDTWS 1.0 is intended to be installed to WSAD 5.0, and LDTWS 1.1 is intended for WSAD 5.1. The discussion in this section applies to either version.

Using the LDTWS

How easy is it to use the LDTWS tag libraries? The straightforward way is to add the Domino tags to a JSP manually using the JSP editor to insert the explicit HTML source for the tag definition. For example, suppose we want to populate a text area in a JSP with the value of a field in a certain document in a Domino database. We could edit the JSP and enter the Domino tag `<domino:item id="somefield"/>` as the HTML source.

There's an even easier way. The LDTWS plug-in provides a special WSAD navigator view, "Domino," which can access and open any Domino database available on your local workstation or via a Domino server. Once a Domino database is made accessible in this view, we can select any Domino design element directly and drop it onto a JSP in the editor, and the Domino tag corresponding to the design element is created in the JSP source. OK, this is easier.

Before we look more closely at how to use it, let's consider the broader implications of the Domino navigator view. Not only does it make adding Domino tags to JSPs easy, it at the same time makes available all of your existing Domino designs (from databases or design templates) to be reused as JSP design elements. But you may be thinking…I can just enable my Domino databases for Web access. True enough. However in many cases, making a Domino server directly accessible on the Internet (as it would have to be to allow the Domino HTTP server task to serve the database pages) causes consternation among the security minded. By using JSPs, the JSP engine can be separated from the HTTP server across a firewall and the JSP engine from the Domino server. In addition, using JSPs allows the entire Web application to be presented from the same server rather than having some pages come from WAS and some from Domino. Finally, it enables Web application developers unfamiliar with Domino design tools to work with Domino design elements at an easy to grasp level.

Now let's look at how to use the Domino navigator. You can access it from the Web perspective within WSAD by selecting the Domino tab at the bottom of the navigator pane as indicated in Figure 8-1.

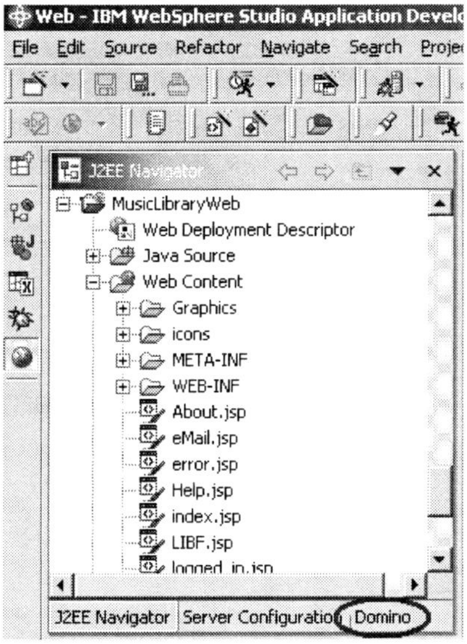

Figure 8-1 Domino tab.

The Domino navigator view is then displayed and initially looks as shown in Figure 8-2. This view is blank the first time you open it.

To open a Domino database in this view, either select the first icon on the title bar, or right-click in the view frame and select "New database connection…" The result is the dialog shown in Figure 8-3.

There are two ways to specify the input fields in this dialog depending on where the Domino database you want to access resides. If the database is on your workstation, then leave the Domino Server and User fields blank and specify the file name of the database in the Path field. WSAD looks for the database under the default Notes path on your workstation—if it's in another directory, you must specify the full path to the database (see Figure 8-3).

Figure 8-2 Domino navigator view.

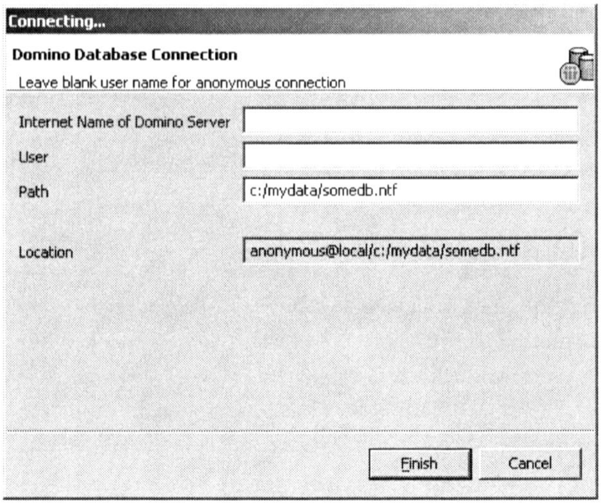

Figure 8-3 Domino Database Connection dialog.

If the database is on a remote Domino server, then specify the TCP/IP host and domain name of the server in the Domino Server field and, optionally, the user name under which to access the database in the User field. For local databases, specify the file name of the database in the Path field.

Press the Finish button to establish the database connection. (If a user name was specified, the user is prompted for the password for that user name.) If the database connection can be established, the database path appears in the Domino view window as shown in Figure 8-4.

Figure 8-4 Domino view with database connection.

Now you can select any of the design elements shown by this view, including forms, individual fields, entire views, or columns within a view and server agents. In Chapter 9, "Accessing JE22 Elements from Domino," we discuss the Domino tag libraries and their usage in more detail.

Testing LDTWS Applications

Now that you've built a Web application that accesses Domino databases using the Domino tag libraries, how does one go about testing? Certainly, you can set up a server system for testing and install a version of WebSphere Application Server and Domino on this server. Then it is a matter of deploying your application to the WAS system (along with the Notes.jar library) and testing from this system. Depending on your organization's development environment, this may be relatively easy, or it may present difficulties.

If your development workstation is capacious (with a capital "C"), another possibility is to test solely on the workstation. WSAD V5 comes with a nearly full-fledged version of WAS V5 itself, which can be run from within the WSAD workbench. This WAS within WSAD is referred to as the WebSphere Test Environment (WTE). The WTE can be configured and started to run servlets, JSPs, EJBs, and even JMS components. With a bit of extra configuring, the WTE can easily run JSPs utilizing the Domino tag libraries. With just this "out of the box" setup, JSPs using the Domino tags can be tested against a remote Domino server system. Going a step further, it is possible to set up a Domino server on a workstation with WSAD to get a self-contained test system.

Here's how: Install the Domino 6 server to the workstation following the standard install and server setup procedure. (The type of server to install, Messaging, Utility or Enterprise, is not important for testing the Domino tags under WTE.) Once the server is installed, add the top level path to the server to the `PATH` environment variable. For example, the usual server path is `/opt/lotus/domino` on a Unix system or `C:\Lotus\Domino` on a Windows system. (If a Domino client is also installed on the workstation, it is important that the Domino server path precede the client path in the `PATH` string—this order will have no effect on the client execution.) With this `PATH` string setting, the Domino Java classes (in Notes.jar), which are invoked by the Domino JavaScript tags, find the notes.ini file for the Domino server and then are able to access Domino databases on the workstation itself. This is precisely how the Domino tags within JSP classes can access Domino databases under the WTE. Notice that nothing was mentioned about starting the Domino server—accessing Domino databases via the Notes.jar does not require a running Domino server.

There it is—WSAD/WTE and Domino server installed on a single system such that JSPs utilizing the Domino tag libraries can be written and tested.

Domino Designer Features for J2EE Development

Domino Designer is the development tool component of the Lotus Notes/Domino product suite. For many developers, Designer is and has been the primary tool for application development in their enterprise. It provides the necessary ease of use features for rapid application development (RAD) around the Notes database model, and allows for the addition of sophisticated programming control to provide almost any degree of automation required. Developers can program Notes database function using the Lotus Formula language, a set of parameterized functions; LotusScript, a Visual Basic-like script language with object-oriented features; or Java and JavaScript.

Designer is a mature product, having its origins with the first Notes client/server offerings from Lotus and growing with the Domino family under the IBM/Lotus unbrella to its current version with Domino R6. This growth, from a tool to code Notes @Formula language, to LotusScript, to HTML, to Java, and now with support for XML and JSP tags, follows the path of application evolution, from Notes client-centric, to Web accessible, to J2EE and Web service-oriented.

A Brief Look at Domino Designer

To get a feel for what Designer offers Web application building, let's take a brief look at how it is used. Needless to say, an exhaustive presentation of Designer function and features is beyond the scope of our book, and there are many references you can go to for this level of detail. (One in particular is the IBM Redbook, "Domino Designer 6: A Developer's Handbook," SG246854, available at `http://www.redbooks.ibm.com/redbooks/SG246854.html`.)

Looking at a typical Designer layout in Figure 8-5, we find the following features:

- A Bookmarks View pane (the left-most frame) shows the Domino databases recently opened in Designer. The active database expands to a Navigator view, which shows the the Domino design elements within the database. For example, under the Forms branch in the MusicLib.nsf database, the Library Item form is shown.
- The Editor pane (the blue shaded pane containing the form contents) displays the contents of the design element being edited.

- The Programmer's pane (the two frames below the editor pane) allows the entry and editing of code associated with the design element. The left side of this pane shows the objects or events and reference information available for the language being used. The right side is where code is entered and edited. The code can be Lotus @Formula language, LotusScript, JavaScript, or Java (depending on the type of event or where the element is intended to run—for example, in the Notes client or in a Web browser).

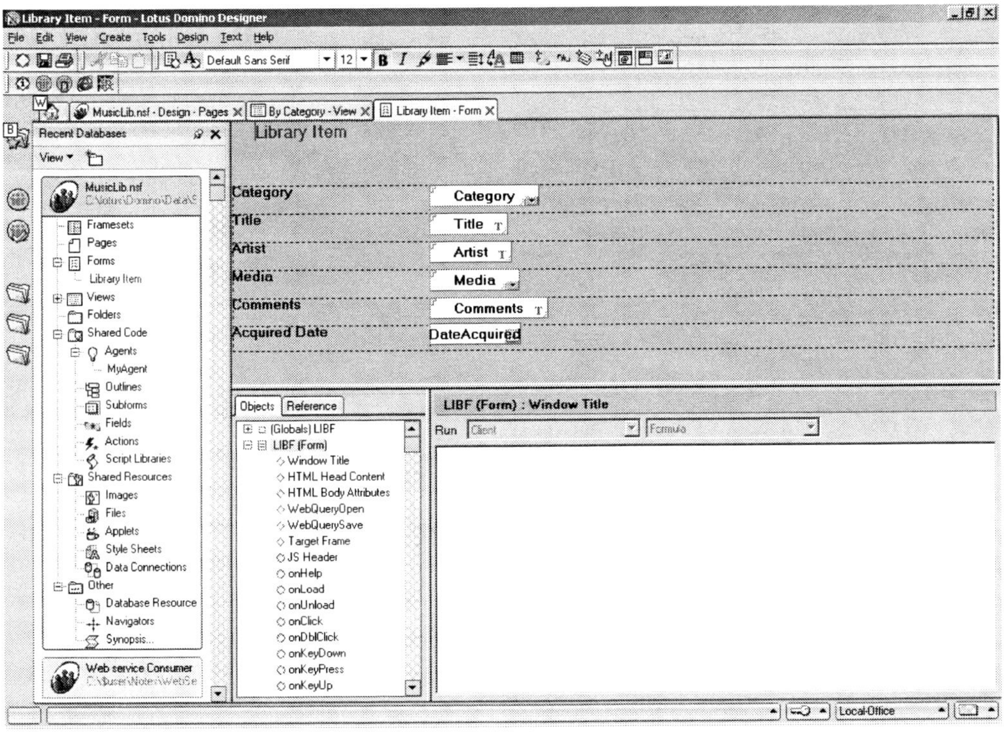

Figure 8-5 Domino Designer R6.

Using these views and the actions and tools available in Designer, a developer builds a Domino application by constructing one or more databases with the appropriate forms, views, agents, resources, and so on—and possibly code associated with the elements and their events. Some of the Domino design elements may be familiar to a Web developer, such as Framesets, Pages, Applets, and Style Sheets, whereas others will be more familiar to a Domino developer such as Forms, Views, Agents, and so on.

Features for J2EE Development in Designer

What can we find in Domino Designer that will help us build applications in the J2EE environment? Among the recently added features to the Domino R6 version of Designer, there are three that provide such aid: XML support for Domino database elements, JavaScript tag libraries for Domino elements and, although not an explicit Designer feature, the ability to implement Web service providers and consumers.

As XML is used as the basic data format syntax for many Web enabled applications and especially for Web services, it is important that the Domino application development environment provides the means to work with Domino elements in terms of XML. Beginning with Domino R5 and continuing with Domino R6, Lotus has provided XML support in the following ways:

- Lotus has specified a Document Type Definition (DTD) for Domino database elements such as views, forms, and documents. The XML described by this DTD is referred to as "Domino XML" or DXL.
- Domino Designer provides tools (via the menu bar) to export design elements as DXL to files, view the DXL, or apply XSL transforms to DXL.
- The Domino Java class library includes support for generating DXL for parsing and transforming Domino objects such as documents.
- Domino includes the XML4J parser and the LotusXSL transform processor, which enable agent code to parse and transfrom XML (and DXL). (The XML4J and LotusXSL code were contributed by IBM/Lotus to the Apache open source Xerces parser and Xalan XSL transformer projects.)
- Domino Designer provides LotusScript classes, which allow generating, parsing, and transforming XML and DXL.

Let's examine the Designer XML related features a bit more closely. We'll look at how Designer enables XML to be manipulated in Java and in LotusScript. For Java, which is primarily used for agent and applet code, Designer only provides explicit support for those XML related classes defined as part of the lotus.domino package and a small subset of the XML4J and XSLT classes (see Figure 8-7). By "explicit" we mean that the classes are listed in Designer's reference pane and help viewer. This largely reflects Designer's focus on the Domino model and is more a lack of convenience than a restriction of function since any class made available via the Domino classpath can be used within Domino Java code. Figure 8-6 shows some of the XML related classes in the lotus.domino package.

The `lotus.domino.Document.generateXML()` methods shown in Figure 8-6 can be used to generate an XML (DXL) stream from a Document object either as a String, as input to a Writer object, or as the result of an XSLT transform (the `generateXML(Object, XSLTResultTarget) signature`). The agent code shown is an example of how the generateXML method is used to generate the DXL for all documents in a database to a file.

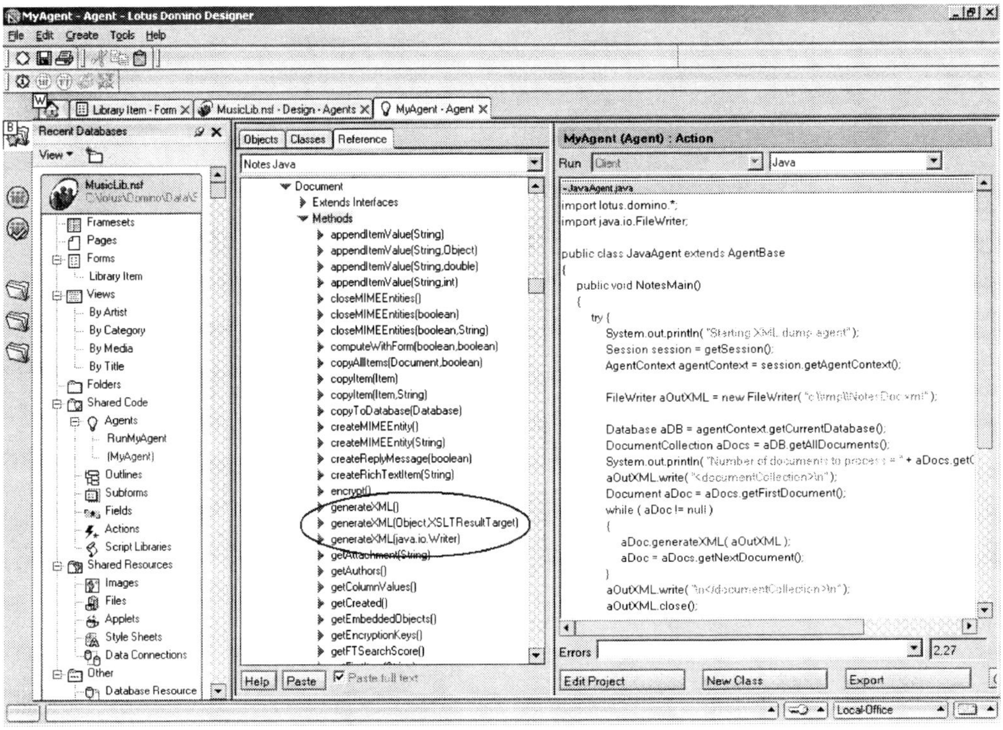

Figure 8-6 Lotus.domino XML classes.

> **NOTE**
> You might be wondering if this same function can be accomplished via the DXL utility menu items mentioned here. The answer is no; the Tools>DXL Utilities menu functions apply only to Designer (design) documents—you must write an agent like the one shown to generate or transform XML for regular documents in a Domino database.

Figure 8-7 shows the subset of XML classes from the `org.w3c.dom` and `org.xml.sax` packages, which are provided explicitly by Designer. The `org.w3c.dom` classes can be used to traverse an XML DOM tree generated by a parse of a Domino document element (such as an attachment, rich text field, or item text). The three `org.xml.sax` classes are provided to enable input to the XSL transform function and to handle exceptions from the XML transformer (which uses a SAX parser for the XSL stylesheet). Note that these packages appear under the Third-Party Java drop-down item.

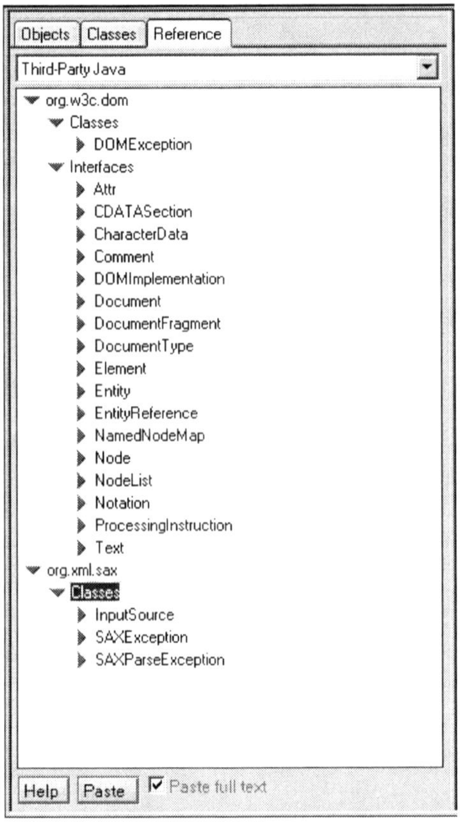

Figure 8-7 Designer XML-related classes.

Lotus Web Services Enablement Toolkit

The Lotus WSTK was developed and offered by IBM/Lotus in May of 2001 and contains several early examples of contructing Web services in the Domino environment. Specific examples illustrate how to define Web service providers to Domino applications and how to have Domino applications consume Web services. Being such an early offering in the Web service evolution, the WSTK takes a more or less brute force approach to constructing the Web service interfaces, using the level of Java and LotusScript function available at that time. For example, there are samples provided for constructing a Web service using the `org.apache.soap` Java packages, where the instructions describe how to manually package and deploy the Web service Java application to a SOAP-enabled Web server. Several sample applications are provided with the toolkit, which demonstrate interfaces to various Lotus products such as Discovery Server, Learning Space, and Sametime Connect.

Although it does not show off the latest development techniques, the WSTK still does provide useful examples of integrating Web services with Domino applications.

SOAPConnect

SOAPConnect is a more recent toolkit offered by IBM/Lotus and enables LotusScript applications to invoke Web services. It is available either from the Lotus Developer Domain (`http://www-10.lotus.com/ldd`) as a stand-alone package (in the Sandbox area of the LDD site) or as part of the IBM Web Services Toolkit package from the IBM AlphaWorks site. To be precise, SOAPConnect is a partial implementation of the SOAP V1.1 standard for the LotusScript language. It provides several new LotusScript classes for formating and sending SOAP requests and receiving and accessing the responses.

Note that SOAPConnect provides for invoking or "consuming" Web services. It cannot be used to expose Domino functions as Web services. As we will show in subsequent chapters, Domino Web agents (LotusScript) or Domino Objects for Java can be used to provide Web service access to Domino function.

Future Direction for WebSphere/Domino Development Tools

One future direction for WebSphere/Domino development tools is in the Eclipse area. As mentioned at the beginning of this chapter, the core of the WebSphere Studio Application Development IDE platform was released as an open-source software project, Eclipse (www.eclipse.org). The Eclipse project originated as an open consortium of software development tool vendors interested in creating a development environment that fosters product integration through plug-and-play, componentized tools. As of mid 2003, the organization Web site had delivered more than seven million download requests, incorporated more than 30 member companies, and recognized more than 250 Eclipse projects. WebSphere Studio 5.0 is based on Eclipse 2.0.

Eclipse 2.1

The most recent version of Eclipse—2.1—makes it even easier for developers to build tools that will mesh with the Eclipse platform. Eclipse 2.1 adds support for MacOS to the list of many operating systems it already supports (AIX, HP/UX, QNX, Linux, Windows, and Solaris). The release also adds Ant utility support and more flexible project layouts. The Java development tools received editor and debugger improvements, refactorings, and other enhancements.

Eclipse 3.0

Even more and bigger changes will come with the next major Eclipse release. Eclipse 3.0 is targeted for release in the second quarter of 2004. With that version, the organization is focusing on three themes: the user experience, scalability, and ease of use. Because the number of plug-ins has gone from a handful at Eclipse's conception to hundreds today, the organization wants to simplify the platform's ease of use. The organization also would like to increase the scalability of the platform. And finally, the group intends to broaden the user experience so that developers can do more with the platform.

Taking Eclipse into yet another area, IBM and others have organized Project Hyades to take the tools developing into the testing arena. Hyades implements an Object Management Group (OMG)-defined Unified Modeling Language (UML) testing profile, including Test Case, Test Trace, and Test Objective and Verification artifacts. The tools will let developers evaluate objectives related to performance and scalability for a range of real-world deployment environments, including alternate mixes of server and network interconnection technologies. Hyades will make it easier to integrate a broad range of functional verification, quality assessment, and load testing tools with the Eclipse platform's workbench and other tools so that developers will have access to tools that span a full lifecycle of project development, from design to development to testing and deployment.

Eclipse has expanded its global reach with translated versions in nine languages, including French, German, Italian, Brazilian Portuguese, Spanish, Simplified Chinese, Korean, Japanese, and Taiwanese. Development platforms include IBM AIX, Sun Solaris, and HP-UX. Eclipse also supports Windows and Linux on Intel-based computers. There is an extensive set of plug-in tools built on the Eclipse platform and included in WebSphere Studio to integrate IBM's entire middleware platform. The result is a portal-like environment for developing, testing, and debugging.

The portal-like programming environment is similar to a Web portal in that it lets developers customize their environment using WebSphere Studio to select the tools, data, and assets needed for individual projects. The tools signal a maturing e-business development environment. In using an open application foundation such as Eclipse, vendors no longer compete on providing the basic developer tools, but rather on providing specialized, higher-level technology. This can only be good for developers as they will have only a single framework to learn, but the ability to enhance it in many ways.

Beyond the IDE

One of the interesting ways people might be able to use future versions of Eclipse is to build applications and not just integrated development environments. Some have already started working toward that concept. Seen in this light, Eclipse is a natural framework for "rich client" types of applications.

IBM in particular has announced an Eclipse-based client for Lotus workplace functions. It is hard to over-emphasize the import of the phrase "Eclipse-based" in this last statement. By being based on Eclipse, the Lotus client immediately addresses several requirements that Lotus Notes customers have been interested in for a long time. An Eclipse-based client would be inherently cross-platform, tailorable, and extendable (even by third parties).

CHAPTER 9

Accessing J2EE Elements from Domino

"Then came the Internet. Microsoft CEO Bill Gates called it the 'Internet tidal wave.' After the initial hype by the popular press, it soon became clear that the Internet's biggest immediate impact would be on the corporate workplace. Internet technology makes it easy to publish and share information with colleagues, whether in the same office or in branches overseas. Intranets, as these internal networks are called, seem to promise all the capabilities of Lotus Notes, and they are easier to set up and cheaper to maintain."

"But in fact, the Internet may have been what Notes needed all along. Since it launched Notes in 1989, Lotus's biggest challenge has been getting people to understand its purpose. The Internet, the World Wide Web, and intranets give people a window into the world of collaboration across a network. Eric Schmidt of Sun (at that time) says: 'Notes will offer more of the pieces for intranets than any other solution.'"

—*Fortune*, July 8, 1996

When Lotus introduced Workplace, its open-standards-based, modular collaboration platform built on J2EE, there became two Lotus options for J2EE: Workplace and the traditional Lotus Notes/Domino products combined with WebSphere. Lotus has produced a Domino upgrade road map through 2005 that shows the path to J2EE. The portal-like interface offered in Notes/Domino R6.5 is designed to bring Domino users to "the same ultimate destination" as Workplace.

Notes/Domino R7, due out late in 2004, will move even closer to Workplace. It will give businesses the ability to use DB2, rather than the traditional NSF database, as the Domino data store. (Workplace runs atop DB2 and WebSphere.) Ultimately, the option of using a relational database with Domino should mean a faster e-mail client, letting end users benefit from improvements in maintaining views, and clearing caches. Notes/Domino R7 will also feature changes to Domino Designer and WebSphere Studio, including better rapid application development capabilities. Plus, it will deliver disconnected mail support for managing e-mail data off-line.

Notes/Domino R8, not due for delivery until 2005, will feature the ability to integrate third-party J2EE-based applications, such as customer relationship management software in Domino. In the meantime, we still have many ways to build Domino applications using J2EE-based functions.

This chapter deals with the currently available J2EE access from Domino R6 to the WebSphere Application Server (WAS) V5 product.

WebSphere Web Server Plug-In for Domino

The WebSphere product versions provide Web server "plug-ins" for various Web servers enabling the Web servers to forward requests for J2EE Web elements, namely servlets and JSPs, to the WebSphere Application Server. (As with most J2EE application servers, serving HTML is not a primary focus for WAS. WAS V5 does come with an embedded HTTP server based on the IBM HTTP Server, but it is provided mainly to handle specific features such as WAS administration and to offer HTTP itself as a means of communication between external components and WAS.) In particular, a Web server plug-in is provided for the Domino HTTP Server, which allows requests for servlets and JSPs directed to Domino to be handled by a WAS node. This plug-in mechanism was one of the first integration points between the Domino and WAS products.

The WebSphere Domino HTTP Server plug-in can be viewed as a basic means of enabling J2EE access from Domino. With the plug-in installed, access to J2EE elements is via explicit URL links. For example, sending a URL of the form `http://A.Domino.Host/servlet/SomeServlet` to a Domino Server will result in the Domino HTTP task passing the request via the plug-in to a WAS node, which passes the response back via Domino. Figure 9-1 depicts Domino and WAS using the WAS HTTP plug-in.

Let's examine more closely how this WebSphere plug-in is configured and how it operates. The plug-in itself is provided as a shared library (DLL for Windows, .a for AIX, .so for Solaris, and as of this writing no plug-in exists for Domino on Linux) and a default configuration file: plugin-cfg.xml. The plug-in files are packaged with both the Domino and WAS product installations and can be automatically installed during a WAS installation. The plug-in also can be installed manually. The plug-in files can be found under the Domino install path: Domino/Data/domino/plugins/was5.

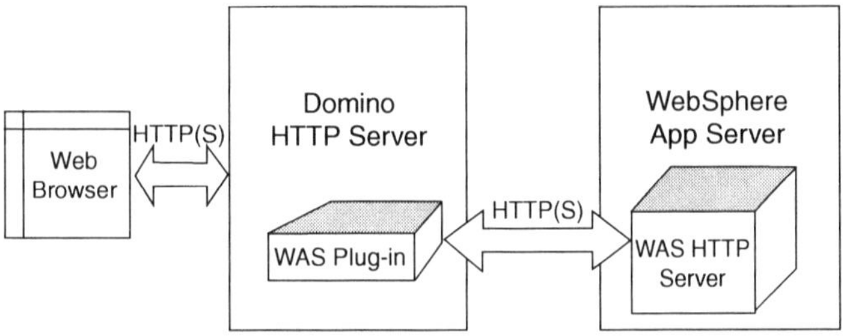

Figure 9-1 WAS plug-in for Domino HTTP.

There are two essential steps to installing the plug-in: 1) specifying the plug-in library file to the Domino HTTP Server and 2) specifying the location of the plug-in configuration file to the plug-in itself.

The first step is straightforward and simply requires a change to the Domino server document in the Domino directory. You specify the plug-in library file as a DSAPI filter file on the Internet Protocols/HTTP page of the server document as indicated in Figure 9-2.

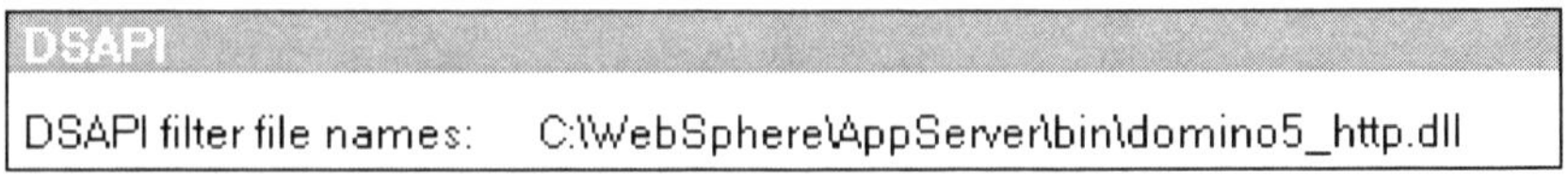

Figure 9-2 Domino Server plug-in setup.

The second step is not as straightforward and is operating system-dependent. The plug-in library itself looks for the plug-in configuration file via an operating system environment variable (registry key on Windows). Unfortunately, these variable names are not well advertised in the WebSphere or Domino documentation. For Windows systems, a registry key value needs to be set as shown in Figure 9-3 (you can use the `regedit` utility program to create the required key path if it is not already defined). For Unix systems, the plug-in looks for the configuration file in the `config` sub-directory under the path defined by the `WAS_HOME` environment variable, that is, `$WAS_HOME/config`. For IBM iSeries systems, the configuration file location is specified in the Domino notes.ini file.

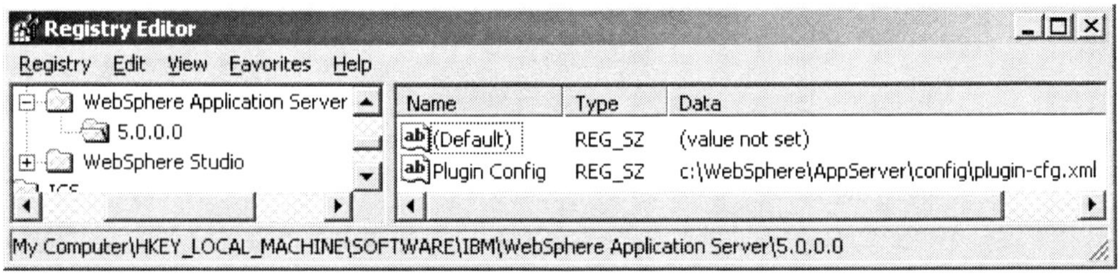

Figure 9-3 Registry setting for the WAS plug-in.

Once these settings have been made, you can restart the Domino HTTP Server task to have it load the WebSphere plug-in as a DSAPI filter. If the settings are correct, you will see the following messages at the Domino console:

```
HTTP Server: Java Virtual Machine loaded
WebSphere HTTP DSAPI filter loaded
HTTP Server: DSAPI WebSphere HTTP DSAPI Filter Loaded successfully
HTTP Server: Started
```

Next, let's take a closer look at the plug-in configuration file, plugin-cfg.xml, to see what it controls and how it can be tailored. This configuration file specifies three key aspects of how the plug-in operates: which URL hostnames and which URL types get routed to WAS and to which WAS servers they get routed. It also allows specification of the log file written by the plug-in and the level of logging. The following shows the configuration file that comes with Domino R6:

```
<Config>
   <!-- The LogLevel controls the amount of information that gets
        ➥written to the plug-in log file. Possible values are
        ➥Error,Warn, and Trace. -->
   <Log Name="C:/WebSphere/AppServer/logs/native.log" LogLevel=
   ➥"Error"/>

   <!-- Server groups provide a mechanism of grouping servers
        ➥together. -->
   <ServerGroup Name="default_group">
      <Server Name="default_server">
         <!-- The transport defines the hostname and port value
              ➥that the web serverplugin will use to communicate
              ➥with the application server. -->
        <Transport Hostname="localhost" Port="9080" Protocol=
        ➥"http"/>
      </Server>
   </ServerGroup>

   <!-- Virtual host groups provide a mechanism of grouping
        ➥virtual hosts together. -->
   <VirtualHostGroup Name="default_host">
      <VirtualHost Name="*:*"/>
   </VirtualHostGroup>

   <!-- URI groups provide a mechanism of grouping URIs together.
        ➥Only the context root of a web application needs to be
        ➥specified unless you want to restrict the request URIs
        ➥that get passed to the application server.  -->
```

```
    <UriGroup Name="default_host_URIs">
       <Uri Name="/servlet/*"/>
       <Uri Name="/webapp/examples/*"/>
       <Uri Name="*.jsp"/>
       <Uri Name="*.jsv"/>
       <Uri Name="*.jsw"/>
       <Uri Name="/ErrorReporter"/>
       <Uri Name="/j_security_check"/>
       <Uri Name="/tradetheme/*"/>
       <Uri Name="/theme/*"/>
       <Uri Name="/WebSphereSamples/*"/>
       <Uri Name="/estore/*"/>
    </UriGroup>

    <!-- A route ties together each of the above components. -->
    <Route ServerGroup="default_group" UriGroup="default_host_URIs"
         VirtualHostGroup="default_host"/>
</Config>
```

As with all WAS configuration files, the plug-in configuration file has XML syntax and thus can be edited with any text editor. The `Log Name` element specifies the file name of the log file written by the plug-in and the level of logging desired. The `ServerGroup` element specifies the WAS server(s) to which URLs are routed and the protocol to be used. The default is to route URLs to the local WAS server at port 9080 (the embedded WAS HTTP Server) over HTTP. Additional servers can be specified here, as well as HTTPS instead of HTTP. If more than one server is specified within a server group, the plug-in routes to the servers in a round-robin fashion. If HTTPS is used, additional system environment variables must be set pointing to the SSL libraries and the certificate and encryption key files required by the plug-in for the HTTPS protocol.

One or more `VirtualHostGroup` elements can be specified to define the hostname patterns, which the URLs must match to be considered part of that host group. Similarly, one or more `UriGroup` elements can be specified to define the types of URLs belonging to the URI group. Finally, the `Route` element specifies a single routing definition in terms of the virtual host and `Urigroup` elements. That is, a `Route` element defines that URLs belonging to the specified `VirtualHostGroup` and `UriGroup` should be routed to the specified `ServerGroup`.

The operation of the plug-in is straightforward. Every URL sent to the Domino HTTP task is intercepted by the plug-in and compared to the set of `Route` definitions. If a match is found, the URL is routed to a server specified in the specified `ServerGroup` for the Route definition. Otherwise, the URL is passed back to the Domino HTTP task for processing. This is straightforward indeed, but some care should be taken. Since every URL, including requests for images, goes through the plug-in, and each URL is compared against every `VirtualHost` and `Uri` element within each `VirtualHostGroup` and `UriGroup`, the `VirtualHost` and `Uri` elements should be kept as minimal as possible. If a URL pattern is not supported by the WAS node, you should remove it from the `UriGroup`. (The fact that every URL is passed to the plug-in is due to the nature of the Domino DSAPI used by the plug-in—and can't be avoided.)

The WAS HTTP Server plug-in periodically rereads the plug-in configuration file, so any changes made to it are picked up by the plug-in without having to restart the Domino HTTP task.

Invoking J2EE Elements from Domino Applications

Now we'll discuss techniques for invoking J2EE elements, such as servlets, JSPs, and EJBs from Domino applications presented over the Web. In this context, we are assuming that a Domino server is handling requests for Domino database elements (documents, forms, views, etc.) and that the J2EE elements are being invoked from the Domino elements—for example, invoking a servlet to process a Domino form submission. In fact, let's start with just that scenario.

Servlets/JSPs on Form Submission

Since a JSP is a type of servlet, assume that you can replace "servlet" with "JSP" in what follows. To get a Domino form processed by a servlet, you must somehow invoke the servlet when the user hits the Submit button (or equivalent) on the form displayed in the user's browser. There are two commonly used techniques: overriding the `$$Return field` or overriding the HTML form.

The `$$Return field` override method makes use of the fact that the Domino server, in response to a form submit action (`document.forms[0].submit()`), sets the HTML response output as the contents of the `$$Return field` in the Domino form if this field exists. By setting the `$$Return field` to a URL string enclosed in brackets, the Domino server will invoke this URL. If the URL string is set to specify a servlet, the servlet is invoked as a result of the form submission. The servlet URL string can be set such that any of the form's field contents appear as parameters to the servlet. But note that these parameter values will be carried as clear text (not encrypted) in the servlet request URL string, so avoid passing sensitive data this way.

The HTML form override method involves modifying the HTML generated by the Domino server for the form. The form HTML is modified to specify a secondary HTML form element with its `METHOD` and `ACTION` parameters set to the servlet to invoke on the submit action. To use this method, the programmer creates the Domino form with "pass-through HTML" text at the top of the form as in the following:

```
</Form>
<Form METHOD=POST  ACTION="/servlet/MyServlet">
```

This HTML, when processed by the browser, ends the 0th form object generated by Domino and starts a new form object (`forms[1]`) with the specified method and action parameters. Then, on a `document.forms[1].submit()` action, the servlet specified by the `ACTION` parameter is invoked. Since the fields on the Domino form are rendered by the browser as HTML input fields, they get passed to the servlet in the request header by the `POST` method (unless the fields had the HTML attribute "`Type=Hidden`" specified). Note that like the `$$Return field` method, the servlet parameter data is passed in "clear" text, so be careful what you pass as parameters.

EJBs from Domino

As discussed in a previous chapter on J2EE features, Enterprise Java Bean (EJB) components possess architectural features that make them well-suited for use in certain areas of application design, namely what's often referred to as "business logic" and persistent data storage. The two types of EJBs largely correspond to these uses: Session EJBs are used for business logic, and entity EJBs are used for persisting data. Given that EJBs are designed to be invoked in a distributed object, client/server fashion via a J2EE application server, can they be invoked via WAS from within Domino elements?

The answer for Domino is similar to the answer for other applications—it is often not easy nor flexible to invoke EJBs from outside of the J2EE application server, especially server to server. The reason lies with the communication protocol used for EJB client/server interaction. This EJB distributed object protocol is Internet Inter-ORB Protocol (IIOP), which is based on the Common Object Request Broker Architecture (CORBA). For Java, the CORBA/IIOP communication is encapsulated by the Remote Method Interaction (RMI) package. Thus, to utilize EJBs in a distributed fashion, you must communicate via Java RMI. The use of RMI often results in design issues and complexities. For example, RMI implementations entail certain restrictions on thread usage, which conflicts with that of the Domino agent manager JVM, and IIOP uses non-fixed TCP/IP ports, which are difficult to handle through firewalls. So even though it is possible to invoke EJBs directly from Domino Java code, it is usually not worth the design difficulties that ensue.

As with other J2EE applications and frameworks, the simpler approach to utilizing EJBs from Domino is indirectly via servlets (JSPs), or more recently, by using Web services or JMS message queues (for message-driven EJBs). With these approaches, the code that needs to utilize the EJB invokes a servlet or Web service or puts a message to a queue, which, in turn, invokes the EJB. This way, the Domino-to-WAS communication is via HTTP, which is straightforward to code and passes through firewalls. Also the EJB interaction is entirely within the application server and can be made very efficient by the use of local EJB interfaces.

Let's look at a specific example of how an EJB can be invoked via a servlet by a Domino agent and from a Domino form action. In this example, we make use of an entity EJB, `BookBean`, representing a single book published by a fictitious publishing house. Our EJB was developed within WSAD using the EJB creation and deployment wizards (see Chapter 8, "Development Tools," for references on how to develop EJBs within WSAD). The `BookBean` EJB is defined with a local home interface and with Container Managed Persistance (CMP). We deploy the EJB module containing BookBean (and its interface components) to a WTE server using the CloudScape RDBMS engine for the CMP implementation. Deploying the EJB module this way allows us to test the EJB under the WTE on our workstation without requiring access to an external database. Figure 9-4 shows the `BookBean` definition as viewed via the WSAD EJB Deployment Descriptor "Beans" page.

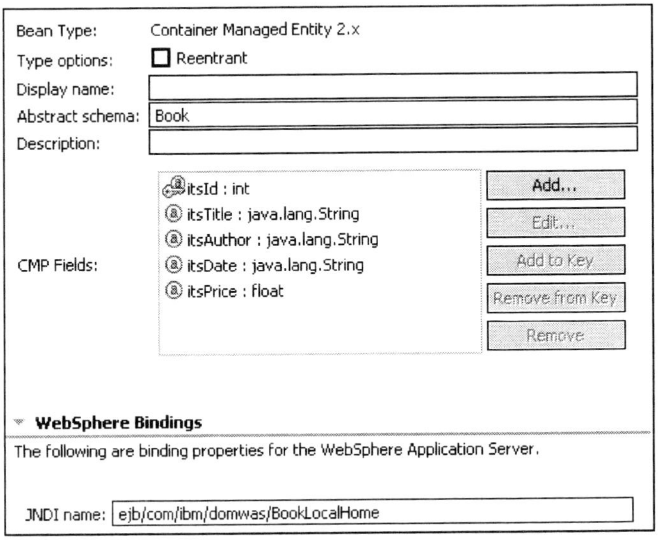

Figure 9-4 BookBean EJB.

Besides the basic EJB components generated by the WSAD wizards (those needed for the EJB to conform to the EJB 2.0 CMP specification), we add another finder method, `findByTitle`, to the local home interface via the EJB Query Language (EJBQL) feature of the EJB 2.0 CMP. This again is a wizard-driven procedure that results in the required Java code being added to the EJB components without any explicit Java coding from us.

Next we add a servlet to the WSAD J2EE project containing our `BookBean` EJB module, again using a wizard. (We tell our children that being a modern-day software engineer is akin to being Harry Potter.) The servlet is named `GetBook` and provides the interface to the `BookBean` EJB via the `GET` request function (i.e., via the `HTTPServlet.doGet() method`). Since we are working within the same J2EE project under WSAD, it is a simple matter to reference the EJB from the servlet: First, we add the BookBean EJB reference to the Web deployment descriptor. Then we add the Java code to the servlet to access the `BookBean` EJB via its local home interface. Figure 9-5 shows the References page from the Web deployment descriptor after adding the EJB reference.

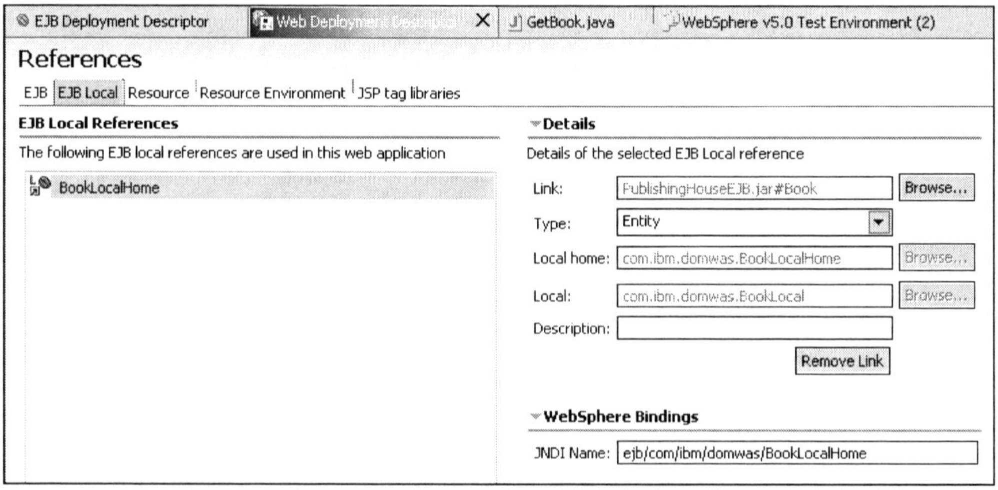

Figure 9-5 EJB reference from Web Deployment Descriptor.

The following listing shows the servlet code. We chose to lookup the JNDI initial context and the EJB local home interface object in the servlet's initialization method, which is called only when the servlet is (re)loaded. Note that the JNDI name used for the initial context lookup must be of the form: `java:comp/env/<EJB reference name>`.

```
import java.io.*;
import javax.servlet.ServletException;

import javax.servlet.http.HttpServlet;
import javax.servlet.http.HttpServletRequest;
import javax.servlet.http.HttpServletResponse;

import java.util.*;
import javax.naming.*;

/**
 * GetBook Servlet - Obtains data from Book EJB.
 * @version   1.0
 */
public class GetBook extends HttpServlet
{
   // Holds EJB Home interface
   BookLocalHome itsBookLH;
```

```
   /**
    * Handle the Get request - look for the Book EJB with the
      ➥given id.
    *
    * @see javax.servlet.http.HttpServlet#void
(javax.servlet.http.HttpServletRequest,
javax.servlet.http.HttpServletResponse)
    */
   public void doGet( HttpServletRequest req, HttpServletResponse
   ➥resp )
      throws ServletException, IOException
   {
      PrintWriter aPW = resp.getWriter();
      aPW.println( "<HTML>\n<HEAD>"
         + "\n<META http-equiv=\"Content-Type\" content=\"text/
         ➥html; charset=ISO-8859-1\">"
         + "\n<META http-equiv=\"Content-Style-Type\" content=\
         ➥"text/css\">"
         + "\n<TITLE>GetBook Response</TITLE>"
         + "\n</HEAD> \n<BODY>\n"
      );

      try {
         // Lookup the Book EJB via one of its finder parameters.
         // We look for the finder parameters as an argument of
            ➥the GET
         // request.
         // 'id' = find by primary key.
         // 'title' = find by title.
         //
         BookLocal aBook = null;
         int aId;
         String aFinder = req.getParameter( "id" );
         if ( aFinder != null )
         {
            aId = Integer.parseInt( aFinder );
            aBook = itsBookLH.findByPrimaryKey( new BookKey
            ➥( aId ) );
         }
         else if ( (aFinder = req.getParameter( "title" )) != null
)
```

```
        {
           Collection aMatches = itsBookLH.findByTitle
           ➥( aFinder );
           aBook = (BookLocal)aMatches.toArray()[0];
        }
        else
           aPW.println( "<BR><B> Book finder argument not passed
           ➥on request! </B><BR>" );

        // Format Book EJB fields as HTML response.
        aId = ((BookKey)aBook.getPrimaryKey()).getItsId();
        aPW.println( "<FONT color=\"#0000ff\" face=\"Arial\"
        ➥size=\"+2\"><b>Book Details</b></FONT>"
           + "\n<HR> <BR>   ID Number: " + aId + "<BR>"
           + "<BR> Title: " + aBook.getItsTitle() + "<BR>"
           + "<BR> Author: " + aBook.getItsAuthor() + "<BR>"
           + "<BR> Date: " + aBook.getItsDate() + "<BR>"
        );
        aPW.println( "<BR> Price: " + Float.toString(
        ➥aBook.getItsPrice() ) + "<BR>" );
     }
     catch ( Exception e )
     {
        System.out.println( "doGet error: " + e );
        aPW.println( "<br> Requested book not found. " );
        aPW.println( "<br><br> Error: " + e );
     }

     aPW.println( "</BODY>" );
     aPW.close();
  }

  /**
   * Initialize the servlet - set up the EJB home interface.
   *
   * @see javax.servlet.GenericServlet#void ()
   */
  public void init() throws ServletException
  {
     InitialContext initContext = null;
     String aJNDIName = "java:comp/env/BookLocalHome";
```

```
        super.init();

        // Set up EJB home interface for use by servlet threads.
        // Get the initial context
        try {
           System.out.println( "Getting the InitialContext..." );
           initContext = new InitialContext();
        }
        catch ( javax.naming.NamingException e )
        {
           System.err.println("NamingException: "  + e);
        //   e.printStackTrace();
          }
        // Perform the lookup
        try {
              System.out.println( "Performing the lookup.." );
              Object obj = initContext.lookup( aJNDIName );
              itsBookLH = (BookLocalHome)obj;
        }
        catch ( javax.naming.NamingException e )
        {
             System.err.println( "Error retrieving the home
             ➥interface: " + e );
         //   e.printStackTrace();
        }
    }

}
```

Finally, we show an example of a Domino Java agent, which invokes the `GetBook` servlet. This code is a straightforward use of the `java.net.URLConnection` method to invoke a URL via an HTTP connection. After the Domino session and agent context objects are obtained, the agent accesses the document associated with the agent. If the agent is invoked from a document via a form action, hot-spot, or button, that would be the document itself. A `java.net.URL` object is constructed to represent the URL of the servlet to be invoked and its parameters. In this example, we invoke the servlet via the WTE, that is, via a URL specifying a host name of `"localhost"` and port 9080, and use the value of a field in the document as the servlet parameter. The servlet response text is read from the `URLConnection` object and can be processed by the agent code or, in the case where the agent was invoked via a browser, can be returned to the browser (via the `PrintWriter` object obtained from `AgentBase.getAgentOutput()`).

```
import lotus.domino.*;
import java.io.*;
import java.net.*;

public class JavaAgent extends AgentBase
{
   public void NotesMain()
   {
      // Get writer to agent's output (back to browser)
      PrintWriter aPW = getAgentOutput();

      try {
         Session session = getSession();
         AgentContext agentContext = session.getAgentContext();
         Document aDoc = agentContext.getDocumentContext();

         // Open URL for GetBook servlet.  Use doc booktitle as
            ➥argument to servlet.
         URL aURL = new URL( "http", "localhost", 9080,
            URLEncoder.encode(
"/PublishingHouseWeb/GetBook?title=" +
aDoc.getItemValue( "booktitle" ) ) );
         URLConnection aCnctn = aURL.openConnection();
         aCnctn.connect();

         // Copy servlet output to StringBuffer
         StringWriter aSW = new StringWriter();
         PrintWriter aPSW = new PrintWriter( aSW );

         BufferedReader aRdr = new BufferedReader(
            new InputStreamReader( aCnctn.getInputStream() ) );
         String aLine = aRdr.readLine();
         while ( aLine != null )
         {
            aPSW.println( aLine );
            aLine = aRdr.readLine();
         }

         // Here we could process the servlet's response...
         // For example, setting document fields with data from
            ➥the response.
```

```
            aDoc.save();
        }
        catch( Exception e )
        {
            aPW.println( "Agent error: " + e );
        }
    }
}
```

This same technique of invoking an EJB via a servlet also can be used anywhere JavaScript is written as part of a Domino design element. For example, the `onClick` event of a button may be written in JavaScript to invoke the EJB servlet. Using the `GetBook` servlet we created above, Figure 9-6 shows what JavaScript code looks like to invoke this servlet as the result of a button click. In this example, we obtain the servlet parameter text, the book's title, from the document (form) object. We replace any blank characters with "+'s" so that it can be passed as part of a URL and then set the `"location"` DOM object `"href"` property to the servlet URL. The servlet response is taken as the new page to display. Note that we specified `"Common JavaScript"` so that this code will work in both the Notes client as well as a browser.

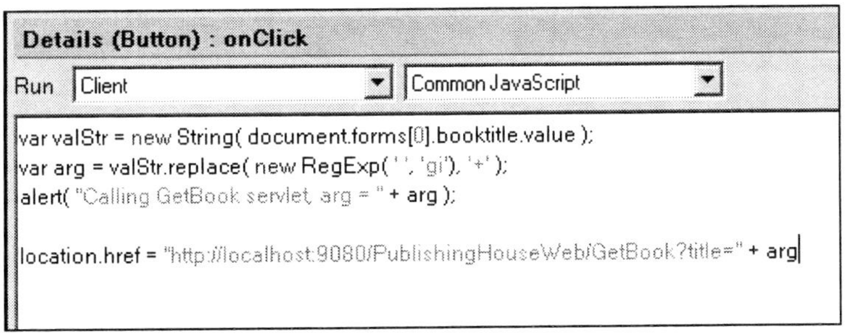

Figure 9-6 Invoking a servlet from JavaScript.

Java from LotusScript

As long-time Domino developers are aware, LotusScript is the interpreted, object-oriented programming language provided by Lotus across its product family, including Domino, Lotus123, WordPro, and Freelance. In order to accommodate the many Domino programmers with primarily LotusScript skills, Lotus has provided a LotusScript extension (LSX) called LS2J, which allows Java classes to be invoked from within LotusScript code. Using the LS2J LSX, it is possible, at least in theory, to implement many of the techniques we discussed here within LotusScript.

The LS2J LSX, however, allows only primitive arguments to be passed to or returned from the Java classes. This limitation makes it difficult to work with basic items such as strings or date/time values since they are encapsulated by classes in Java. In addition, since Domino agents can be coded in Java, and `LotusScript` programs can easily invoke Java agents, one can simply invoke the Java specific code for J2EE processing as agents from LotusScript. Given that, we feel that using the LS2J LSX to invoke J2EE components from LotusScript provides no real advantage, except perhaps to pure LotusScript programmers who lack Java skills.

Using XML, XSLT from Domino

In this section, we discuss how you can work with XML data in Domino applications. The importance of XML in today's application development environments and especially with Domino and WebSphere products is hard to overstate. From data exchange formats to rendering formats (XHTML) to configuration file formats, XML is becoming the pervasive syntax for structured text data. An understanding of how to work with XML in Domino will be essential when implementing Web services in Domino.

With any discussion of XML, a discussion of the Extensible Stylesheet Language (XSL) and XSL-based XML transform processing is not far behind. As with natural languages of a common syntax and structure—for example, English and French—XSL transform processing provides the translation function between XML-based data formats. XSL provides a powerful means for recasting one XML format into another.

Our discussion of XML processing in Domino will cover the following topics:

- **Parsing and building XML data within a Domino database.** Use of the W3C Document Object Model (DOM) and Simple API for XML (SAX) parsers supplied with Domino.
- **The Domino XML (DXL) schema for design elements.** Working with DXL to format Domino elements in XML.
- **XSL transform support in Domino.** Using the built-in Domino XSL transform engine.

The Domino server product contains the libraries for parsing XML via the W3C DOM and SAX APIs. Agents or event actions can be written within a Domino database to parse and then process XML data. XML formatted data itself can be accessed in various ways within a Domino application. XML data can be read from an external file or from a database document field. XML data can also be obtained as the result of the Domino HTTP Server rendering a form or page or from request content sent to a Web agent. The DOM or SAX APIs are available to Java or LotusScript code within Domino Designer.

Most often we find ourselves needing to parse XML contained in document fields or in the request content directed to a Web agent (as when handling a Web service request). Let's examine how to use the parsing APIs, both DOM and SAX, within Domino agents in detail. The basic steps for parsing XML text are:

1. Set up access to the XML input.
2. Invoke the parser.
3. Process the parsed data. For this discussion, we assume some familiarity with the W3C DOM and SAX APIs (see the Bibliography for references).

Briefly, the DOM API parses XML and creates a tree structure with nodes representing the parsed XML elements, text, etc.; whereas, the SAX API parses XML and invokes callback functions for the different XML constructs. When using the SAX API, step three above is done while the data is being parsed. The choice between using the DOM or SAX API is often based on what the application needs to do with the parsed data. If the XML data must be modified or added to and then rewritten, the DOM API is often used. If the need is to search for certain data within the XML, this is more efficiently done using the SAX API. Note that the DOM API requires enough memory for the entire DOM parse tree, whereas the SAX API is a single pass parser and can be ended as soon as the required data is parsed.

The Domino, Java, and LotusScript APIs have been extended to interoperate with the DOM and SAX APIs. In the Java API, a `parseXML()` method, which parses XML into a DOM tree, has been added to the various Domino classes (interfaces) that can hold XML text, such as `Item`, `RichTextItem`, `MIMEEntity`, and `EmbeddedObject`. Another new Java method is the `getInputSource()` method, which returns a `org.xml.sax.InputSource` object for use by a SAX parser. In the LotusScript API, there are now `NotesDOMParser` and `NotesSAXParser` classes. Let's look at a specific example of Java agent code to perform both DOM and SAX parsing of XML text within a Domino database.

The following Java code illustrates how to parse XML text held within a field of a Domino database document using the DOM parser. It also shows how the parsed data can be extracted from the DOM tree built by the parser. In this example, we extract the top level XML elements, which contain only text, and build a new Domino document with fields (items) corresponding to these extracted elements. The root element name is used to set the form associated with the Domino document.

```
// Parses given field (Item) contents as XML using DOM parser.
// Returns root element name.
//
private String ParseXMLviaDOM( Item theField )
    throws java.io.IOException, NotesException,
    ➥org.xml.sax.SAXException
{
    // Parse field contents (false means no validation against
      ➥DTD).
```

```
    org.w3c.dom.Document aDOMtree = theField.parseXML( false );

    // Create a new Notes document, set its form as root element
       ➥name,
    // then build document from elements.
    Document aNewDoc =
       itsAgentContext.getCurrentDatabase().createDocument();
    org.w3c.dom.Node aRoot = aDOMtree.getFirstChild();
    aNewDoc.appendItemValue( "Form", aRoot.getNodeName() );
    XMLUtils.DOM2Doc( aRoot, aNewDoc );
    return( aRoot.getNodeName() );
}

// Convert an XML DOM tree into a set of items in a document.
// Each child of the given Node is turned into an item named
// after the element.
public static void DOM2Doc( Node theDOM, lotus.domino.Document
    theDoc )
   throws lotus.domino.NotesException
{
   NodeList aNodes = theDOM.getChildNodes();

   for( int i = 0;  i < aNodes.getLength();  i++ )
   {
      Node aNode = aNodes.item( i );

      if ( aNode.getNodeType() == Node.ELEMENT_NODE )
      {
         // Look for text in child node, then append to Notes
            ➥document
         Node aChild = aNode.getFirstChild();
         if ( aChild != null && (aChild.getNodeType() ==
         ➥Node.TEXT_NODE
              || aChild.getNodeType() == Node.CDATA_SECTION_NODE) )
           theDoc.appendItemValue( aNode.getNodeName(),
                                   aChild.getNodeValue() );
      }
   }
   if ( aNodes.getLength() > 0 )  theDoc.save();
}
```

(The DOM2Doc method can be made part of a Java script library within Designer so that it could be reused among other databases.)

The next example illustrates how to parse XML text using a SAX parser. The Java code shown next performs the same function as the previous example, parsing the XML text from a Domino document field and creating a new Domino document with fields corresponding to the XML element names and values.

```
// Parses given field (Item) contents as XML using SAX parser.
// Returns root element name.
//
private String ParseXMLviaSAX( Item theField )
   throws java.io.IOException, NotesException,
      org.xml.sax.SAXException, ClassNotFoundException,
      IllegalAccessException, InstantiationException
{
   // Get a SAX parser via its factory
   org.xml.sax.Parser aParser =
      org.xml.sax.helpers.ParserFactory.makeParser(
         "com.ibm.xml.parsers.SAXParser" );

   // Set the XML document handler as our own SAXHandler.
   SAXHandler aHandler = new SAXHandler();
   aParser.setDocumentHandler( aHandler );

   // Create a new Notes document as the target for the parsed
      ➥data
   // and tell our handler about it.
   Document aNewDoc =
      itsContext.getCurrentDatabase().createDocument();
   aHandler.setNotesDoc( aNewDoc );

   // Parse the given field contents
   aParser.parse( theField.getInputSource() );
   return( aHandler.getRootName() );
}

import org.xml.sax.*;

public class SAXHandler extends HandlerBase
{
   // Notes Document object used as target for parsed data.
```

```
lotus.domino.Document itsNotesDoc;

// Holds root element name, also set as form for itsNotesDoc.
String itsRootName;

// Holds current element, value being parsed.
String itsCurElt;
StringBuffer itsValue = new StringBuffer();

public void setNotesDoc( lotus.domino.Document theDoc )
{
   itsNotesDoc = theDoc;
}

public void startDocument()
{
   System.out.println( "Start SAX API parse of document." );
}

public void endDocument()
{
   // Set root element name as Notes document Form, save it.
   if ( itsNotesDoc != null && itsRootName != null )
   {
      try {
         itsNotesDoc.appendItemValue( "Form", itsRootName );
         itsNotesDoc.save();
      } catch( lotus.domino.NotesException e ) {
         System.out.println( "Parse error: " + e );
      }
   }
   System.out.println( "End SAX parse." );
}

public void startElement( String theName, AttributeList
➥theAttrs )
   throws SAXException
{
   // If root element name not set, assume this is it.
   if ( itsRootName == null )
```

```
            itsRootName = theName;
        itsCurElt = theName;
    }

    public void endElement( String theName )
        throws SAXException
    {
        // If current element has a text value, add it to the Notes
        // document as an Item.
        if ( itsNotesDoc != null && itsCurElt != null &&
              itsValue.length() > 0 )
        {
            try {
                itsNotesDoc.appendItemValue(
                    itsCurElt, itsValue.toString() );
            } catch( lotus.domino.NotesException e ) {
                System.out.println( "Parse error: " + e );
            }
        }
        itsCurElt = null;
        itsValue.setLength( 0 );
    }

    public void characters( char theChars[],
        int theStart, int theLength ) throws SAXException
    {
        if ( itsCurElt != null )
            itsValue.append( theChars, theStart, theLength );
    }

    public String getRootName()
    {
        return itsRootName;
    }
}
```

For the SAX parse, the `SAXHandler` class overrides the set of "callback" methods invoked by the parser (via the `HandlerBase` interface). Note that you must supply the class members and logic required to hold element data across the callback methods. In our example, we used the class members, `itsCurElt` and `itsValue`, to hold the element name and value across the `startElement`, characters, and `endElement` methods.

A couple of final notes about the XML parsing examples:

- The code can be tested easily within Domino Designer or the Notes client. The System.out stream writes to the Java Debug Console window on the client and to the Notes log database when run on a server.
- The Designer Reference tab shows the DOM and SAX Java API methods under the "Third-Party Java" topic. Only three of the SAX classes appear in the Reference, although the entire SAX API can be written and compiled.
- The `parseXML()` method can invoke a validating DOM parser by setting its argument to `"true."` In this case, the parser looks for any DTDs referenced in the XML as page elements within the Domino database having the DTD filename. At this time, it is not possible to perform XML Schema validation.

Next, we look at how to construct XML formatted text from data within a Domino database. Of course, with programming, anything is possible, so one can take the "brute force" approach and generate the XML text character by character (or substring by substring) using Java or LotusScript code. For simple XML, this approach may be appropriate but quickly becomes time-consuming and inflexible for more complex XML. A more flexible approach can be taken by making use of the application "programming" features of Domino, namely forms and views.

Although we assume general familiarity with Domino application development, we'll take a couple of paragraphs here to review the features of Domino forms and views. Not only are forms and views useful as a means of displaying and listing database documents to a Notes client or Web browser, but they can be and often are used as programming functions. Forms can be used in a similar way as XSL stylesheets are used to format XML data, that is, to format the contents (items and their values) of a Domino database document in a certain way. A single document can be presented in different ways by associating it with different forms. Views can be thought of as forms with a query function—they can determine a collection of documents (via the view selection formula) and format the collection into a list using a subset of document fields.

Using Domino database forms and views, where the form text or view column text surrounding the field data is XML, we can "format" the document data into XML. Let's look at an example. Suppose we have a Domino database used to hold information about books, say for a publishing house application, and the documents in this database contain fields as shown by the "book" form in Figure 9-7. (Keep in mind that Domino database documents only contain field items and values; the formatting stuff—the table, text, etc.—are part of the form.) Also suppose that we want to generate XML documents for each of these Domino documents. We can simply create another form, say "bookXML" and specify the XML text as the form text as shown in Figure 9-8.

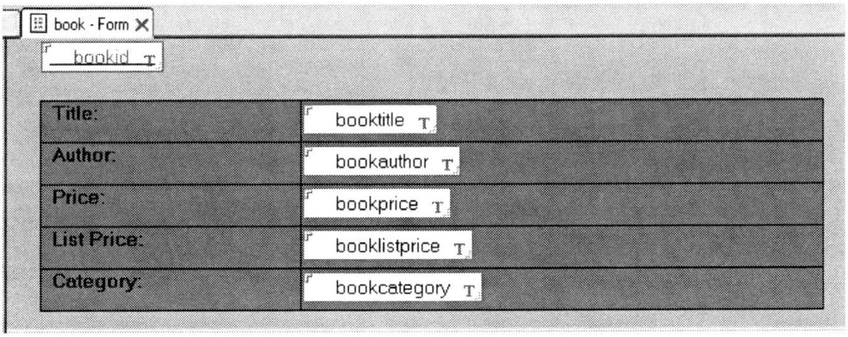

Figure 9-7 Book form for display.

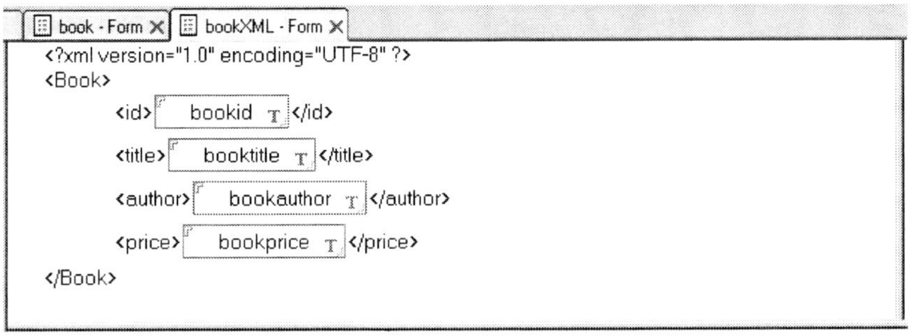

Figure 9-8 BookXML form for generating XML.

We can apply the "bookXML" form to database documents containing the fields originally created by the "book" form, and then we can write, send, or process the resulting content, which will be in XML format. Note that the fields in the "bookXML" form can be a subset of the original set of document fields or, using lookup formulas, could come from other documents or even other databases.

Hopefully, at this point it should be clear how the "bookXML" form can be used to display XML content for a "book" document within a browser or Notes client, but perhaps not at all clear is how to obtain this XML content within agent or event code. Let's look at one technique for obtaining the XML content via a form within agent code. Basically, we need to have the Domino server "apply" the XML form to a specific document and "return" the XML content to our agent code. Having Domino "apply" the form can be done by creating a view in the database having a "form formula" defined specifying the form name. This view can be defined to select all documents for which we want to generate XML (in our case, documents originally created with the "book" form) and to have a sorted column with the document's unique Domino identifier as the column value. Having Domino "return" the XML content (that is, the document with the XML form applied) can then be done by simply requesting the document from this view via a Domino URL. Figure 9-9 shows a "BooksAsXML" view created as we just described.

Figure 9-9 "BooksAsXML" view.

The following Java agent code uses the "BooksAsXML" view to generate the "bookXML" content for a specific "book" document and write this XML content to a file.

```
import lotus.domino.*;
import java.net.*;
import java.io.*;

public class JavaAgent extends AgentBase
{
   public void NotesMain()
   {
      try {
         Session session = getSession();
         AgentContext agentContext = session.getAgentContext();

         // Get selected document - should be a 'book' document.
         DocumentCollection aDocs =
           agentContext.getUnprocessedDocuments();
         Document aDoc = aDocs.getFirstDocument();

         if ( aDoc != null )
         {
            // Set URL to get document with XML form applied.
            URL aURL = new URL( "http", "localhost", 80,
               "/booklist.nsf/BooksAsXML/" + aDoc.
               ➥getUniversalID() );
            URLConnection aCnctn = aURL.openConnection();
            aCnctn.connect();
```

```
                    // Copy XML output to file.
                    PrintWriter aWtr = new PrintWriter(
                       new FileWriter( "c:\\temp\\document.xml" ) );
                    BufferedReader aRdr = new BufferedReader(
                       new InputStreamReader( aCnctn.getInputStream() ) );
                    String aLine = aRdr.readLine();
                    while ( aLine != null )
                    {
                       aWtr.println( aLine );
                       aLine = aRdr.readLine();
                    }
                    aWtr.close();
                 }
              }
              catch( Exception e )
              {
                 System.out.println( "GenerateXML error: " + e );
              }
           }
        }
```

The URL used to obtain the document from our "BooksAsXML" view has the following format: `http://<hostname>/<database-name>/<view-name>/<doc-name>[?OpenDocument]` where the <doc-name> value is used to lookup the document in the first sorted column of the specified view <view-name>. In our example, we used the Domino Universal Document ID as this value. Some final points about this example:

- The "XML" form must have the Content Type option set to "Other" with "XML" specified in the adjacent input field, as shown in Figure 9-10. (The Content Type setting is located in the Form properties dialog under the Advanced tab.) This setting tells the Domino HTTP Server to set the content type of the generated form text as XML; otherwise, it would be generated with HTML headers that would appear before the XML header.
- The agent (or event) code, because it invokes a URL, must be run against a Domino server or under a Notes client with the "Local Web Preview" process running. You can simply select the Preview In Web Browser action in the Notes client to start this process.

If XML content needs to be created that must contain data from multiple documents, for example, a "Category" XML document which contains multiple "Book" elements, it can be generated using a similar techniques as described previously. We can use a page design element to specify the top-level XML (header and root element) and an embedded view to format a set of documents as child elements. In this case, the view would define column formulas to specify the XML tag data around the document fields (column values). See the IBM Redbook, "XML Powered by Domino," for details on how to use this technique.

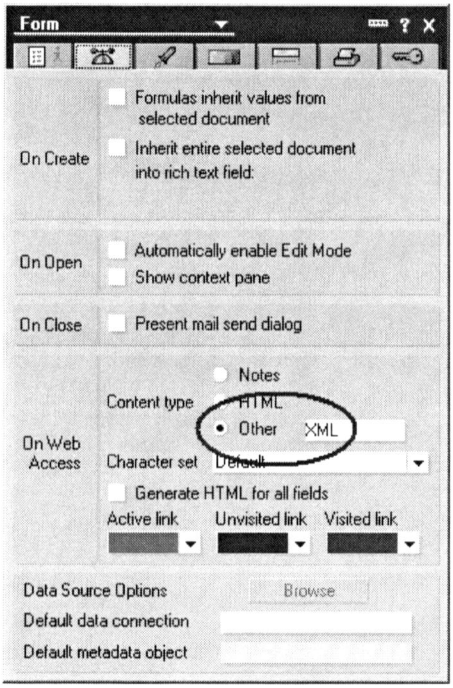

Figure 9-10 Form properties for XML.

A third approach to generating XML content from Domino data is to make use of the Domino programming functions, which generate DXL from Domino database, design elements, and Domino's built-in ability to transform XML via XSL. Recall that DXL is the XML format defined by the IBM/Lotus-supplied Domino DTD and is expressed in terms of Domino design elements such as documents, items, forms, views, etc. With the ability to perform XSL transformation, we can use DXL as the starting point from which to generate other XML formats or other formats entirely, such as HTML or plain text. Domino provides API functions to generate DXL and to perform XSL-based transforms from stylesheets incorporated into the database or external to it. If you are proficient with XSL, this approach may be the most flexible way to produce sophisticated XML content from Domino data.

Using Web Services from Domino

The ability to work with XML within Domino provides the foundation on which we can enable Domino applications for Web services, both as Web service providers and Web service consumers. As discussed in earlier chapters, enabling an application for Web services involves working with three major formats and protocols: SOAP for message exchange, WSDL for service definition, and UDDI for service awareness. Each of these formats/protocols is underpinned by XML; thus, enabling Web services in Domino is largely a matter of working with XML in Domino applications. In this section, we focus on techniques to provide or consume Web services from Domino applications. The focus is definitely on technique. You won't find much about "why Web services should be used" here—just "how is it done."

Providing a Web Service Within a Domino Application

We'll look at providing Web services first. Although not required, the SOAP protocol for Web service invocation is most often implemented "on top" of HTTP, and we will discuss SOAP over HTTP exclusively here. A SOAP request for a Web service is not much more than a HTTP request (POST) with a specific XML formatted message. Hence the invocation takes the form of a URL pointing to a Web service handler. A Web service handler can be anything that responds to an HTTP request. For Domino applications, the handler is naturally implemented as a "Web agent," that is, a Domino database agent that can be invoked by a URL to the Domino server. (The "Web agent" distinction is in contrast to the more traditional uses for Domino agents as scheduled or event-driven server tasks.)

Let us digress a bit about Web service handlers. Since a handler is usually a Web application server that is running the application that provides the Web service, you may have multiple handlers (servers) for the set of Web services you provide. You then have the issue of to what extent you need provide access to these Web service handlers. Does each handler need to be open to service requests, or should the requests be routed through a single access point? These are indeed important design questions. However, since the request for a particular service must ultimately be handled at the application providing the service, we focus on the handler serving the Web service.

So we can define a Web agent in our Domino application to act as the Web service handler. What is required of this agent/handler? Primarily it parses the SOAP (XML) request, performs the service, then formats and returns the SOAP response. Let's look at Java agent code that does just that. We'll call the agent `WSHandler`. Figure 9-11 shows the agent properties we set for our `WSHandler` Web agent. Note that we selected "Agent list selection" and a Target of "None" for the Runtime settings, as these are the appropriate choices for a Web agent. (Recall that a Domino Web agent is simply a Domino agent that is exclusively invoked via a URL from a browser.)

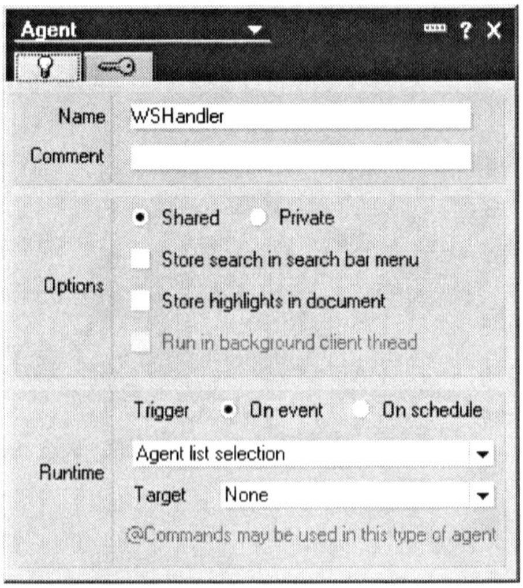

Figure 9-11 `WSHandler` agent properties.

Here is the `WSHandler` agent code:

```
import lotus.domino.*;
import java.io.*;
import java.net.*;

public class JavaAgent extends AgentBase
{
   public void NotesMain()
   {
      Document aRespDoc = null;   // Holds SOAP response
      try {
         Session session = getSession();
         AgentContext agentContext = session.getAgentContext();
         Database aDB = agentContext.getCurrentDatabase();

         // Get parameters, request content from the HTTP call
(1)      Document aDoc = agentContext.getDocumentContext();
         if ( aDoc != null )
         {
            // 'Log' the request as DXL
            BufferedWriter aBW = new BufferedWriter(
               new FileWriter( "c:\\temp\\book-wsreq.xml" ) );
(2)         aDoc.generateXML( aBW );
```

```
            aBW.close();

            // Process the SOAP request.
(3)         Item aReqContent = aDoc.getFirstItem(
            ➥"REQUEST_CONTENT" );

            // Get a SAX parser via its factory
            org.xml.sax.Parser aParser =
               org.xml.sax.helpers.ParserFactory.makeParser(
                  "com.ibm.xml.parsers.SAXParser" );

            // Set the SAX document handler as our own
               ➥SOAPHandler.
            SOAPHandler aHandler = new SOAPHandler();
            aParser.setDocumentHandler( aHandler );
            aParser.parse( aReqContent.getInputSource() );

(4)         System.out.println( "SOAP-ENV: " + aHandler.itsEnv );
            System.out.println( "Service: " + aHandler.
            ➥itsService );
            System.out.println( "Method: " + aHandler.itsMethod );
            System.out.println( "Encoding: " +
            ➥aHandler.itsEncodingStyle );

            // Process the web service request according to the
               ➥method
            // and arguments… (not shown)

            // Prepare the SOAP response
            aRespDoc = aDB.createDocument();
            aRespDoc.appendItemValue( "Form", "SOAPRespXML" );
(5)         aRespDoc.appendItemValue( "SOAPRespBody",
               "<SOAP-ENV:Fault>" +
               "<faultcode>SOAP-ENV:Client</faultcode>" +
               "<faultstring>Service not supported.</faultstring>" +
               "</SOAP-ENV:Fault>" );
            aRespDoc.save();
            PutResponse( aRespDoc.getUniversalID() );
            aRespDoc.remove( true );
        }
    }
```

```
        catch( Exception e )
        {
            System.out.println( "WSHandler error: " + e );
        }
    }

    // Generates SOAP response from SOAPRespXML document with given
    // UNID, copies response to agent output (HTTP response).
    //
    private void PutResponse( String theUNID )
        throws MalformedURLException, IOException
    {
        // Set URL to get document with XML form applied.
(6)     URL aURL = new URL( "http", "localhost", 80,
            "/booklist.nsf/0/" + theUNID );
        URLConnection aCnctn = aURL.openConnection();
        aCnctn.connect();

        // Copy SOAP response XML to agent output (HTTP response).
        PrintWriter aWtr = getAgentOutput();
(7)     aWtr.println( "Content-Type: text/xml " );
        aWtr.println( "  " );
        BufferedReader aRdr = new BufferedReader(
            new InputStreamReader( aCnctn.getInputStream() ) );
        String aLine = aRdr.readLine();
        while ( aLine != null )
        {
            aWtr.println( aLine );
            aLine = aRdr.readLine();
        }
        aWtr.close();
    }
}
```

The agent code is worth a step-by-step examination. We'll use the (n) annotations as reference points.

1. The Domino database document object returned by `AgentContext.getDocumentContext()` contains all of the HTTP request data, including the `SOAPAction` verb and request content.

2. A nice way to view the HTTP request data is to format the context document as DXL and write it to a file.
3. The SOAP request XML is obtained from the HTTP request content field.
4. We use a SAX parser to parse the SOAP request into its constituent elements, especially the service name, method name, and encoding style. We do not show the actual service processing, as we are most interested here in how to implement the Web service request and response. But this is the point at which the service would be performed. Also there are other options to using a SAX parser for the SOAP request. We present a more in-depth discussion of the parsing options later in this section.
5. For the SOAP response XML, we use the "XML form" technique discussed in the previous section. We defined a `SOAPRespXML` form with the static SOAP envelope and body XML and a single field to contain the variable response XML itself (see Figure 9-12). To build the SOAP response, we create a document using this form and set the field to the response XML.
6. The `PutResponse` method obtains the formatted SOAP response via an `OpenDocument` URL specifying the UNID of the document built in step five. (Note that since we built this document with the desired form name, we can use the default view identifier of `0` for the view. There is no need to have a different form applied in this case.)
7. Finally, we copy the response text to the Web agent's output stream, which is the HTTP response stream. We explicitly set the content type as "text/xml." We have found that the response XML text (written in the `SOAPRespXML` form) did not need to contain the XML header (`<?xml ... ?>`) declaration, as this is generated by Domino. When we did include it as part of the response text, our SOAP client reported a SOAP parsing error that there was a processing instruction in the response XML that was not permitted by the SOAP standard.

There it is—a fairly generic Domino Web agent for processing Web service requests. (Look Ma, no WebSphere!) In order to invoke the Web service, you create a URL to invoke the Domino Web agent. Using the WSHandler Web agent example, the URL would look like:

```
http://hostname/database.nsf/WSHandler?OpenAgent
```

If you are more interested in a LotusScript version of a Web service agent, the Designer help documentation shows a detailed example of one. It takes a more direct approach to parsing the request and formatting the response, but the previously described techniques can be coded just as well in LotusScript.

As mentioned here, there are other options for parsing the SOAP request and for building the SOAP response XML. We have investigated using some of the SOAP processing libraries (in Java), such as those from Apache (`org.apache.soap`) and WAS itself. Unfortunately, this proved to be problematic. The primary source of problems was the somewhat out-of-date level of the Domino R6 XML libraries. The level of SAX and DOM parser code provided by Domino

(SAX 1.0 and DOM Level 1) is a major version behind that used by the SOAP libraries (SAX 2.0 and DOM Level 2). At least with the Java libraries, it is possible for Domino's XML parsing libraries to be upgraded by simply replacing the XML-related .jar files with newer versions, but this is not supported by IBM and could affect existing Domino code. At this stage of product evolution, we must make do with the somewhat outdated XML support in Domino.

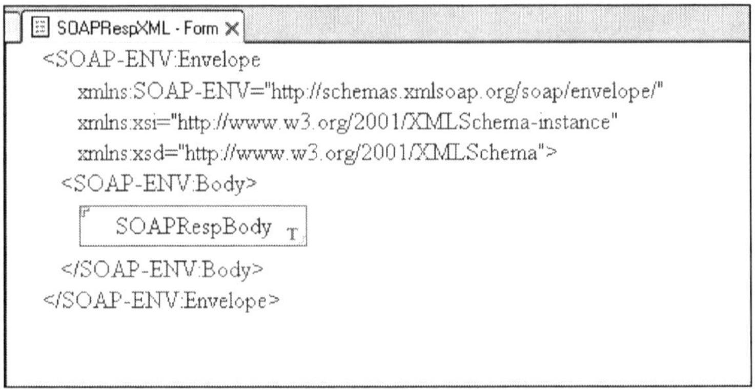

Figure 9-12 XML form for SOAP response.

Even so, it is still possible to write a SOAP parser using the SAX version available with Domino R6. (See the last section for the SAX handler code we wrote to parse the SOAP request for the Web agent.)

Invoking a Web Service Within a Domino Application

Next we look at how to invoke a Web service from within a Domino application. There are several options available, each with its ensuing set of advantages and disadvantages.

One approach is to incorporate an external SOAP package into Domino and utilize it within application code. For example, there are the Java-based SOAP implementations from the Apache project, Apache SOAP (which originated in IBM), and Apache Axis. There is also the IBM/Lotus provided SOAPConnect toolkit, which makes it possible to invoke Web services from LotusScript code. The biggest disadvantage to this approach is the potential for incompatibility between these packages and Domino itself. As we saw in the previous section, the incompatibilities are most likely to arise within the XML libraries. In fact, the SOAPConnect toolkit requires that the standard Domino XML libraries be replaced with those in the toolkit and yet warns that this should not be done in a production environment.

Another approach is to provide the basic SOAP function using standard Domino features. These features, along with the XML processing techniques that we have discussed previously, are sufficient to create and parse the SOAP messages required to invoke a Web service. For Web services that do not involve complex requests or responses, this approach is perhaps the safest from a product support viewpoint and likely the most efficient. The disadvantage with this approach is the amount of code required. We will look at an example of invoking a Web service using this approach, and you can be the judge.

Basically, the code required to invoke a Web service is similar to that for invoking a servlet or EJB, except that we need to provide the SOAP formatted request message as the HTTP request content and, of course, process the SOAP response message. The main steps are as follows:

1. Build the SOAP request message as XML, including the service-specific request data.
2. Make the HTTP request to the Web service host with the SOAP request XML as the HTTP request content.
3. Parse the HTTP response content containing the SOAP response message.

Here is an agent, WSRequest, which implements these steps in Java:

```
import lotus.domino.*;
import java.io.*;
import java.net.*;

public class JavaAgent extends AgentBase
{
   public void NotesMain()
   {
      try {
         Session session = getSession();
         AgentContext agentContext = session.getAgentContext();

         // Get selected document from which to get request
            ➥argument
         DocumentCollection aDocs =
            agentContext.getUnprocessedDocuments();
(1)      Document aDoc = aDocs.getFirstDocument();
         if ( aDoc == null )
         {
            System.out.println(
               "WSRequest error: No selected document!" );
            return;
         }

         // Build SOAP request - set parameter in a SOAPReqXML
            ➥document
(2)      Document aReqDoc =
            agentContext.getCurrentDatabase().createDocument();
```

```
            aReqDoc.replaceItemValue( "Form", "SOAPReqXML" );
            aReqDoc.replaceItemValue( "WSbooktitle",
               aDoc.getFirstItem( "booktitle" ) );
            aReqDoc.save();

            // Get SOAPReqXML document with form applied.
(3)         URL aURL = new URL( "http", "localhost", 80,
               "/booklist.nsf/0/" + aReqDoc.getUniversalID() );
            URLConnection aCnctn = aURL.openConnection();
            aCnctn.connect();

            // Issue the SOAP request
            StringWriter aSOAPResp = new StringWriter();
(4)         MakeSOAPRequest(
               "http://localhost/booklist.nsf/WSHandler?OpenAgent",
               aCnctn.getInputStream(), aSOAPResp );

            // Get a SAX parser to parse the response
            org.xml.sax.Parser aParser =
               org.xml.sax.helpers.ParserFactory.makeParser(
                  "com.ibm.xml.parsers.SAXParser" );

            // Set the SAX document handler as our own SOAPHandler.
(5)         SOAPHandler aHandler = new SOAPHandler();
            aParser.setDocumentHandler( aHandler );
            aParser.parse( new org.xml.sax.InputSource(
               new StringReader( aSOAPResp.toString() ) ) );
            aHandler.Log();

            // Do something with the response value…
         }
         catch( Exception e )
         {
            System.out.println( "WSRequest error: " + e );
         }
      }
```

Let's examine this agent code in detail—using the (n) annotations in the listing above:

1. Our agent is set up to work on selected Domino documents or from a simple form action or button. We get the document here. We'll use this document to get the Web service request parameter(s).

2. We create a Domino document and set it to use a form we have created to contain the SOAP request XML, `SOAPReqXML`. We copy the request parameter field from the selected document to this `SOAPReqXML` document.
3. The SOAP request XML is obtained by requesting the combined document and form we created in step two.
4. The SOAP request is sent to the Web service host. We'll examine the `MakeSOAPRequest` method that follows.
5. The SOAP response returned in the `StringWriter` object is parsed using a SAX parser and our `SAXHandler` class as described in the previous section. The response value can then be gotten from the `SAXHandler` and used in some way.

Here is the code to perform a SOAP request and obtain the response via HTTP:

```
    // Makes SOAP request given request XML as InputStream
    //  theTarget - Web service target URL
    //  theIS - SOAP request content
    //  theOutput - On exit, contains SOAP response content
    //
    private void MakeSOAPRequest(
       String theTarget, InputStream theIS, StringWriter theOutput
)
    {
       try {
          // Set up the URL object for the HTTP request and get the
          // HttpURLConnection object which will control the
             ➥request.
          //
(1)       URL aUrl = new URL( theTarget );
          HttpURLConnection aUC =
             (HttpURLConnection)aUrl.openConnection();

          // Here we set up a HTTP POST request as required by
             ➥SOAP.
          aUC.setRequestMethod( "POST" );
          aUC.setDoOutput( true );

          // Set the content type as XML, and as a SOAPAction
(2)       aUC.setRequestProperty( "Content-type", "text/xml" );
          aUC.setRequestProperty( "SOAPAction", "" );
```

```
            PrintWriter aWtr = new PrintWriter( aUC.
            ➥getOutputStream() );
            BufferedReader aRdr = new BufferedReader(
(3)            new InputStreamReader( theIS ) );
            String aLine = aRdr.readLine();
            while ( aLine != null )
            {
               aWtr.println( aLine );
               aLine = aRdr.readLine();
            }
            aWtr.close();

            // Now make the connection with the request header we
               ➥set up,
            // get the response code/message, then disconnect.
            //
(4)         aUC.connect();
            int aRspCode = aUC.getResponseCode();
            String aRspMsg = aUC.getResponseMessage();

            // Copy response content (SOAP response) to output object
            aWtr = new PrintWriter( theOutput );
            aRdr = new BufferedReader(
(5)            new InputStreamReader( aUC.getInputStream() ) );
            aLine = aRdr.readLine();
            while ( aLine != null )
            {
               aWtr.println( aLine );
               aLine = aRdr.readLine();
            }
            aWtr.close();

            aUC.disconnect();
         }
         catch ( Exception e )
         {
            System.out.println( "SOAPRequest error: " + e );
         }
      }
```

The details:

1. We first set up a `HttpURLConnection` object to use for the SOAP request. Note that `openConnection()` does not actually connect to the server, it just returns the `URLConnection` object associated with the URL object. This allows the request header to be set up prior to the request send.
2. Here we set the necessary HTTP request header items for SOAP.
3. Next, the SOAP request that XML passed to this method is copied to the HTTP request content.
4. Now the network connection is made and the request sent. On return from this method, the response content is available.
5. Here we copy the response content, the SOAP response XML, as text to the output object.

As you can see, the required code is not too complex (thanks to the Java libraries used). Well...almost. The `SAXHandler` we used to parse the SOAP XML here and in the previous section is admittedly oversimplified. It will handle request and response formats containing the basic set of SOAP parameter encoding types, but it needs to be enhanced to handle more complex data encodings such as arrays and structures. Here is where a full-function SOAP package would be worth its price.

Additional Code Used for Domino Web Services

We include the code here that we used in the previous sections to parse XML. The first listing is code to perform SAX parsing of XML text. The code provides an implementation of the `org.xml.sax.HandlerBase` interface.

```
import org.xml.sax.*;

public class SAXHandler extends HandlerBase
{
    // Notes Document object used as target for parsed data.
    lotus.domino.Document itsNotesDoc;

    // Holds root element name, also set as form for itsNotesDoc.
    String itsRootName;

    // Holds current element, value being parsed.
    String itsCurElt;
    StringBuffer itsValue = new StringBuffer();

    public void setNotesDoc( lotus.domino.Document theDoc )
```

```
{
    itsNotesDoc = theDoc;
}

public void startDocument()
{
    System.out.println( "Start SAX API parse of document." );
}

public void endDocument()
{
    // Set root element name as Notes document Form, save it.
    if ( itsNotesDoc != null && itsRootName != null )
    {
        try {
        itsNotesDoc.appendItemValue( "Form", itsRootName );
            itsNotesDoc.save();
        } catch( lotus.domino.NotesException e ) {
            System.out.println( "Parse error: " + e );
        }
    }
    System.out.println( "End SAX parse." );
}

public void startElement( String theName, AttributeList
➥theAttrs )
    throws SAXException
{
    // If root element name not set, assume this is it.
    if ( itsRootName == null )
        itsRootName = theName;

    itsCurElt = theName;
}

public void endElement( String theName )
    throws SAXException
{
    // If current element has a text value, add it to the
       ➥Notes
```

```
// document as an Item.
        if ( itsNotesDoc != null && itsCurElt != null
        ➥&& itsValue.length() > 0 )
        {
            try {
                itsNotesDoc.appendItemValue( itsCurElt,
                ➥itsValue.toString() );
            } catch( lotus.domino.NotesException e ) {
                System.out.println( "Parse error: " + e );
            }
        }
        itsCurElt = null;
        itsValue.setLength( 0 );
    }

    public void processingInstruction(
String theTarget, String theDescr ) throws SAXException
    {}

    public void characters(
char theChars[], int theStart, int theLength )    throws
    SAXException
    {
        if ( itsCurElt != null )
            itsValue.append( theChars, theStart, theLength );
    }

    public String getRootName()
    {
        return itsRootName;
    }
}
```

The next code listing implements a SAX parser to specifically parse SOAP requests.

```
import org.xml.sax.*;

public class SOAPHandler  extends HandlerBase
{
    // SOAP Request attributes
    public String itsEnv;
    public String itsService;
    public String itsMethod;
```

```
    public String itsEncodingStyle;

    // State of parse processing
    int itsState;

    // Parse SOAP Envelope
    static final int kStateParseEnvelope = 0;
    // Indicates next element is SOAP Body
    static final int kStateParseBody = 1;
    // Look for request parameters
// (we take any text element as a parameter).
    static final int kStateParseParms = 2;
    static final int kStateEnd = 999;

    // Inner class to capture parameters
    public class SOAPParm
    {
        public String itsName;
        public String itsType;
        public String itsValue;
    }
    // Holds parsed parameters as SOAPParm objects
    public java.util.Vector itsParameters;

    // Holds current element, type (if set), value being parsed.
    String itsCurElt;
    String itsCurType;
    StringBuffer itsCurValue = new StringBuffer();

    public void startDocument()
    {
        System.out.println( "Start parse of SOAP request." );
        itsState = kStateParseEnvelope;
        itsParameters = new java.util.Vector();
    }

    public void endDocument()
    {
        System.out.println( "End SOAP parse." );
    }
```

```
    public void startElement( String theName, AttributeList
    ➥theAttrs )
        throws SAXException
    {
        itsCurElt = theName;

        // If we're expecting the Body element, assume this is
           ➥it...
        switch ( itsState )
        {
        case kStateParseEnvelope:
            // If SOAP Envelope, look for SOAP-ENV attribute
            if ( theName.equals( "SOAP-ENV:Envelope" ) )
                itsEnv = theAttrs.getValue( "xmlns:SOAP-ENV" );
            else if ( theName.equals( "SOAP-ENV:Body" ) )
                itsState = kStateParseBody;
            break;

        case kStateParseBody:
            itsService = theAttrs.getValue( "xmlns:ns1" );
            if ( itsService == null )
                itsService = theAttrs.getValue( "xmlns" );
            itsMethod = theName;
            itsEncodingStyle = theAttrs.getValue(
"SOAP-ENV:encodingStyle" );
            itsState = kStateParseParms;
            break;

        case kStateParseParms:
            // Remember type if specified
            itsCurType = theAttrs.getValue( "xsi:type" );
            break;

        default:
        }
    }

    public void endElement( String theName )
        throws SAXException
    {
        if ( itsState == kStateParseParms && itsCurValue.length()
```

```
> 0 )
        {
            // Create new SOAPParm object to hold this parameter
               ➥element
            SOAPParm aParm = new SOAPParm();
            aParm.itsName = itsCurElt;
            aParm.itsType = itsCurType;
            aParm.itsValue = itsCurValue.toString();
            itsParameters.addElement( aParm );
        }
        itsCurElt = null;
        itsCurValue.setLength( 0 );
    }

    public void characters(
char theChars[], int theStart, int theLength )    throws
    SAXException
    {
        if ( itsCurElt != null )
            itsCurValue.append( theChars, theStart, theLength );
    }

    // Logs SOAPHandler contents to System.out
    public void Log()
    {
        System.out.println( "SOAP-ENV: " + itsEnv );
        System.out.println( "Service: " + itsService );
        System.out.println( "Method: " + itsMethod );
        System.out.println( "Encoding: " + itsEncodingStyle );
        System.out.println( "Parameters: ". + itsParameters.size()
);

        for ( int i = 0;  i < itsParameters.size();  i++ )
        {
            SOAPHandler.SOAPParm aParm =
                (SOAPHandler.SOAPParm)itsParameters.elementAt( i
);
            System.out.println( "   " + aParm.itsName + " = " +
       aParm.itsValue + " (" + aParm.itsType + ")" );
        }
    }
}
```

CHAPTER 10

Accessing Domino from J2EE

> **"In addition to Lotus's unmatched collection of collaboration features and customizability, setting up our test installation of Domino was a far less headache-inducing prospect than setting up Exchange. Add to this a superior security record and strategy, and Domino is the clear winner of our Editors' Choice."**
>
> **—*PC Magazine*, Editors' Choice Awards, February 2, 2004**

IBM/Lotus has long provided a Java package (a bundle of Java classes and interfaces) that provides access to Domino objects. This Java package, `lotus.domino`, is the basis for direct access to and manipulation of Domino objects from J2EE applications. In this section, we examine how the `lotus.domino` package can be used from J2EE application elements running in the WebSphere Application Server. We'll see in concrete terms how this access to Domino from WebSphere extends the range of functions that can be incorporated into Web and enterprise applications.

Before we delve into using the `lotus.domino` package from J2EE, we highlight some points about programming with the package itself. We're not attempting to provide thorough exposition on the `lotus.domino` package in what follows, just some of the key aspects of the package that we hope will clarify how to use it from J2EE programs. The Domino Designer Help pages are a good source of detailed information about the `lotus.domino` package.

Java Domino Classes

The Java classes and interfaces in the `lotus.domino` package provide access to Domino objects as if the Domino server were performing the action. The `lotus.domino` objects are used to access and manipulate actual Domino databases and data elements within those databases. However, though the access is directly to Domino data, it does not bypass the Domino access control mechanism. The `lotus.domino` objects must be used in the context of a client session so that the access is still subject to the access controls defined for the session's user identity. The session access can be made in two ways: "local," where the `lotus.domino` objects access Domino data on the same system as the objects are running, and "remote," where the access is from the program's system to a remote Domino server. (Remote sessions are established using a `CORBA` mechanism via the Internet Inter-Object Protocol (IIOP). The HTTP and DIIOP server tasks must be running on the Domino server that is the target of a remote session.) The type of access for a `lotus.domino` session is determined by the signature of the `createSession` method of the Session class. See Table 10-2 for how the signatures of the `createSession` method map to the type of session access.

In addition to the two types of client access, the `lotus.domino` sessions can be established in three major contexts. You can think of session context as the type of JVM environment under which the `lotus.domino` program runs. The three context types are Application, Agent, and Applet. Agent session context is where the Java program is written as a Domino database agent and runs under the JVM established by the Domino server (or Notes client). We saw examples of Agent context in the previous chapter on accessing J2EE elements from Domino applications. Applet session context naturally applies to Java programs coded as applets and which run under a Web browser JVM. Application session context is for all other types of Java programs that run under a JVM outside of Domino or a Web browser. One most often writes J2EE programs that invoke Domino objects using Application context. The client session context is established by the type of lotus.domino base class under which the session object is created or obtained. Application context is set by using the `NotesFactory` class, Agent context via `AgentBase`, and Applet context via `AppletBase`.

To compile a Java program using the `lotus.domino` package, the Notes.jar Java archive file must be specified to the Java compiler. (In WSAD, you can specify Notes.jar as an external jar file in the project's Java build path.) To run the program, you must make either the Notes.jar (for local sessions), NCSO.jar (for remote sessions), or both accessible via the JVM's class path, the system path, or the `java.library.path` property value. If the program uses local session access, then the Domino server (or client) platform-specific library files must also be accessible. We've found that the only means of specifying the runtime libraries that works on all platforms is via the `java.library.path` property. For example, on a Unix/Linux system, you would specify it in this way:

```
java -Djava.library.path=$PATH:/opt/lotus/notes/latest/linux/:/
➥local/notesdata <class>
```

A program using local session access will use the Domino installation corresponding to the first Domino configuration file (notes.ini) found on the system path. Thus, if a Domino server and Notes client are installed on the same system, a local session will be established to the server if its notes.ini file is found first in the system path and vice versa. If the local session is established without a user identity specified, the user identity defaults to the Domino ID file pointed to by the notes.ini file (usually the server ID file for a server installation and a client ID file for the client).

Remote session access is made via the Internet Inter-Object Protocol (IIOP) and requires the Domino server DIIOP task be running on the target server (as well as the HTTP server task). With Domino R6, it is now possible to specify the server port on which the IIOP communication is conducted, making it easier to accommodate firewall restrictions. Keep in mind that the DIIOP task will terminate the remote Domino session due to inactivity. Your Java code should check the status of the Domino session object (via `Session.IsValid()`) before accessing any Domino objects, or handle the CORBA/IIOP exception thrown if the session is terminated.

Let's now look at the specific classes and interfaces available in the `lotus.domino` package. The classes can be categorized into three main groups: session-related, Domino objects, and utilities. Table 10-1 shows the classes and interfaces in this grouping.

Table 10-1 `lotus.domino` Class Categories

Category	Classes/Interfaces
Session Related	`NotesFactory, NotesThread, AgentBase, AgentContext, AppletBase, Session`
Domino Object Model	`ACL, ACLEntry, Agent, Database, DateRange, DateTime, DbDirectory, Document, DocumentCollection, EmbeddedObject, Form, International, Item, MIMEEntity, MIMEHeader, Name, Newsletter, Outline, OutlineEntry, Registration, Replication, RichTextItem, RichTextParagraphStyle, RichTextStyle, View, ViewColumn, ViewEntry, ViewEntryCollection, ViewNavigator`
Utility	`AgentLauncher, AgentLoader, AgentRunner, Log, Parser, Processor, Stream, XSLTResultTarget`

When using the Domino Java classes from J2EE applications, you will most often use the session-related classes of `NotesThread`, `NotesFactory`, and `Session`. The `NotesThread` class serves to establish the thread context under which Domino objects are managed. You must establish the `NotesThread` context for your application if it will access Domino objects via a local session (it is not required for remote session access). For J2EE applications running under the control of a Web or EJB container, the thread context setup is done by using the `NotesThread` static methods `sinitThread()` and `stermThread()`, since the container controls the thread setup for the servlet, JSP, or EJB. We will see an example of using the `NotesThread` static methods for servlets and EJBs in the following sections.

The `NotesFactory` class is used to establish the Domino session context for the Domino Java classes. Specifically, the `createSession()` method is used for session setup. The `createSession` method is multi-faceted, each method signature resulting in a particular type of session and user identity being established. Table 10-2 shows the various signatures of `createSession` and the resulting session type and user identity.

Table 10-2 `lotus.domino.Session.createSession()` Signatures

`createSession` Signature	Type of Session/User Identity
`()` `( null, null, null )`	Local session, user identity set from ID file pointed to by first notes.ini file in system path.
`( null, null, String Pw )`	Same as above, except String must match password of ID file.
`createSessionWithFullAccess()` `" ( String Pw )`	Same as above, except document access (on server) is not restricted by Readers fields. String must match password of ID file.
`( null, "", "" )`	Local session, user identity is Anonymous if Server record in local Domino directory permits Anonymous access.
`( String Host, "", "" )`	Remote (IIOP) session where Host is TCP/IP host name of the remote Domino server (Domino 6 allows host names suffixed with a port number, such as "`host:port`"). User identity is Anonymous if permitted as above.
`( null, String User, String Pw )`	Local session, user identity is that of given user name if it matches a Person record in the local Domino directory and password matches Internet password.
`( String Host, String User, String Pw )`	Remote session, Host as above. User identity is that of given user name if it matches a Person record in remote Domino directory and password matches Internet password.
`( null, String Token )`	Local session, user identity is that associated with the Token string. The Token string must be in LTPA token format as used for Single Sign-On. (See Chapter 12 for a discussion of SSO.)
`( String Host, String Token )`	Remote session, user identity set as above.

`createSession` Signature	Type of Session/User Identity ()
`( null, Credentials Cr )`	Local session, user identity is that associated with the `org.omg.SecurityLevel2.` Credentials object passed as the `Cr` parameter. This Credentials object is used in the `loginHelper` servlet supplied with WAS.
`( String Host, Credentials Cr )`	Remote session, user identity set as above.
`( null, null )`	Local session, user identity is set from the default—Credentials object within the JVM environment. This signature can be used within an EJB to establish identity based on the default Credentials associated with the EJB container.
`( String Host, null )`	Remote session, user identity set as above.
`( null, HttpServletRequest Req )`	Local session, user identity is set by the Domino HTTP server from the HTTP client authentication data in the request header.
`( String Host, HttpServletRequest Req )`	Remote session, user identity set as above.

For application processing, the most often used interfaces from the Domino Object Model are Database, Document, and Item. The Database interface represents a single Domino database file (.nsf or .ntf) and can be obtained from either the `AgentContext` (if running as a Domino agent), `DbDirectory`, or `Session` object (if not running as an agent). The Document interface represents a single document within a database and can be obtained from the `Database`, `DocumentCollection`, `Newsletter`, or `View` objects. (Used within a program, a `View` object acts as a document search or query function.) The `Item` interface represents a single data element within a document. Using these three interfaces, you can access and manipulate most any data within Domino databases, including the databases used for address book, mail, calendar, and discussion functions.

If you are using a local session for Domino access, you are responsible for managing the "garbage collection" of the Domino Java objects. This is because the Domino Java objects in a local session use the Java Native Interface (JNI) to invoke Domino libraries via the Domino C API. Each Domino Java class makes a `recycle()` method available for freeing the JNI resources. You should invoke the `recycle()` method on an object when it is no longer used to avoid using up system memory. Any Domino object obtained should be freed via the `recycle()` method, especially objects acquired in a collection class. For parent/child structures, recycling the parent recycles the children. Objects should be recycled in the same thread in which they are obtained. Recycling of objects is not needed for remote sessions since the Domino access is done at the Domino server.

Using Domino Classes from J2EE Servlets

Next, we'll look specifically at how to use the Domino Java Classes from J2EE servlet applications. For servlets, we have choices as to where in the servlet life cycle to establish the Domino session and, if required for local sessions, the `NotesThread` context. Recall that the Web container manages a servlet through four primary stages: loading, initialization (`init()`), service handling (`doGet()`, `doPost()`), and destruction (`destroy()`). We could establish the Domino session and thread context in the servlet's initialization stage (and undo it in the destruction stage) or establish them for each service request. Naturally, it would be more efficient to establish the session once for the servlet in the initialization method. However, this implies that the session would be established using single user identity for all servlet requests. If a single user identity is appropriate for the servlet application, then create the Domino session in the servlet's `init()` method. If the Domino access needs to be made under a user identity associated with the servlet requester, then the session will have to be created within the session handling methods.

As for establishing the `NotesThread` context for a servlet, recall that this is only required for local Domino sessions. If needed, the `NotesThread` context must be established for each thread invoking Domino Java objects. Since a Web container will invoke the service handling methods from multiple JVM threads (unless the servlet is defined to be single threaded), we really have no choice but to establish the `NotesThread` context within the service handling methods. This can be done via the `NotesThread` static methods, `sinitThread()` and `stermThread()`. The `stermThread()` method should be put in a `finally` clause since it must be invoked if the `sinitThread()` method is successful.

Let's look at a sample servlet, `GetBookDomino`, which accesses a Domino database, "booklist.nsf," via the Domino Java classes and accesses a view within this database. We will focus on the use of the Domino classes rather than the function of the servlet itself.

`GetBookDomino.java:`

```
import java.io.PrintWriter;
import java.io.IOException;
import javax.servlet.ServletException;

import javax.servlet.http.HttpServlet;
import javax.servlet.http.HttpServletRequest;
import javax.servlet.http.HttpServletResponse;

import java.util.Date;
import java.util.Enumeration;
import java.text.SimpleDateFormat;

import lotus.domino.*;

/**
```

```
 * GetBookDomino - Demonstrates using Domino Java classes from
➥WAS.
 */
public class GetBookDomino extends HttpServlet
{
   // Holds single Domino session, if SingleDominoSession=yes
   Session itsDominoSession;

   public void init() throws ServletException
   {
      super.init();

      // See if single Domino session to be set up
      itsDominoSession = null;
      String aSessParm = getInitParameter( "SingleDominoSession" );
      if ( aSessParm != null && aSessParm.equalsIgnoreCase( "yes"
      ➥) )
      {
(1)      NotesThread.sinitThread();
         try {
            itsDominoSession = NotesFactory.createSession();
         }
         catch ( NotesException e )
         {
            System.out.println( "Error creating Domino session!" );
            itsDominoSession = null;
         }
      }
   }

   public void doGet( HttpServletRequest theReq,
   ➥HttpServletResponse theResp )
      throws ServletException, IOException
   {
      // Set up writer for servlet response content, write HTML
         ➥heading.
      theResp.setContentType( "text/html" );
      PrintWriter aPW = theResp.getWriter();
      aPW.println( "<HTML>\n<HEAD>"
         + "\n<META http-equiv=\"Content-Type\""
```

```
        + " content=\"text/html; charset=ISO-8859-1\">"
        + "\n<META http-equiv=\"Content-Style-Type\"
        ➥content=\"text/css\">"
        + "\n<TITLE>GetBook Response</TITLE>"
        + "\n</HEAD> \n<BODY>\n"
    );

    // Save the servlet request parameters.
    // We look for: host, user, pw, auth (these are used for
    ➥createSession)
    String aHost = null;  String aUser = null;  String aPass =
    ➥null;
    Enumeration aReqParms = theReq.getParameterNames();
    while ( aReqParms.hasMoreElements() )
    {
        String aParm = (String)aReqParms.nextElement();
        if ( aParm.equalsIgnoreCase( "host" ) )
            aHost = theReq.getParameter( aParm );
        else if ( aParm.equalsIgnoreCase( "user" ) )
            aUser = theReq.getParameter( aParm );
        else if ( aParm.equalsIgnoreCase( "pw" ) )
            aPass = theReq.getParameter( aParm );
    }
    aPW.println( "<br>Session parameters: Host = " + aHost +
        ", User = " + aUser + ", Passwd = " + aPass );

    try {
        // Set up Domino Java thread and local session.
        Date aStart = new Date();
        SimpleDateFormat aSDF = new SimpleDateFormat("MM/dd/yyyy
        ➥HH:mm:ss");
        aPW.println( "<p>Servlet started: " + aSDF.format( aStart
) );
        String aSessType;
        Session aSession = itsDominoSession;
        if ( aSession == null )
        {
(2)         if ( aHost == null )  NotesThread.sinitThread();
            aSession = NotesFactory.createSession( aHost, aUser,
            ➥aPass );
            aSessType = "created by servlet";
```

```
          }
          else
             aSessType = "from servlet init";
          aPW.println( "<br>Domino session " + aSessType +
             ", set up time (ms): " +
             ( (new Date()).getTime() - aStart.getTime() ) +
             "<br>Session server: " + aSession.getServerName() +
             "<br>Connected as: " + aSession.getUserName() );

          // Access Domino database we want to display, fail if not
          ➥found.
          String aDbName = getInitParameter( "DbName" );
          if ( aDbName == null )  aDbName = "booklist.nsf";
(3)       Database aDb = aSession.getDatabase( null, aDbName,
          ➥false );
          if ( aDb != null )
          {
             View aBooks = aDb.getView( "Books" );
             if ( aBooks != null )
             {
                aPW.println( "<p><TABLE border=\"1\">" );
                aPW.println( "<TR><TH>Title</TH><TH>Author</TH><TH>
                ➥Price</TH>" );
                Document aDoc = aBooks.getFirstDocument();
(4)             while ( aDoc != null )
                {
                   aPW.println( "<TR> <TD>" +
                   ➥aDoc.getItemValueString( "booktitle" )
                      + "</TD><TD>" + aDoc.getItemValueString(
                      ➥"bookauthor" )
                      + "</TD><TD>" + aDoc.getItemValueString(
                      ➥"bookprice" )
                      + "</TD>" );
                   aDoc = aBooks.getNextDocument( aDoc );
                }
                aPW.println( "</TABLE>" );
(5)             aBooks.recycle();
             }
             else
```

```
                aPW.println( "<p><b>Error: Domino database view not
                ➥found!</b>" );

            aDb.recycle();
         }
         else
            aPW.println( "<p><b>Error: Domino database not
            ➥found!</b>" );

         // Recycle session only if created by this servlet
         if ( itsDominoSession == null )
            aSession.recycle();
      }
      catch ( Exception e )
      {
         System.out.println( "GetBookDomino error: " + e );
         aPW.println( "<p><b>Servlet error: </b>" + e );
      }
      finally
      {
         // Terminate NotesThread only if established by this
         ➥servlet
(6)      if ( itsDominoSession == null && aHost == null )
            NotesThread.stermThread();
         aPW.println( "</BODY>" );
         aPW.close();
      }
   }

   public void destroy()
   {
      if ( itsDominoSession != null )
      {
         try {
            itsDominoSession.recycle();
         }
         catch ( NotesException e )
         {
```

```
                System.out.println( "Error recycling single Domino
                ➥session!" );
            }
(7)         NotesThread.stermThread();
        }
        super.destroy();
    }

}
```

Here are the specific points about the servlet code. The numbered list that follows refers to the (n) annotations in the preceding code.

1. The servlet is coded to set up the Domino session in the initialization method based on a servlet parameter. Here, since the session is local, we must also establish a `NotesThread` context.
2. If a session has not been established already, the `doGet` method creates the session based on the host, user, and password parameters passed via the request. If a remote host is not specified, then the `NotesThread` context is established. Even if the `NotesThread` context is established during servlet initialization, it may need to be established in `doGet` (or the other service-handling methods) because the method may be running under a different thread than the one under which initialization was performed. We found no way to determine if `NotesThread` is established other than attempting a Domino object access and catching a `NotesException` if the access fails.
3. A specific Domino database object is obtained.
4. A `View` object is used to iterate through a set of documents in the database. Views are used in this way to act as queries on the database content.
5. Domino objects should be recycled after they are done being used. The recycle method performs additional resource cleanup beyond Java garbage collection.
6. The `NotesThread` context must be terminated if established by the service method.
7. The `NotesThread` context must be terminated if established by servlet initialization.

In order to run a servlet utilizing the Domino Java classes, the application server must have access via its Java classpath to the jar files containing the Domino Java support. For local session access, the required file is Notes.jar and for remote access it is NCSO.jar. Both of these files can be found under the Domino R6 installation path. When running the servlet under WSAD in the WebSphere Test Environment (WTE) server, you must specify the Notes and NCSO .jars to the WTE server environment file, as shown in Figure 10-1. Since our servlet can use both local and remote sessions, we specified both the Notes and NCSO .jar files.

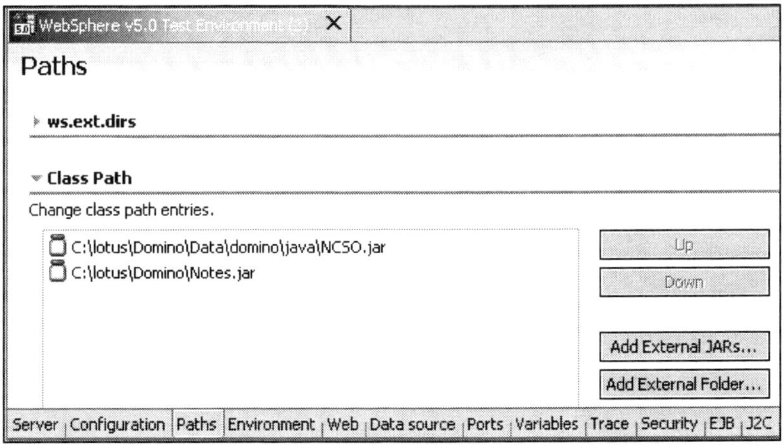

Figure 10-1 WTE path specification for Domino .jars.

Using Domino Classes from J2EE EJBs

Now we'll look at using the Domino Java classes within an EJB. Access via an EJB may be appropriate if the Domino resources are accessed in the same way across different servlets or JSPs or the access is part of a larger transaction. The first question to address is what type of EJB to use. Out of the three EJB types now available—session, entity, or message-driven—we have only two practical choices: session or message-driven EJBs. Using Domino classes from entity EJBs would require us to provide Bean Managed Persistence using Domino databases, and would be difficult at best since the Domino database is document-oriented and not relational.

We first focus on accessing Domino from a session EJB. To avoid confusion with the use of the term "session," we will always qualify it as either "session EJB" or "Domino session."

There are two types of session EJBs, different in the way they are managed by the EJB container. We'll need to understand these differences so that we can incorporate the Domino Java classes appropriately. A session EJB can be either stateless or stateful. Architecturally, the difference between stateless and stateful session EJBs is the extent to which the session is tied to the client. Stateless session EJBs have no ties to the client and can be reused across many clients. A stateless session EJB usually offers only a single function such as a query or update function. Stateful session EJBs can be tied to a client and maintain a conversation state across EJB method invocations. A shopping cart function is often implemented using a stateful session EJB. Stateless session EJBs are created in pools by the EJB container and allocated to clients out of the pool. Stateful session EJBs are created and destroyed in response to client actions.

For stateless session EJBs, we can create the Domino session in the `ejbCreate()` method and attempt to reuse it in the business method(s) if the Domino session can be established under a single user identity not associated with the client (such as the server's identity or some identity representing all clients). There are two concerns with this approach. The first is that if a local

Domino session is to be established via the `NotesThread` static methods, there is a serious issue with managing the `NotesThread` context across the threads used by the EJB container—see the following section on `NotesThread` issues. The second concern applies to remote Domino sessions. The remote session might be timed out by the Domino server and so should be checked each time it is referenced (via `lotus.domino.Session.isValid()`) to see if it needs to be reestablished. Avoiding these concerns implies using the less efficient approach of establishing the Domino session in every business method invocation. This approach also would be our only choice if the Domino session had to be established under the identity of the EJB client.

With stateful session EJBs, the concerns around establishing the Domino session would, at first glance, seem to be simplified because of the facts that the stateful EJB is created, used, and destroyed for each client and is single-threaded relative to the client. It should be a matter of just creating the Domino session during `ejbCreate()` and terminating it in `ejbRemove()`. Unfortunately, although the EJB container is required to "single-thread" the EJB methods from the client point of view, it may (and, in the case of WAS, does) run separate EJB method invocations on different threads within the application server. This makes it virtually impossible to establish a local Domino session for reuse across the EJB business methods. A remote Domino session can be established in `ejbCreate()` and terminated in `ejbRemove()`, but its use must be validated in case it is dropped by the Domino server.

In addition, establishing the Domino session within stateful session EJBs brings another set of concerns. Besides the two states of "Does Not Exist" and "Method Ready," which apply to both types of session EJBs, the stateful session EJB can be put into a "Passive" state via the `ejbPassivate()` method and returned to the "Method Ready" state via `ejbActivate()`. Because client state must be preserved across passivation/activation, client data within the EJB must be serialized. Since we can't serialize a Domino session object, it must be terminated before passivation and restored during activation.

After determining where and when we need to establish the Domino session, we next must consider the user identity under which to establish the session. The simple cases are those where we can establish the Domino session either anonymously under the Domino server's identity or under a single user identity for all clients. The not so simple case is where we want or need to establish the Domino session under a specific user identity, say for a stateful session EJB representing a shopping cart. Unfortunately for this case, we find that the J2EE and Domino client/user identity functions don't make it a simple matter of passing the EJB client identity to Domino. A full explanation of why this is so would take us far afield of our topic here, but basically the J2EE EJB container is architected to handle client identity authentication and authorization so that the EJB itself doesn't need to (and thus can be kept generic). Programatic authorization within the EJB is provided for but must be based on the rather minimal set of authenticated client information available to the EJB (namely, the authenticated caller name and role). We can't get the credentials used to authenticate the EJB client in order to pass them to Domino. Nevertheless, the situation is not hopeless, and we can take the following approach.

We'll call this technique the "handoff" approach: We can pass the user identity parameters needed to establish the Domino session directly to the EJB via a separate "business" method. Referring back to Table 10-2, we see that there are four ways to establish a specific user identity for the Domino session—userid/password, LTPA token, `org.omg.SecurityLevel2.Credentials`, or `HttpServletRequest` objects. If the EJB is being invoked from a servlet, then any of these four can be readily used. The LTPA token string can be extracted from the `HttpSession` object (it is kept as a cookie on the client and we discuss it in detail in Chapter 12); the Credentials object can be obtained from the `LoginHelper` servlet provided by WAS; and the `HttpServletRequest` object can be passed directly (it contains the name and password information from an HTTP basic authentication). Unfortunately, if the EJB is invoked from a Java application or an EJB client container, then only the userid/password is easily passed.

`NotesThread` Issues with EJBs

For those interested in the details of the "serious" `NotesThread` issue with establishing local Domino session within EJBs, we attempt an explanation here. Otherwise, this section can be skipped.

From the Domino Designer Help database, we find the following caveats concerning multithreading the Domino Java classes (we have numbered the points for reference). You should avoid multithreading unless you have good reason to use it, such as proceeding while file input/output and while Web requests are processing. Observe the following guidelines:

1. Within a session, Domino objects are shared, synchronized, and recycled across threads. By using a different session on each thread, you lose these capabilities; you must explicitly manage synchronization and recycle on a per-thread basis.
2. Do not use `DbDirectory` across threads.
3. Accessing an existing document on multiple threads is permissible, but accessing it on just one thread simplifies memory management. Restricting accesses to one thread allows you to recycle without checking the other threads. Creating documents across threads is always safe, and these objects can be recycled without reference to the other threads.
4. Profile documents are cached on a per-thread basis. In the event of an update contention, the last thread update wins.
5. Take care not to delete a document needed for navigation by a view or collection on another thread.
6. When child objects are used on threads rather than the parent, keep the parent thread alive until all child threads terminate. This is particularly important when using Domino objects in AWT event handlers.

The first five points could be managed with some careful programming. It is the sixth caveat that makes it virtually impossible to manage a `NotesThread` context within an EJB instance being run by an EJB container. If the `NotesThread` context is initialized and terminated around the Domino calls within every EJB method, we can't keep a "parent" thread context established—and any Domino objects created in the method disappear when the `NotesThread` context is terminated. If we try (and we have) to keep a "parent" thread context active in `ejbCreate()` (at least for stateful EJBs), we won't be able to terminate that same thread context because there is no guarantee that the `ejbRemove()` is invoked on the same thread as the `ejbCreate()`.

Accessing Domino from Web Services

With the ever-expanding support for Web services within WebSphere Application Server and Developer, wrapping almost any server-based function as a Web service is becoming easy. If appropriate, a Domino application can be opened to intranet or Internet clients by exposing key function points as Web services, say using SOAP-XML over HTTP. If the Domino application functions are already accessible via the Domino Classes for Java, turning these functions into Web services is nearly automatic using the tools available in WSAD.

As with servlets and EJBs, the major consideration with using the Domino Classes for Java within Web services is Domino session setup and management, especially for local Domino sessions. In this section, we discuss these considerations in detail. We take the scenario of converting an existing Java class, which utilizes the Domino Classes for Java into a set of Web services. The Web service conversion is done within WSAD and tested using the WebSphere Test Environment (WTE) application server.

Before delving into the details, it may be helpful to briefly review how a J2EE application server, especially WAS, handles Web service requests and responses. The Web service requests are handled by servlets within the application server. There are two Web service handling servlets: `messagerouter`, which handles "document style" requests, and `rpcrouter`, which handles SOAP style requests. The `rpcrouter` servlet receives the HTTP request data and parses the SOAP-RPC XML request. It then takes the parsed request and passes it to an invoker servlet, which invokes the defined service class. The service class can be a Java bean or an EJB (WebSphere allows only stateless session EJBs as Web service providers). The key points to remember here are that the Web service provider class is invoked as a servlet by the application server's servlet engine and is subject to the multi-threading used by the servlet engine.

Now let's look at the process of creating a Web service from a Java class using WSAD. As one would expect, a WSAD wizard guides us through the process. We'll number the next paragraphs to highlight the steps in the process and their order.

1. Import or create the Java class within a Web project. If needed, you may need to create the Web project first. At this step, ensure that the appropriate Notes .jar file is specified in the project's Java build path and in the WTE classpath.

2. Right-click the imported Java class source file and then select **WebServices>Deploy as Web Service** (menu item), which starts the Wizard. The following dialog appears:

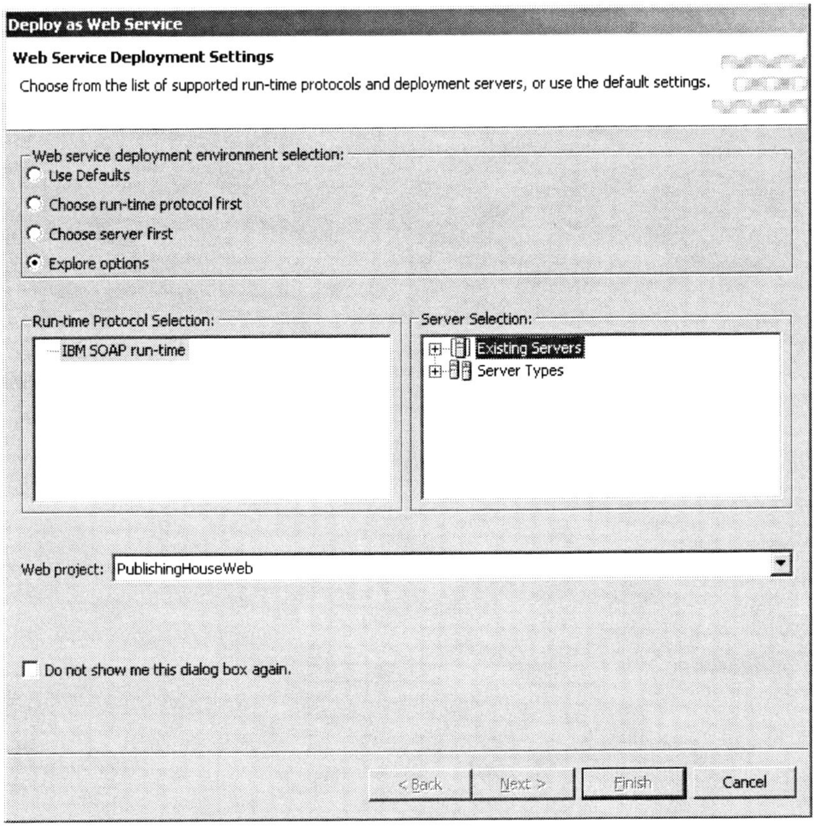

Figure 10-2 Web Service Deployment Dialog 1.

3. Select **Explore options** and choose an existing server WTE (see Figure 10-2). Select the **Next>** button. (The button becomes selectable once a server is chosen.)
4. Specify the Java bean (class) containing the service methods. Select the **Next>** button. The dialog shown in Figure 10-3 appears.
5. The Web Service Java Bean Identity dialog is where the Java class is configured as a Web service (see Figure 10-3). Using this dialog, we can specify the service URI, how it should be invoked by the application server, and where the various WSDL documents defining the service should be placed. We'll look at the Scope field in detail, since it affects how we can manage the Domino session within the service methods for our service that utilizes Domino objects.

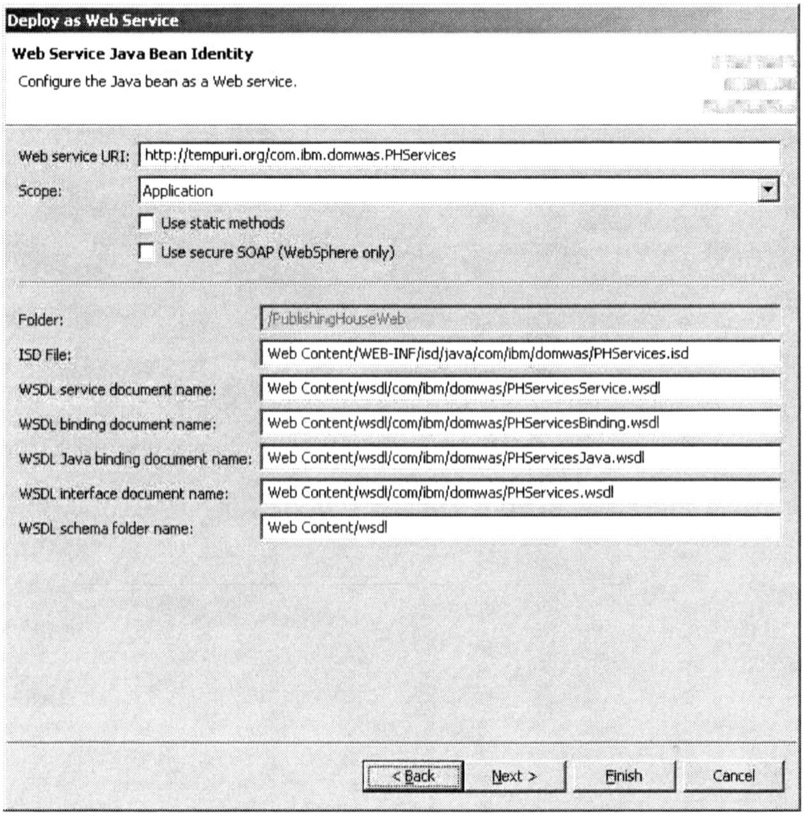

Figure 10-3 Web Service Deployment Dialog 3.

The Scope field specifies the lifetime of the Java class object created by the application server to handle the Web service request. This object is referred to as the "service object." The Scope can be specified as "Application" (the default), "Session," or "Request." If "Application" scope is specified, the server creates a single instance of the service object for the life of the server and uses this single object to handle the Web service requests from all clients. "Session" scope results in the server creating a separate service object for each SOAP session (established via client-side cookies). For "Request" scope, the server creates an instance of the service object for each service request. (The Scope field applies only to Web services generated from Java classes or beans and not to those generated from EJBs.)

We selected a Scope of Application for our services and selected the **Next>** button.

6. The next dialog allows selection of which methods in the Java class to expose as Web services. It also allows selection of the type of encoding for the service request and response. This dialog is shown in Figure 10-4.

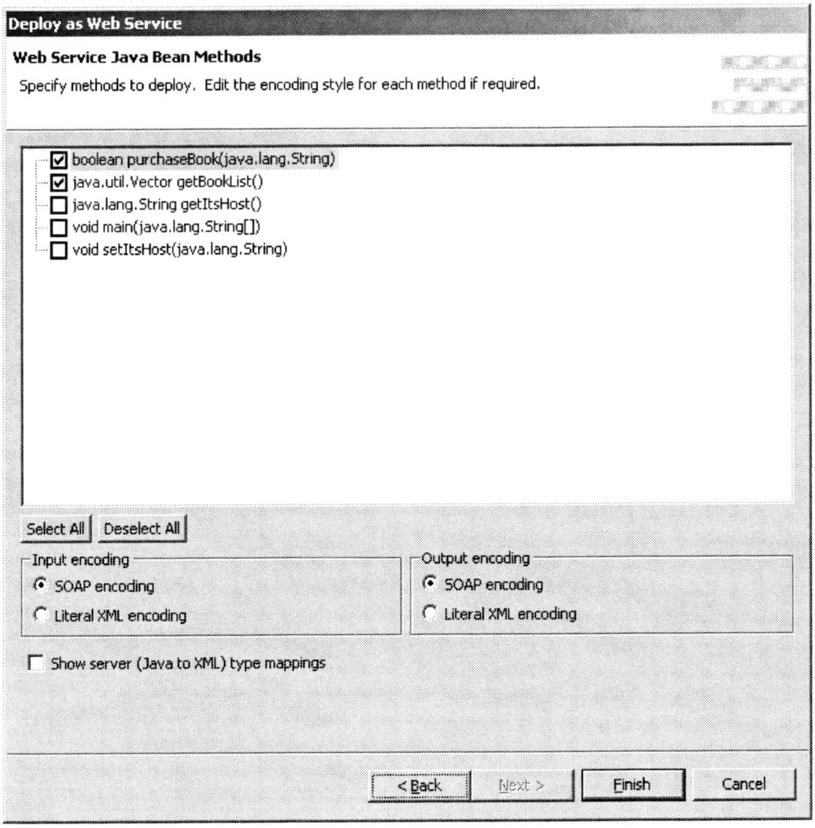

Figure 10-4 Web Services Deployment Dialog 4.

7. Selecting the **Finish** button on the dialog results in the generation of the WSDL documents within the Web project. The WSDL documents are used by the application server Web service tasks (namely the `messagerouter` and `rpcrouter` servlets) to associate the Web service request with the corresponding Java class.

For our service, which is using Domino objects, only Application or Session scope gives us the opportunity to share a Domino session across service invocations. We can implement the Java class containing the service methods to hold the Domino session as a class member and allocate the Domino session in the class constructor and terminate (recycle) the session in the finalizer. We then can use the procedure above to generate Web services from individual methods in our Java class. By specifying a scope of Application or Session, the Domino session will span the service method invocations. In this way, we can avoid the overhead of establishing the Domino session for each service invocation. We can recycle individual Domino objects within the service methods where they are not shared or need to be reacquired before each use because they frequently change (such as database views).

As with servlets and EJBs, there are complications to maintaining local Domino sessions across Web service invocations. Again, the complications are due to the need to establish a `NotesThread` context for local Domino sessions. Even though the same Java object is reused for the service under Application or Session scope, the service methods may be invoked on separate threads. It is possible with careful programming to maintain the `NotesThread` context when sharing local Domino sessions across service invocations. A "parent" `NotesThread` context can be established in the service class constructor with the establishment of the Domino session and the thread identifier saved as a class member. Each service method can then check to see if they are running under this "parent" thread, and if not, can establish another `NotesThread` context for the duration of the method. The "parent" context can be terminated in the service object finalizer.

If remote Domino sessions are used, then the above complications involving thread context are not a factor. In many cases, because of the usual separation of the J2EE application server and Domino servers into separate tiers, remote rather than local Domino sessions are used.

What about establishing the Domino sessions under the identity of the Web service requestor? Unfortunately, at this point in product and architecture development, the story is a bit bleak. There is no easy way to obtain client credentials from the SOAP-XML Web service request in the service object itself. This makes it difficult to establish the Domino session under the identity of the requesting client. Any client credentials would have to be passed as parameters to the individual service methods. Consequently, client-specific Domino sessions would have to be established within each service method on every service invocation.

Lotus Domino Toolkit for WebSphere Studio

The Lotus Domino Toolkit (LDT) for WSAD provides a set of JSP tag libraries that allow creation of JSP-based Web applications from existing Domino databases. The toolkit contains the Domino JSP tag libraries introduced with Domino R6 and installs into WSAD as a feature plug-in, thus providing the WSAD developer with the means to easily create Domino-based Web applications without having to leave the WSAD environment.

In this section, we will cover how to install the LDT and make use of the Domino JSP tag libraries to create Web applications. We take the point of view of an experienced J2EE and WSAD developer who may be less familiar with Domino function. We highlight the considerations relating to Domino that the J2EE developer may need to take into account when writing a Web application using the LDT.

Installation and Setup

The LDT install file is shipped as part of the Domino product beginning with the Domino 6.02 release. It can be found in the Applications directory of the Domino Designer install CD. The Domino 6.02 version of the LDT only works with WSAD 5.0. You must use the LDT version shipped with Domino 6.5 with WSAD 5.1. In case you may be wondering, the LDT does not work with any version of the basic Eclipse IDE, since it requires certain support provided only with WSAD.

The LDT installs to the WSAD root directory. It installs the Domino JSP tag libraries, the Domino Java jar files, and a set of documentation files that integrate into the WSAD help system. The LDT files are installed to subdirectories under the [WSAD_Root]/wstools/eclipse/plugins directory. In particular, the Domino Java jar files are placed in the com.ibm.lotus.lib subdirectory. Note that the 6.02 version of the LDT contains the NCSOW.jar for remote Domino sessions. This jar file is an older version of the CORBA/IIOP support that is not compatible with Domino R6 servers and should not be used. You should use the NCSO.jar file, which can be found in the R6 versions of either the Notes client or Domino server installations. As with the Domino Classes for Java, the Notes.jar and NCSO.jar files must be made available to the application server on which the JSPs using the Domino tag libraries are run.

Once the LDT has been installed, the help documentation is available via the main WSAD help window and a new "Domino" view is shown as part of the Web perspective (see Figure 10-5).

Figure 10-5 Domino view in WSAD.

Using this Domino view, the WSAS programmer can access any local or remote Domino database (to which they are authorized) and construct JSP elements from the Domino database design elements such as forms, views, or actions. We'll examine just how to use this view to access Domino databases and use their design elements, but, first, let's look at how to set up a Web project to use the Domino JSP tag libraries and how to create a JSP with Domino tags.

Using the LDT Within WSAD

Assuming we have a Web project already created, we can set up the project to use the Domino JSP tag libraries via the properties dialog for the project. Open the properties dialog for the Web project, select the **Web** property page, and then select the **Include Domino Custom Tags Library** feature, as shown in Figure 10-6.

The Domino tags feature results in WSAD copying the Notes.jar, NCSO.jar, and domtags.jar files to the Web project's WEB-INF library path for deployment and to its build library path for compilation, as shown in Figure 10-7.

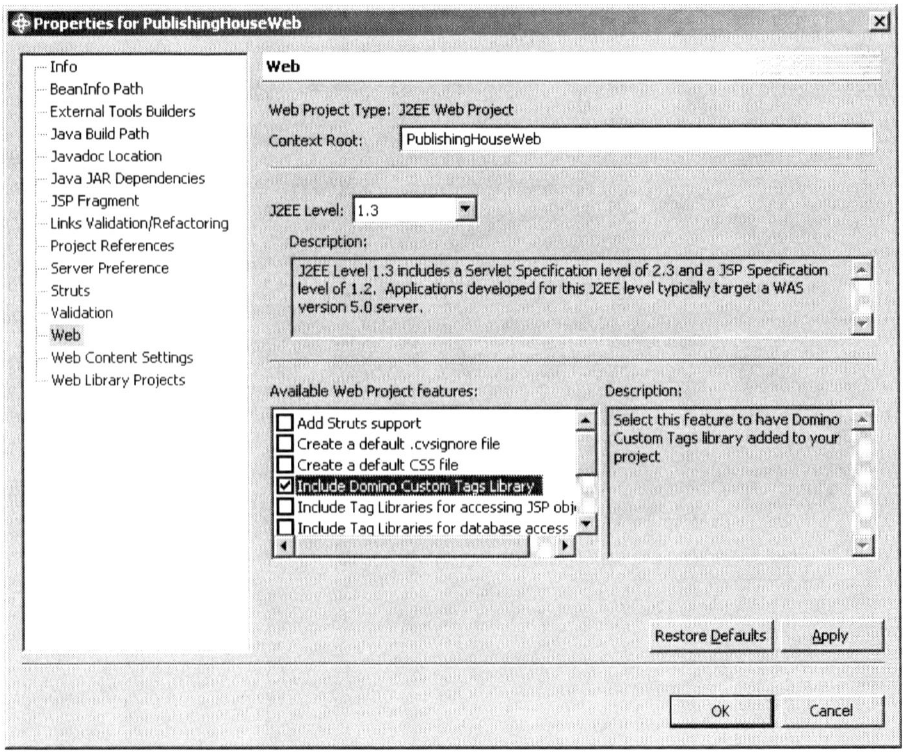

Figure 10-6 Web Project properties.

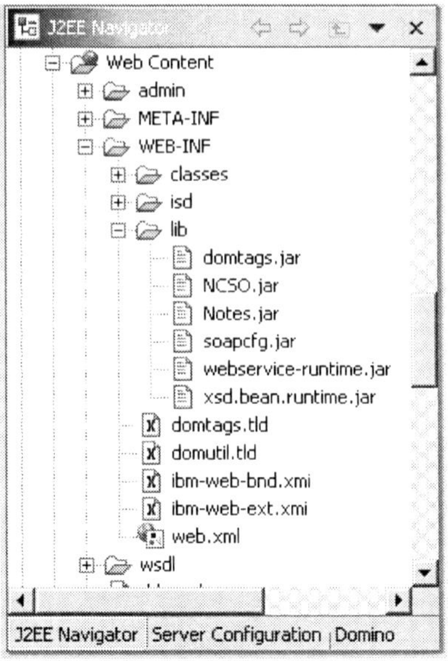

Figure 10-7 Web Project library files.

If the Web project is created, the creation wizard allows selection of the Domino Custom Tags Library feature in the same way.

Note that the Notes.jar or NCSO.jar files may already have been added to the project's build library path if they were required by other project components such as other servlets. In this case, the duplicate versions can be removed. Also, the Notes.jar or NCSO.jar files can be removed from the WEB-INF library path if the target WAS server has these files in its runtime classpath. As with the Domino Java objects, the Notes.jar file can be removed if only remote Domino sessions will be established and vice versa for the NCSO.jar file.

On creating a JSP, the WSAD wizard allows selection of JSP tag libraries to be included in the page. Selecting **Add Tag Library** in the dialog shown in Figure 10-8 brings up the selection dialog shown in Figure 10-9.

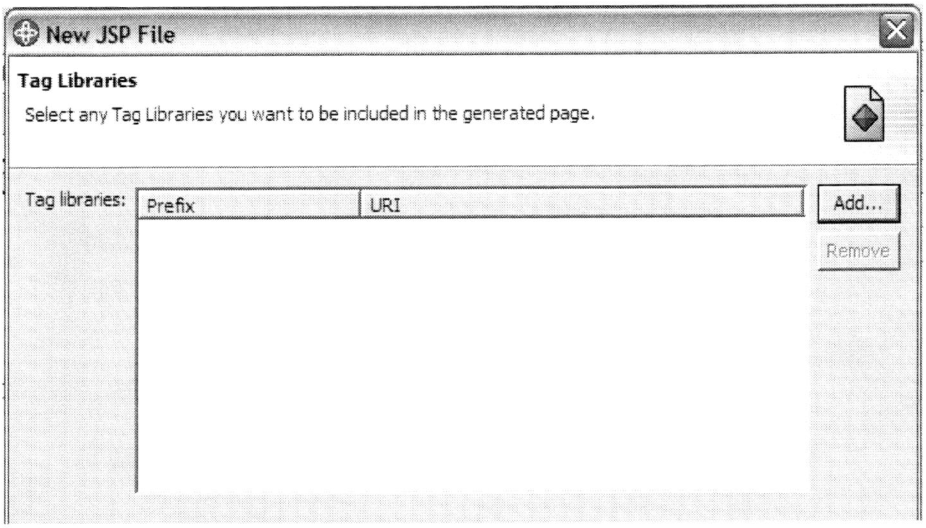

Figure 10-8 Selection of JSP tag libraries.

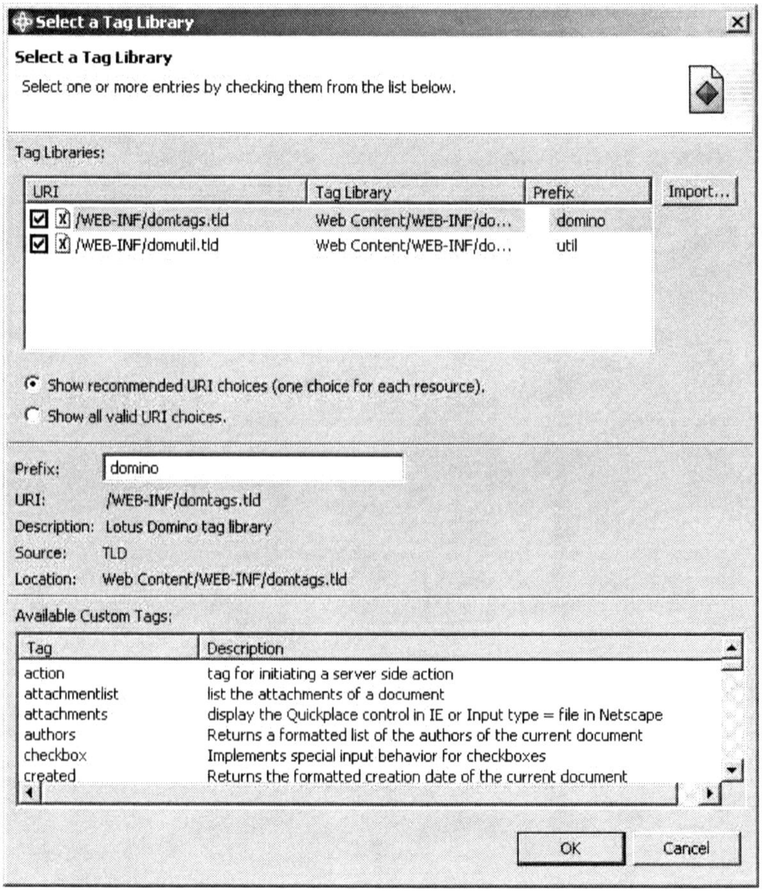

Figure 10-9 Adding JSP tag libraries.

Since the Domino JSP tag libraries were made available to our project, they appear in this dialog and we can select them. We can specify a tag library prefix for each library by highlighting the library, then entering the prefix name in the Prefix: input field. The prefix names associate the JSP tags with a tag library definition statement in the JSP page. We will use a prefix of "domino" to refer to the tags contained in the domtags.tld and "util" for the domutil.tld. When finished specifying the prefix names, selecting the **OK** button takes us back to the previous dialog from which we can select the tag libraries for our JSP, as shown in Figure 10-10.

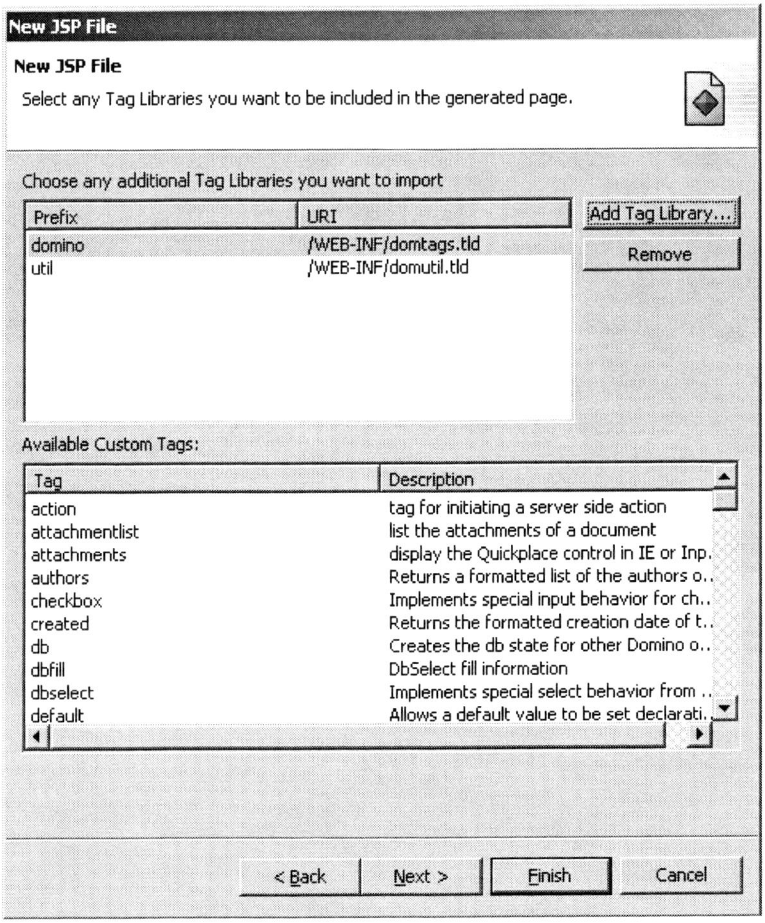

Figure 10-10 Selecting Domino tag libraries.

Now that we have our JSP created with the ability to use the Domino JSP tag libraries, we can access the Domino database from which we can obtain the design elements to include in our page. Display the Domino view (refer back to Figure 10-5) and either select the left-most icon on the right side of the title bar or the menu drop-down next to it, or right click in the view window and select **New database connection**. In this dialog, you specify the Domino server TCP/IP host name on which the database resides (or leave this field blank for local database access), the user under which to access the database (blank to use the default user ID file), and the path to the database. Selecting the **Finish** button results in the Domino database being shown in the WSAD Domino view as an expandable tree with the database's forms, views, and agents as leaves (see Figure 10-11). Now the Domino design elements can be simply dragged and dropped on any JSP set up to use the Domino tag libraries.

Note that a JSP can be set up to use Domino tag libraries either via the creation wizard as shown above, or by explicitly specifying the `taglib` JSP directives for the Domino tag libraries. This setup is not done by dragging/dropping the Domino elements to a JSP page created without the Domino tag libraries, and JSP compilation errors will result until the directives are added.

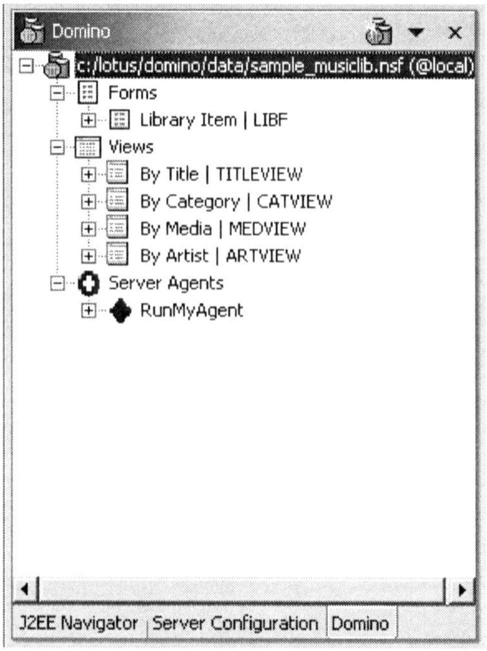

Figure 10-11 WSAD Domino view with database.

(The capitalized names shown next to the form and view names are alias names for the forms and views. They are used in Domino application code to avoid having to change code when a display name is changed.)

Let's now create two JSPs, one based on the `By Title` view from the sample_musiclib.nsf database and the other from the `Library Item` form. This will give us a feel for working with the Domino JSP tags. We create the JSPs as above using the New>JSP Wizard and select the domtags.tld and domutil.tld tag libraries to be added to the page. We create the first JSP, domview.jsp, in this way, then select the `By Title` view and drag it into the body of the JSP. Looking at the JSP source at this point (see the following listing), we see that the `domino:view`, `domino:viewloop`, and domino`:viewitem` tags have been added by the drag and drop of the view element (in bold below):

```
<!DOCTYPE HTML PUBLIC "-//W3C//DTD HTML 4.01 Transitional//EN">
<HTML>
<HEAD>
```

```
<%@ taglib uri="/WEB-INF/domtags.tld" prefix="domino" %>
<%@ taglib uri="/WEB-INF/domutil.tld" prefix="util" %>
<%@ page
language="java"
contentType="text/html; charset=ISO-8859-1"
pageEncoding="ISO-8859-1"
%>
<META http-equiv="Content-Type" content="text/html; charset=ISO-
➥8859-1">
<META name="GENERATOR" content="IBM WebSphere Studio">
<TITLE>domview.jsp</TITLE>
</HEAD>
<BODY>
<P><B>All Documents by Title:<BR>

<domino:view viewname="By Title" dbserver=""
   dbname="c:/lotus/domino/data/sample_musiclib.nsf"
   user=""  password="" >
   <table>
      <tr> <td>Title</td> </tr>
   <domino:viewloop>
      <tr>
         <td>
            <domino:viewitem col="1"/>
         </td>
      </tr>
   </domino:viewloop>
   </table>
</domino:view>

</B></P>
</BODY>
</HTML>
```

Those familiar with Domino views will see that the `viewloop` tag iterates over the contents of the first column of the view, which in this case consists of the `Title` field of the documents using the `Library Item` form. Note also that, by default, the `domino:view` tag is generated with scripting variables for the user and password attribute values as such:

```
user="<%username%>" password="<%password%>"
```

We can remove these variables to make it easier to test against a local Domino database via the WTE. With the `dbserver`, `user`, and `password` attributes all set to empty strings, a local Domino session is established under the default Notes ID file for the local server configuration.

The second JSP, domform.jsp, is created in the same way, except we drag and drop the `Library Item` form element to the page. This results in the following JSP source:

```
<!DOCTYPE HTML PUBLIC "-//W3C//DTD HTML 4.01 Transitional//EN">
<HTML>
<HEAD>
<%@ taglib uri="/WEB-INF/domtags.tld" prefix="domino" %>
<%@ taglib uri="/WEB-INF/domutil.tld" prefix="util" %>
<%@ page
language="java"
contentType="text/html; charset=ISO-8859-1"
pageEncoding="ISO-8859-1"
%>
<META http-equiv="Content-Type" content="text/html; charset=ISO-
➥8859-1">
<META name="GENERATOR" content="IBM WebSphere Studio">
<TITLE>domform.jsp</TITLE>
</HEAD>
<BODY>
<P><B>Library Item</B>

<domino:form name="formname"
dbname="c:/lotus/domino/data/sample_musiclib.nsf"
   dbserver=""  user="" password="" schema="Library Item">
   <table>
      <tr>
      <td>Date Acquired</td>
      <td><domino:input name="DateAcquired" type="date"/></td>
      </tr>
      <tr>
      <td>Comments</td>
      <td><domino:input name="Comments" type="text"/></td>
      </tr>
      <tr>
      <td>Media</td>
      <td><domino:input name="Media" type="text"/></td>
      </tr>
      <tr>
```

```
            <td>Artist</td>
            <td><domino:input name="Artist" type="text"/></td>
            </tr>
            <tr>
            <td>Title</td>
            <td><domino:input name="Title" type="text"/></td>
            </tr>
            <tr>
            <td>Category</td>
            <td><domino:input name="Category" type="text"/></td>
            </tr>
        </table>
        <BR>
        <domino:savedoc showdisabled="false" text="Submit"/>
    </domino:form>

    </P>
    </BODY>
    </HTML>
```

Dragging and dropping the form design element adds the `domino:form` and `domino:input` tags to the JSP. The `domino:form` tag acts as a container for the `domino:input` tags and establishes the Domino session. To provide the submit function, we manually added the `domino:savedoc` tag (as shown in bold above), which will result in either the creation of a new document or the update of an existing document in the Domino database. We'll see how this same JSP can be used for both document creation or update by enhancing the domview.jsp.

As we saw, domview.jsp displays a simple static text list of the `Title` fields of the documents in the database. We embellish the list by changing the text rows into active links, which will display the document contents using the domform.jsp. We also add a "Create New Document" link at the bottom of the table. These changes are shown in bold:

```
<domino:view viewname="By Title" dbserver=""
   dbname="c:/lotus/domino/data/sample_musiclib.nsf"
   user="" password="" >
   <table>
      <tr> <td>Title</td> </tr>
   <domino:viewloop>
      <tr>
         <td>
            <domino:formlink href="/PublishingHouseWeb/
            ➥domform.jsp">
               <domino:viewitem col="1"/>
```

```
            </domino:formlink>
          </td>
        </tr>
      </domino:viewloop>
      </table>
    </domino:view>

    </B></P>
    <P><A href="/PublishingHouseWeb/domform.jsp" target="_self">
    Create New Document
    </A></P>
    </BODY>
```

The domino:formlink tag creates an HTML link to an existing Domino document by passing the document's unique identifier (UNID) to the specified URL reference. The URL reference is assumed to contain either a domino:document or domino:form tag, which uses the UNID to access the document. In our JSP, the domino:formlink tag passes the UNID of the documents in the view to our domform.jsp, which then displays the document's contents. We used a simple hyperlink (<A>) referencing the same domform.jsp to present a page for creating a new document. Since no UNID is passed to the domform.jsp in this link, the domino:form (and domino:input) tag generates the field elements as HTML input tags, allowing the field values to be input by the user. Note that apart from adding a submit action, we did not have to modify the domform.jsp to provide this dual function.

Structure of the Domino JSP Tag Library

Now that we have seen a basic example of using the Domino JSP tags, we'll examine the overall structure of the tags and their function. By understanding the structure, we will be better able to use the tags to create efficient JSPs to access Domino databases and functions.

The set of Domino JSP tags can be divided into the following categories: the core tags, data tags, action tags, navigation tags, and control tags. The core tags are those that provide access to the basic Domino database or server functions by establishing an underlying Domino Java Objects session from the JSP container to the Domino server. The data tags are those that allow access to and manipulation of data items in the Domino databases. Action tags provide means to invoke Domino server-side processing. Navigation tags enable the JSP user to navigate to other Web pages, especially other Domino JSPs. The control tags simplify adding conditional processing to the JSP, where the conditions are based on Domino attributes. Apart from the core tags, these categories are our own definitions and are not used in the Domino Help documentation.

As JSP tags, they have XML structure so that the tags can be written hierarchically—where the sub-element tags can take on the attributes of the parent element tag. A parent tag that contains sub-element tags is referred to as the context of the sub-element tags. The function of several of the Domino JSP tags often depends on their context.

Since they are most important, let's look at the core tags more closely. As mentioned earlier, they establish the underlying Domino Java session from the JSP to the Domino server. Each of the core tags—`view`, `document`, `form`, `ftsearch`, `mailto`, and `runagent`—defines attributes allowing specification of the host, user, and password, which, as we saw in the section on Domino Java Classes, are the primary parameters for creating a Domino Java session. Since creating a Domino session is relatively expensive, we want to create only a single Domino session per JSP. If more than one of the core tags is required on a single page, the session tag can be used as their parent and the Domino session established only once. Similarly, if a single Domino database needs to be accessed via separate core tags (say from `view/viewloop` and `ftsearch` tags), then the `db` tag can be used to wrap the tags referring to the database so that the database open is done only once.

The following tables list the Domino JSP tags in the categories we defined here.

Table 10-3 Domino JSP Core Tags

Core Tags	Context	Description
`view`	`[db, session]`	Displays a Domino view, which is a set of documents in a database and a set of fields for each document.
`document`	`[db, session]`	Displays, edits, or creates a single Domino database document. Contains item or formula tags that display or edit document data.
`form`	`[db, session]`	Displays, edits, or creates a single Domino database document with optional client-side field validation.
`ftsearch`	`[db, session]`	Performs full-text search of a database.
`mailto`	`[db, session]`	Sends enclosed text as a Domino mail message. Sender is the user owning the current session.
`runagent`	`[db, session]`	Runs a database agent on the server.

Table 10-4 Domino JSP Data Tags

Data Tags	Context	Description
`attachmentlist`	`form`	Displays attachments as list. Intended for Netscape Navigator/Mozilla.
`Attachments`	`form`	Displays attachments as list. Intended for Internet Explorer; supports upload via ActiveX control.
`authors`	`viewloop, docloop, document, form`	Displays Authors field of document. Document must have been saved.

Data Tags	Context	Description
checkbox	form	Displays or edits list of item contents as checkbox.
created	viewloop, docloop, document, form	Displays document creation date.
dbfill	dbselect	Sets controls in page to values from document fields specified via `dbselect`.
dbselect	form	Specifies a set of document values from a database view or of `option` or `dbfill` tags. Displays as a drop-down list.
docloop	db, ftsearch, responses, page, view	Generates the tag body for each document in the `context` tag (for example, for each document in a view).
fill	option	Sets client controls from specified target and value attributes. Used within `select`, `dbselect` tags.
formula	viewloop, docloop, form, document	Displays results of formula specified in tag body.
input	form	Allows editing of a document field.
item	viewloop, docloop, form, document	Displays the value of a document field. Allows editing of a field in form context.
lastaccessed **lastmodified**	viewloop, docloop, form, document	Displays last accessed/modified date and time of a document. (Forces reload of document.)
option	select, dbselect	Displays single option in a drop-down list control.
profiledocument	form	Obtains fields from user profile documents.
radio	form	Displays or edits list of item values as radio buttons.
richtext	form	Allows editing of a rich text item via Java RichText applet or Internet Explorer rich text control.
select	form	Specifies set of list values via option tags. Displays as drop-down list.
selectedloop	[view]	Generates the tag body for each document in the `context` tag (for example, for each document in a view).
selectentry	viewloop	Displays loop item with checkbox.
setitem	document, form	Sets item in document. Item not saved until document is saved.
textarea	form	Displays text item as HTML `textarea` control.

Data Tags	Context	Description
unid **user**	`viewloop, docloop, form, document`	Displays document UNID/current user.
viewitem	`viewloop, selectentry`	Displays item from specified view column.
viewloop	`view, page`	Generates the tag body for each row in a view.

Table 10-5 Domino JSP Action Tags

Action Tags	Context	Description
action	`form, view`	Generates action link causing action performed on server. Action specified via `ifserverevent` tag.
delete-attachment	`attachmentlist, form`	Deletes selected attachment from document.
deletedoc	`form`	Deletes current document; displays next document.
editdoc	`form`	Places form in edit mode (displayed only if in read mode).
newdoc	`form`	Creates new document using current form.
nextdoc	`form`	Displays next document in current document collection.
saveclosedoc	`form`	Saves and closes the current document.
savedoc	`form`	Saves current document (with validation).
savenow	`form`	Saves current document immediately.

Table 10-6 Domino JSP Navigation Tags

Navigation Tags	Context	Description
docnavimg	`form`	Generates a navigation link from specified images, one showing the link enabled, and the other disabled.
formlink	`viewloop, docloop, form, document`	Creates link between a document reference in a `viewloop` or `docloop` and a form or document page. Link is via UNID.
pextdoc **prevdoc**	`form`	Creates `navigation` tag to display next, previous document in current document collection.

Navigation Tags	Context	Description
pagebar	page	Provides for navigation to a particular page.
pagefirst **pagelast** **pagenext** **pageprev**	page	Generates HTML to navigate to first, last, next, or previous page accessed in a session.
parentlink	viewloop, docloop, form, document	Generates a link to a document's parent.

Table 10-7 Domino JSP Control Tags

Control Tags	Context	Tags / Description
Access Control	db, view, ftsearch, document, form	ifauthor, ifdbrole, ifdepositor, ifdesigner, ifeditor, ifmanager, ifnoaccess Generates body of tag only if session user has corresponding ACL access to the Domino database
	Docloop, viewloop, document, form	ifdocauthor, ifdocreader Generates tag body only if session user has corresponding access to the document (via author, readers fields).
View Control	viewloop	ifcategoryentry, ifconflictentry, ifdocumententry, iftotalentry Generates body of tag only if `ViewItem` corresponds to tag.
Validation Control	form	validatecc, validatedate, validatelength, validatepattern, validaterange, validaterequired, validatesummary, validatewith Performs specified client-side validation on the specified input tag. `clientvalidate` attribute on input tag must be other than `none` and set to `true` on `form` tag.

Hints and Tips for Using the Domino JSP Tags

Here we highlight some hints and tips for using the Domino JSP tags, especially within WSAD. Hopefully these items will be useful to you in becoming comfortable with the Domino tags so that you can quickly start generating quality Domino-based Web applications. The items are in no particular order.

1. **Tags generate client-side as well as server-side code.** Although most Domino tags generate server-side (Java) code, keep in mind that several Domino tags result in purely client-side function (HTML or JavaScript). Examples of client-side tags are `select`, `option`, `fill`, and the set of `validate*` tags.
2. **Certain tags might force a page reload.** Several of the Domino tags require that the current page be refreshed in order to function. Be aware of this possibility when designing the page flow to avoid delays in user interaction.
3. **Not all document item types are fully rendered when using the LDT in WSAD.** When using the Lotus Domino Toolkit plug-in for WSAD, the drag-and-drop feature for creating JSPs from Domino forms and views, all of the Domino item (field) types are generated as simple (Domino) input tags. For certain field types, such as "Dialog list," "Checkbox," and "Radio," this results in JSP fields that are displayed as single text input controls rather than drop-down list selection controls or checkboxes or radio buttons. In the case of "Dialog list" (keyword) fields, rendering them as input tags results in the alias values for the list items being displayed and provides no way to make a list selection in edit mode. You need to enhance the JSP by using the Domino `select`, `option`, and `fill` tags to generate a drop-down list control for these fields. For example, in the domform.jsp we used to show the results of dragging-dropping a Domino form to create a JSP, the `Category` field is defined in the Domino database design as a "Dialog list." We can change the input tag generated for this field to the following source:

```
<tr>
   <td>Category</td>
   <td> <%--domino:input name="Category" type="text"/--%>
      <%-- Drop-down list selections for Category item dialog
      ➥list --%>
      <domino:select name="Category">
         <domino:option value="1" display="Book">
            <domino:fill target="Category" value="1"/>
         </domino:option>
         <domino:option value="2" display="Music">
            <domino:fill target="Category" value="2"/>
         </domino:option>
         <domino:option value="3" display="Film">
            <domino:fill target="Category" value="3"/>
         </domino:option>
         <domino:option value="4" display="Software">
            <domino:fill target="Category" value="4"/>
         </domino:option>
      </domino:select>
   </td>
</tr>
```

The select tag generates the drop-down list and identifies the form item target. The option tags are used to associate the list item display names with the dialog list alias values. The fill tags are used to set the alias value of the item when a list selection is made.

4. **Drag/drop of a Domino form to a JSP page does not preserve form layout.** A JSP page created by dragging and dropping a Domino form via the Domino view often does not match the layout of the Domino form. For example, the order of fields within a table is reversed.
5. **JSPs can be generated from Domino Designer XML (DXL) using XSLT.** One way to overcome the limitations of the LDT with generating JSPs from Domino database designs is to generate the JSPs from the DXL for the design documents via XML stylesheet transforms. Since the DXL captures the exact content and structure of the design document, transforming from DXL will produce a truer rendering of the Domino design. The LDT rendering problems mentioned in tips 3 and 4 can be overcome this way.

For example, here is an XSL template for converting a Domino form field from DXL into the appropriate Domino JSP tags.

```
<xsl:template match="dxl:field[@type]">
   <xsl:choose>
      <xsl:when test="self::node()[@type='keyword']">
         <!-- We have a keyword type field...
              Generate a Domino select tag with option and fill
              ➥tags set to the keywords given by the dxl:keywords
              ➥tag. -->
         <xsl:element name="domino:select">
         <xsl:attribute name="name">
            <xsl:value-of select="@name"/>
         </xsl:attribute>
         <!-- Look for the keyword values in text elements.
         ➥The values are generated as: "display-name|value-name",
         ➥so we need to parse the values using XPath string
         ➥functions as below. -->
         <xsl:for-each select=".//dxl:text">
            <xsl:element name="domino:option">
            <xsl:attribute name="display">
               <xsl:value-of select="substring-before(string(),
               ➥'|')"/>
            </xsl:attribute>
            <xsl:attribute name="value">
               <xsl:value-of select="substring-after(string()
               ➥,'|')"/>
```

```
                </xsl:attribute>
                    <xsl:element name="domino:fill">
                    <xsl:attribute name="target">
                        <xsl:value-of select="ancestor::node()/
                        ➥dxl:field/@name"/>
                    </xsl:attribute>
                    <xsl:attribute name="value">
                        <xsl:value-of select="substring-after(string()
                        ➥,'|')"/>
                    </xsl:attribute>
                    </xsl:element>
                </xsl:element>
            </xsl:for-each>
            </xsl:element>
        </xsl:when>
        <xsl:otherwise>
            <!-- Generate a Domino input tag, with name and type
            ➥attributes taken directly from the corresponding
            ➥dxl:field attributes. -->
            <xsl:element name="domino:input">
                <xsl:attribute name="name">
                    <xsl:value-of select="@name"/>
                </xsl:attribute>
                <xsl:attribute name="type">
                    <xsl:value-of select="@type"/>
                </xsl:attribute>
            </xsl:element>
        </xsl:otherwise>
    </xsl:choose>
</xsl:template>
```

Note that this XSL is general enough to generate the correct Domino tags for most field types in the DXL (it would need to be enhanced to generate the correct tags for checkbox and radio fields). It will generate simple Domino input tags for most field types and will generate the `select`, `option`, and `fill` tags for keyword field types. For example, it would generate the JSP tags shown in hint 3 for a keyword field. Admittedly, this is sophisticated XSL, but would be well worthwhile if there were a need to generate many JSPs from a Domino application. Writing and testing XSL such as this is made even easier by using the XML/XSL development tools offered by WSAD.

Using JavaMail with Domino

The JavaMail package (`javax.mail`) provides a set of interfaces for creating and sending e-mail messages to various types of mail providers. JavaMail can be configured to send e-mail messages to providers that support the Internet standard Simple Mail Transfer Protocol (SMTP). SMTP is one of the e-mail protocols supported by the Domino server. Thus, it is common in combined WebSphere and Domino-based applications to use JavaMail to generate e-mail within J2EE Web applications and have it handled by the Domino server. Fortunately, this approach works with relatively little configuration required for either the JavaMail package or for Domino.

One of the default e-mail provider implementations supplied with JavaMail is for SMTP providers. Since WAS is built on J2EE versions that include JavaMail, WAS is capable of supporting JavaMail applications using SMTP. Moreover, WAS (V5 and above) itself supports the configuration of JavaMail sessions via its Web administration application. With former versions of JavaMail, the provider configurations were made via Java properties files, which were usually built within the application itself. With the newer JavaMail versions, the JavaMail sessions are preconfigured and obtained by the application via JNDI lookup. It's this preconfigured JavaMail session approach that is now used in WAS.

To define a preconfigured JavaMail session via the WAS Web administrative console, expand the **Resources** item in the left-hand pane of the console and select **Mail Providers**. Select **Server** as the resource scope (see Figure 10-12); then click the **Apply** button. Select the **Built-in Mail Provider** preference page and then open the **Mail Sessions** properties page from the link at the bottom of this page (see Figure 10-13).

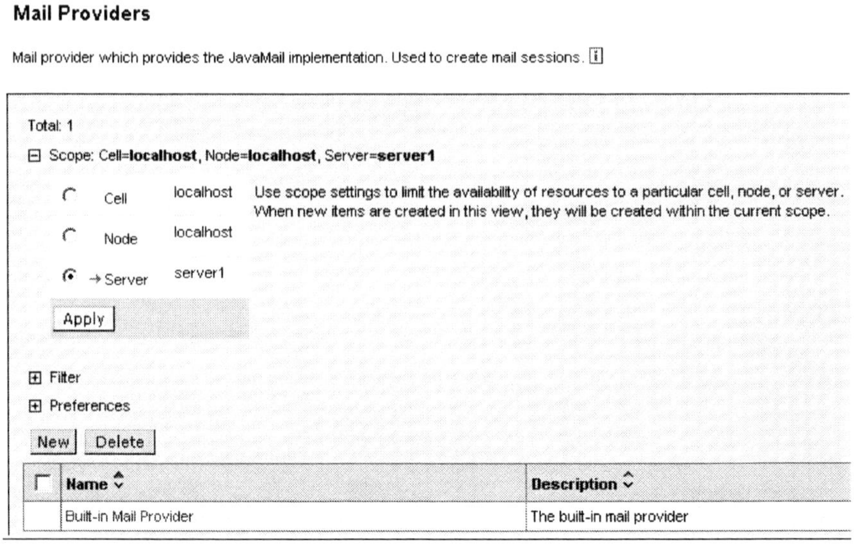

Figure 10-12 WAS Admin Mail Providers page 1.

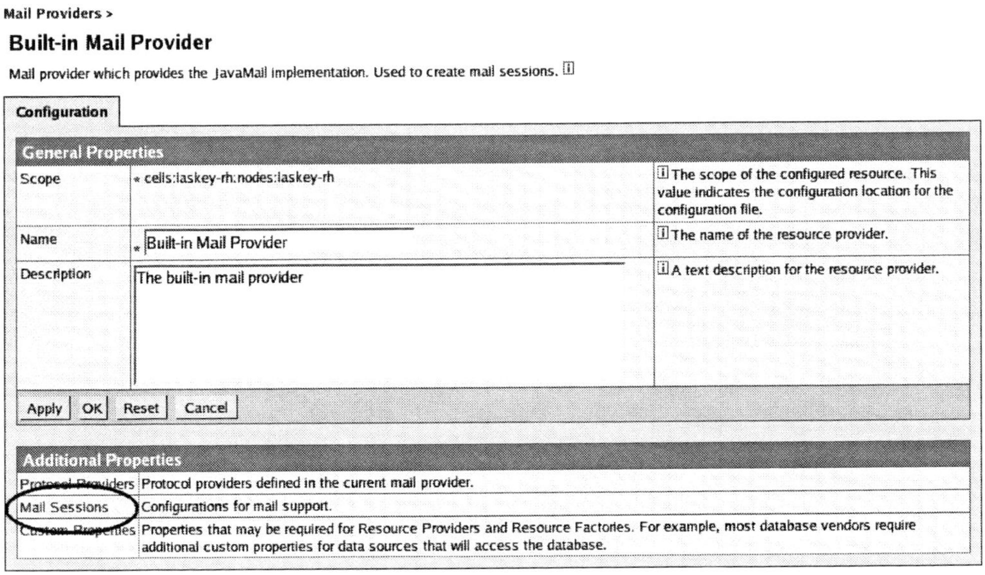

Figure 10-13 WAS Admin Mail Providers page 2.

The **Mail Sessions** properties page allows you to specify a name and JNDI name for this SMTP mail session resource. The JNDI name is used in the application to refer to this resource. You also specify the properties to be used to connect to the SMTP provider on this page, such as the **Mail Transport Host**, **User ID**, and **Password** (see Figure 10-14). Once this resource is saved, it can be used by the Web application.

On the Domino server side, you must configure Domino as an SMTP listener. This is done in the Server document in the Domino directory by enabling the SMTP protocol for the server. If the e-mail messages are directed to addresses within the Domino domain, you must ensure that the recipient addresses exist in the Domino directory (either in the primary names.nsf database or in external directories adjoined via the Directory Assistance feature). Also ensure that the Internet domain name used for these recipient addresses matches the Internet domain name specified for the server in the Server document.

Once WAS and Domino have been configured for SMTP interchange, you can then add JavaMail classes and invocations to your Web application in a straightforward manner. There are many references available on using the JavaMail package from within a J2EE application (for example, java.sun.com), but we will not go into the details of the JavaMail API here.

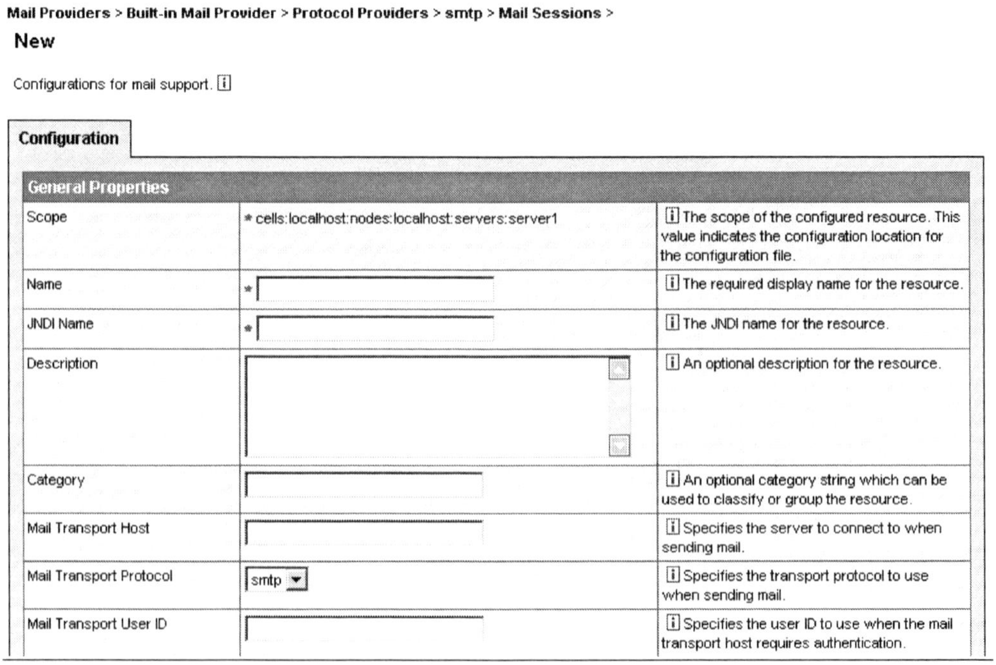

Figure 10-14 WAS Admin Mail Providers page 3.

CHAPTER 11

Issues with WAS and Domino Together

> **"The idea is to get the two brands working together, not make them one. Future versions of Domino will have tighter alignment with DB2, Tivoli, and WebSphere."**
>
> **—Al Zoller, head of Lotus (at that time), at LotuSphere 2002, February 2002**

In this chapter, we examine the issues that often arise when WebSphere Application Server and Domino are used together. Ideally, we want to have the combined system appear as "seamless" as possible to the end user. Some of the issues that need to be addressed to achieve a seamless integration are shared authentication or "Single Sign-On," the use of clustering, performance and high availability issues, and issues with building high-capacity combined systems. The chapter is intended to present the set of integration issues and give an overview of each. Chapters 12 through 15 look at depth into each of these issues, so you can look at this chapter as providing a "lead in" to each of the WAS and Domino together issues that will each be discussed in detail in later chapters.

Single Sign-On (SSO) for WAS and Domino

One of the first areas of integration between WebSphere and Domino application servers was user authentication. It was not long after developers started to build applications requiring secure access to both J2EE and Domino components that the need for shared user authentication, commonly referred to as "Single Sign-On (SSO)" was encountered.

Basic SSO Between WAS and Domino

The need for SSO surfaces when an application is designed with both Domino and J2EE components, which need to be accessed in a secure fashion. Let's consider the case where an application consists of a J2EE servlet and a Domino database, both of which are to be restricted to a certain group of users. If we protect both the servlet and database and set up basic (name and password) authentication for the users, we would end up with a situation where the users would be prompted for their name and password on the first access to either resource. The user is presented with multiple sign-on requests and is left wondering about the cohesiveness of the application. Ideally, we want authentication to be performed only once during a user's HTTP session, no matter how the resource is served—that is, we want a single sign-on.

How do WAS and Domino provide for SSO authentication? The answer is two-fold. First, there must be a way for WAS and Domino to share and convey the fact that a user has been authenticated, no matter which performed the authentication. Second, both WAS and Domino must share the information kept about the users so that authorization can be consistent. The first requirement is met by configuring WAS and Domino to use a special user authentication mechanism called Lightweight Third-Party Authentication (LTPA). The second requirement is met by configuring WAS and Domino to use a shared user directory (or "registry") via the Lightweight Directory Access Protocol (LDAP).

The LTPA mechanism provides a secure means for conveying the fact that a user has already been authenticated. For Web applications, this mechanism must work in the context of a browser to Web server session—and other contexts if possible. LTPA makes use of HTTP session "cookies" to store an encrypted set of user credentials at the browser and have it returned to the server on subsequent requests for protected resources. Both WAS and Domino can be configured to use LTPA, that is, to generate encrypted LTPA cookies and receive and decrypt LTPA cookies and to share the encryption keys needed for encrypting and decrypting. In Chapter 12, "Security and Single Sign-On," we discuss the use and configuration of LTPA by WAS and Domino in detail and discuss some of the issues that arise with LTPA.

The requirement that WAS and Domino share a user directory for SSO is necessary so that authentication and authorization can be consistent. One needs to ensure that a user is authenticated and authorized in the same way no matter which application server performs it. Basing the shared user directory access on LDAP is a natural choice since LDAP is an open Internet standard that broadens the set of possible shared directory solutions. The use of LDAP enabled directories fits nicely with Domino as it has long provided LDAP access into and out from its own user directory. In Chapter 12, we present in detail various ways in which the shared user directory or registry can be configured and discuss some of the issues surrounding these configurations.

SSO and WebSphere Portal Server

With the advent of the WebSphere Portal Server (WPS) product, the need for SSO support is even more critical and, at the same time, more challenging. By design, the portal user interface displays separate applications in separate frames of the Web pages, called "portlets," and each portlet may be driven by a separate application on the server. Without SSO support, a portal would become a maze of different sign-on pop-ups and authentication mechanisms as the user navigates among the portlets. The portal server and portlet architecture make it easier for disparate applications to be integrated at the browser, but, at the same time, make it more challenging for the designer and developer to provide the necessary integration on the server.

A prime example of a portal environment is the WPS-based IBM Lotus Workplace product. The Lotus Workplace portal brings together Domino, Lotus Instant Messaging (formerly known as Sametime), and QuickPlace products, along with J2EE applications, to form an e-mail and collaboration portal. Each of these Lotus products can be run as a stand-alone application with its own user directory and authentication/authorization mechanism. Bringing them all under a single portal requires an even broader scope for the SSO requirements of shared user directories and authentication. In the case of the original release of Lotus Workplace, the fact that all of the applications are Domino-based provides us a common set of tools, namely LTPA and LDAP, to help us achieve SSO. But as we discuss further in Chapter 12, even this common base can lead to difficulties in implementation. For clarification, it should be noted that beginning with Lotus Workplace 1.1, there is no longer the need to connect to a Lotus Sametime/Domino server for Web conferencing and Instant Messaging. In Lotus Workplace 1.1 those components are now J2EE-based applications running on top of the WebSphere server. Appendix F gives details on Lotus Workplace 1.1. At the time this book was written, the latest version of the WebSphere Portal Server (V5.02 Extended Edition) still required separate Sametime and QuickPlace servers.

Other SSO Mechanisms

There are other approaches to providing SSO support than those offered directly by WAS and Domino. These solutions usually involve a dedicated authentication component that can manage user authentication across a wide range of products and plug into the product-specific authentication mechanisms. The dedicated authentication solutions also provide a wider range of authentication mechanisms that can be presented to the user, such as hardware-based authentication, smart cards, certificates, and so on.

IBM offers a product within the Tivoli product suite, which is focused on managing security among a broad range of enterprise applications. The product, Tivoli Access Manager (TAM), provides for enterprise-wide access control, security policy management, authorization, and auditing functions. TAM is most applicable for customers who want to centralize security among heterogeneous Web applications (beyond WPS and Domino) or those who desire flexible authentication alternatives (tokens, certificates, and so on), enhanced access control mechanisms (time-of-day checking, day-of-week-checking, etc.), and for customers who have selected TAM for their overall security and SSO architecture.

TAM contains a component called WebSEAL that provides a Web access proxy function; that is, it intercepts Web requests from the open Internet, authenticates and validates the requests, and then forwards the requests to the application servers on secure network connections. WebSEAL is designed to sit in the network layer exposed to the Internet where it handles user authentication for all Web resource requests.

The WebSEAL proxy provides a module interface, called a "junction," which allows specific implementations to be used for common functions. By plugging in different modules to junctions, WebSEAL can be configured to use different mechanisms to perform the user authentication for the Web request and authentication/authorization to the application servers. WebSEAL provides an LTPA junction so that SSO can be achieved among WebSphere and Domino application servers fronted by a WebSEAL proxy server. At the time of this writing, the LTPA junction is the only means available for achieving SSO with WebSEAL, WebSphere, and Domino. (Refer to IBM Redbook Technote TIPS0331 at `www.redbooks.ibm.com`.)

Clustering in the Combined System

Both Lotus Domino and WebSphere offer what each calls "clustering." However, clustering in Domino is quite different than clustering in WebSphere, even though the basic concept is the same—i.e., to have a group of servers appear as if they are one server. Clustering is a basic way for both Domino and WebSphere to provide high availability (one server in a cluster can crash but the cluster is still available and fully functional) and high scalability (just double the number of servers in the cluster, and you've doubled the capacity of the cluster).

J2EE application server vendors define a cluster as a group of machines working together to transparently provide enterprise services (support for JNDI, EJB, JSP, HttpSession and component failover, and so on). They leave the definition purposely vague, because each vendor implements clustering differently. At one end of the spectrum rest vendors who put a dispatcher in front of a group of independent machines, none of which has knowledge of the other machines in the cluster. In this scheme, the dispatcher receives an initial request from a user and replies with an HTTP redirect header to pin the client to a particular member server of the cluster. At the other end of the spectrum reside vendors who implement a federation of tightly integrated machines, with each machine totally aware of the other machines around it along with the objects on those machines. Actually, most IT devices can be comprised of "clusters" that allow the devices to be redundant and failover-capable. For example, we can cluster devices such as

- **Load balancers:** Single points of entry into the cluster and traffic directors to individual Web or application servers.
- **Web servers.**
- **Gateway routers:** Exit points out of an internal network.
- **Multilayer switches:** Packet and frame filters to ensure that each machine in the cluster receives only information pertinent to that machine.
- **Firewalls:** Cluster protectors from hackers by filtering port-level access to the cluster and internal network.

- **SAN (Storage Area Networking) switches:** Connect the application servers, Web servers, and databases to a backend storage medium; manage which physical disk to write data to; and failover.
- **Databases.**

Regardless of how they are implemented, all clusters provide two main benefits: scalability and high availability (HA).

Scalability

Scalability refers to an application's ability to support increasing numbers of users or application data volume. Clusters allow you to provide extra capacity by adding extra servers, thus ensuring scalability.

High Availability

HA can be summed up in one word: redundancy. A cluster uses many machines to service requests; therefore, if any machine in a cluster fails, another machine can transparently take over.

A cluster only provides HA at the application server tier. For a Web system to exhibit true HA, it must be like Noah's ark in containing at least two of everything, including Web servers, gateway routers, switching infrastructures, and so on.

Cluster Types

J2EE clusters usually come in two flavors: shared-nothing and shared-disk. In a shared-nothing cluster, each application server has its own filesystems with its own copy of applications running in the cluster. Application updates and enhancements require updates in every node in the cluster. With this setup, large clusters become maintenance nightmares when code pushes and updates are released.

In contrast, a shared-disk cluster employs a single storage device that all application servers use to obtain the applications running in the cluster. Updates and enhancements occur in a single file system, and all machines in the cluster can access the changes. Until recently, a downside to this approach was its single point of failure. However, SAN gives a single logical interface into a redundant storage medium to provide failover, failback, and scalability.

We also can implement Application Level Clustering. Domino Clustering falls into that category. Of course, clustered servers also can relate to the Chapter 14 topic of Performance and High Availability. In Chapter 13, "Clustering," we give details on how to implement clustering for both Domino and WebSphere.

Performance and High Availability for Combined Systems

Almost every company needs to have high performance and high availability for their Web servers; however, there are many different aspects of high availability. At one time, the only avenue for achieving high availability in the UNIX realm was through fault tolerant technology. Fault tolerance relies on specialized hardware to detect a hardware fault and instantaneously switch to a redundant hardware component—whether the failed component is a processor, memory board, power supply, I/O subsystem, or storage subsystem. Although this cutover is apparently seamless and offers non-stop service, a high premium is paid in both hardware cost and performance because the redundant components do no processing. More importantly, the fault tolerant model does not address software failures, by far the most common reason for down-time. High availability views availability not as a series of replicated physical components, but rather as a set of system-wide, shared resources that cooperate to guarantee essential services. High availability combines software with industry-standard hardware to minimize down-time by quickly restoring essential services when a system, component, or application fails. While not instantaneous, services are restored rapidly, often in less than a minute. The difference between fault tolerance and high availability, then, is this: A fault tolerant environment has no service interruption, while a highly available environment has a minimal service interruption. Many sites are willing to absorb a small amount of down-time with high availability rather than pay the much higher cost of providing fault tolerance. Additionally, in most highly available configurations, the backup processors are available for use during normal operation. Chapter 14 gives details on the different ways to achieve high availability for your Domino and WebSphere systems.

High-Capacity Combined Systems

Another important aspect of building a combined WebSphere and Domino system is in the area of building high-volume, high-capacity, scalable systems. What are the design aspects of high-volume Web sites—especially those based on both WebSphere and Lotus Collaboration tools? IBM's experience designing the Web sites for the Olympic games, which included Domino for collaboration, will serve as an example of the evolution of these Web site design concepts. Many of the design concepts for high-volume, high-availability, high profile sites that included collaboration features went through the intense proving ground of Web sites for the Olympics and other international sporting events.

Both WebSphere and Lotus Collaboration tools support the concept of horizontal and vertical scaling for high-volume, high-capacity systems. These types of scaling are important for building both high availability and high capacity systems.

- **Horizontal scaling** consists in distributing work among many machines, some or all of which can be less powerful. Failover is a main advantage of horizontal scaling. If a machine becomes unavailable, its work can be routed to other machines containing server clones. Load balancers or IP "sprayers" can be used to distribute requests to identical multiple machines, each running an application server clone. These techniques do not have to interfere with, but can complement the workload management provided automatically by WebSphere and Domino.

- **Vertical scaling** consists in using the product cloning support so as to get more from each machine on which the product runs. A server and several of its clones can share the same machine, enabling you to utilize more of the machine's resources, such as CPU. This works well when the individual machines in your topology are powerful but underutilized.

IBM and other computer services companies have long hosted Web applications on the Internet for corporations as part of "outsourcing" agreements. Many of the high capacity sites are beginning to incorporate function from both WebSphere and Lotus Collaboration. Details are given in Chapter 15, "Implementing High-Volume Web Sites."

J2EE Servlets

What are some of the issues in using servlets from both WebSphere and Domino? This section briefly discusses this aspect of running a combined WebSphere and Domino system. Chapter 9, "Accessing J2EE Elements from Domino," covered aspects of accessing J2EE elements from Domino. Chapter 10, "Accessing Domino from J2EE" covered aspects of accessing Domino from J2EE. Based on those chapters there are few if any "issues" dealing with servlets in a combined WebSphere and Domino system. Here is a short summary of the areas involving servlets discussed in Chapters 9 and 10.

Invoking Servlets from Domino

As described at length in Chapter 9, the WebSphere product versions provide Web server "plug-ins" for various Web servers, enabling them to forward requests for J2EE Web elements, namely servlets and JSPs, to the WebSphere Application Server. In particular, a Web server plug-in is provided for the Domino HTTP server, which allows requests for servlets and JSPs directed to Domino to be handled by a WAS node. This plug-in mechanism was one of the first integration points between the Domino and WAS products. The plug-in itself is provided as a shared library (DLL for Windows, .a for AIX, .so for Solaris, and as of this writing no plug-in exists for Domino on Linux) and a default configuration file: `plugin-cfg.xml`. The plug-in files are packaged with both the Domino and WAS product installations and can be automatically installed during a WAS installation. The plug-in also can be installed manually. The plug-in files can be found under the Domino install path: `Domino/Data/domino/plugins/was5`.

Passing Data to Servlets in the URL

The WebSphere Domino HTTP server plug-in can be viewed as a basic means of enabling J2EE access from Domino. With the plug-in, access to J2EE elements is via explicit URL links. For example, sending a URL of the form `http://A_Domino_Host/servlet/SomeServlet` to a Domino server with the plug-in installed will result in the Domino HTTP task passing the request via the plug-in to a WAS node, which then passes the response back via Domino.

Posting Data to Servlets from Domino Forms

Since a JSP is a type of servlet, assume that you can replace "servlet" with "JSP" in what follows. To get a Domino form processed by a servlet, you must somehow invoke the servlet when the user hits the Submit button (or equivalent) on the form displayed in the user's browser. There are two commonly used techniques: overriding the $$Return field or overriding the HTML Form. These two techniques are discussed in detail in Chapter 9.

IIOP

What are some of the issues in using IIOP with a combined WebSphere and Domino system? As with servlets, IIOP was discussed in both Chapters 9 and 10, and there were few if any issues related to "IIOP vs. XML." This section gives a short summary of the discussions in Chapters 9 and 10 on IIOP and XML.

Connecting to Domino Using IIOP

As described in Chapter 10, remote sessions to Domino are established using a CORBA mechanism via IIOP. The HTTP and DIIOP server tasks must be running on the Domino server, which is the target of a remote session. The type of access for a lotus.domino session is determined by the signature of the `createSession` method of the `Session` class.

SSL Enabling of IIOP Session

As with all WAS configuration files, the plug-in configuration file has XML syntax and thus can be edited with any text editor. The `LogName` element specifies the file name of the log file written by the plug-in and the level of logging desired. The `ServerGroup` element specifies the WAS server(s) to which URLs are routed and the protocol used. The default is to route to the local WAS server at port 9080 (the embedded WAS HTTP server) over HTTP. Additional servers can be specified here, as well as HTTPS instead of HTTP. If more than one server is specified within a server group, the plug-in routes to the servers in a round-robin fashion. If HTTPS is used, additional system environment variables must be set pointing to the SSL libraries and the certificate and encryption key files required by the plug-in for HTTPS.

CHAPTER 12

Security and Single Sign-On

> **"Notes has so many components that the Internet lacks. Authentication [checking the identity of users logging in], for instance, is not built into the Web, but it's handled by Notes. At some point the Web will have all these things, but it will be challenging to integrate them."**
>
> **—Bill Wilson, Senior Vice President of Johnson & Higgins, a New York City insurance brokerage that has 6,000 employees on Notes (*Fortune*, July 8, 1996)**

In this chapter, we present and discuss the security features of the WebSphere and Domino server products, which are used together to secure Web applications. Both products provide support for securing the application components that they serve. They each implement a range of Internet standard security mechanisms for verifying user identity (authentication) and controlling access to protected resources by users (authorization). Some of these mechanisms are common to both products, such as HTTP authentication, Secure Sockets Layer (SSL), and some are not, such as the authorization models. By combining the security features of each, you can appropriately protect your Web applications as if they were running on a single application server.

We discuss the security features most often used to secure Web applications that run on both WebSphere and Domino servers. First, we present a brief review of the WebSphere and Domino security models and then discuss in detail the setup and use of the shared user authentication mechanism, or what's commonly referred to as "Single Sign-On" (SSO) support. We also discuss how to troubleshoot security-related problems and some of the remaining limitations with the current product implementations.

Security Concepts

Let's start with a review of the Domino and WebSphere security models. For our discussion here, we assume you are at least conceptually familiar with the major Internet standard security mechanisms, such as HTTP basic (name and password) authentication, SSL authentication via public key certificates, and user directory protocols such as LDAP and X.500 naming formats. When discussing security configuration, we also assume the reader has some familiarity with both WebSphere and Domino administration functions.

Domino Security

The Domino security model is referred to as a "resource-based" model; that is, the control of access to resources is defined on the resources to be protected. Specifically, the users and the types of actions permitted to users are defined in Access Control Lists (ACLs) stored with any Domino database. The access controls can be further refined at the document or field level. Additional access controls to the Domino server itself can be specified in the server configuration document (the "server" document in the Domino directory).

You specify the set of users and user groups to be granted access to a Domino database and their level of access in the database ACL. Two special user names can be specified, "Default" and "Anonymous." The Default user name controls the level of access granted to users not specified in the ACL. The Anonymous user name controls the level of access granted to unauthenticated users making Web requests to the Domino server. If both the Default and Anonymous users are specified with a level of "No Access," then a requesting user must be authenticated and matched to a user or group name in the ACL. (This assumes that the requesting user has been permitted access to the server itself.)

You specify the type of authentication and session protocol via the port configuration in the Domino server document. For the HTTP port, basic authentication, SSL client-certificate, or session-based (cookie) authentication are supported.

WebSphere Security

As one would expect, the WebSphere security model conforms to the J2EE security model. The WAS security functions are an implementation of the security services specified by the J2EE standard. The J2EE security model is characterized as a "role-based" model. A security role is a logical group of users defined by the J2EE application designer for the purpose of controlling access to application resources. Access controls (referred to as "security constraints") are specified within a J2EE application in terms of the defined security roles. For example, an application may define security roles of "employees" and "customers" and a security constraint, which specifies that the `GetBenefits` servlet may be permitted to only the "employees" role. The type of user authentication to be used by the application server also can be specified within the application. Security functions can only be defined for J2EE *enterprise* applications.

When a J2EE enterprise application is deployed to an application server, the application security roles are mapped to specific users or groups defined in the server's user registry (directory). When the application runs, the application server performs user authentication and applies the security constraints on requests for those resources protected by constraints. To be precise, the container managing the particular J2EE component applies the security function—the Web container applies security for servlets and JSPs, and the EJB container applies security for EJBs.

There are two basic means of specifying access controls within a J2EE application. One way is by defining the security roles and constraints in the application deployment descriptors. This method is referred to as "declarative" security. The other way is to define the security constraints within the application code, referred to as "programmatic" security. Using the declarative security method leads to more portable and flexible applications. Programmatic security can be used when the application security requirements can't be met by declarative security functions. As we will see later, sometimes programmatic security must be used to achieve the desired authentication and access control for an application implemented across WebSphere and Domino.

The Single Sign-On (SSO) Problem

Any security model, whether resource-based or role-based, is predicated on validating that a connecting user is who they say they are, or "user authentication." Both WebSphere and Domino need to authenticate any user attempting access to a protected resource. User authentication is most often done using some sort of credential check or "sign-on" exchange. When a user interacts with a Web application involving protected resources, they naturally expect to be prompted for a sign-on only once for the Web session. Unfortunately, this was not always the case with applications served by both WebSphere and Domino. We had the "SSO problem."

The SSO problem arises when both Domino and WebSphere are configured to use a standard user authentication mechanism, most often HTTP basic authentication. Because both Domino and WebSphere may be serving different components of a Web application separately, they may not be aware that a user has already been authenticated on a previous request and will attempt to reauthenticate the user. If the request URL is different enough from the previous (that is, belongs to a different "realm"), the browser will ask the user for their userid and password again. Thus, the user sees two or more annoying prompts for their userid and password within the same Web application.

In order to provide SSO authentication, both WAS and Domino must generate and understand a common set of authentication information. Indeed WAS and Domino are capable of performing shared authentication with the current product versions and and in versions dating back to WAS 3.5 and Domino R5.05. WAS and Domino are able to recognize previous user authentications, even if they were established by the other product. They employ the HTTP "cookie" mechanism to keep track of user authentications within an HTTP session. Using cookies, which are stored on and returned by the browser, each can recognize when a user has already been authenticated by the other. The use of shared authentication tokens as cookies provides SSO. Next, we'll discuss some of the detail of how SSO works and some of the issues involved.

User Authentication Over the Web

First, let's review the standard user authentication mechanisms for Web applications. There is HTTP Basic Authentication and SSL V3 Client Certificate Authentication.

The basic authentication protocol, also referred to as "name and password authentication," is defined as part of HTTP. It specifies that a server can request authentication (challenge) in response to any request from a client and requires the user agent (browser) to provide a name and password in the subsequent request header. The server's challenge is made relative to a server-specified realm (a text name) and is only valid for that realm. (A realm can be thought of as the set of URLs with the same root path or, in other words, the same host name and path qualifiers as the URL that was challenged.) The user agent may send the name and password in request headers for other URLs in the same realm without being challenged by the server. (In fact, most browsers do this to avoid repeatedly prompting the user for name and password.)

Note that basic authentication should be done over an encrypted connection, using SSL for example. This is because the userid/password is exposed in a simple BASE64 encoding in the HTTP request header.

Client certificate authentication is defined as part of the SSL V3 or TLS protocols. When used with HTTP over SSL/TLS ("HTTPS"), it requires that the HTTPS client send an X.509 certificate representing the client as part of the request header. The X.509 certificate contains the user's name and public key and is signed by a Certificate Authority (CA). The server verifies the client certificate; that is, it checks the CA signature (so that it can trust that the public key belongs to the named user) and then verifies the user's identity by checking that the client system possesses the private key associated with the public key (via an encrypted message exchange using the public/private keys). After this verification process, the server can simply authorize the client's access to the server or its resources based on the authenticated user name. In other words, once the SSL certificate authentication processing has determined that the client is who they say they are, the server simply verifies that the user name is known to it as an authorizied user.

In addition to these authentication mechanisms, there are mechanisms used to persist the user's authentication within an HTTP session. These are referred to as "session based authentication" mechanisms and usually make use of an authentication token, which is kept by the browser. The most common way tokens are maintained between the browser and server is to store the token as part of an HTTP "cookie." Both Domino and WebSphere support session-based authentication.

Authentication in Domino and WebSphere

Domino uses HTTP basic authentication whenever a Web user requests access to a Domino database whose ACL restricts access and client-side SSL authentication is not in effect. A Domino database ACL restricts access when the Default and Anonymous entries are set to "No Access." The name and password supplied by the user is verified against the Domino directory and any external directories configured via a Directory Assistance database. Domino can be configured to match various forms of the user name against the directory entry, including the common name, the last name, the short name, or the full distinguished name ("canonical name" in Notes terminology). The password is compared against the Internet password specified in a person document in the Domino directory or, by default, the `userPassword` attribute of the `inetOrgPerson` schema in an LDAP directory.

You can specify the realm name Domino will return on the basic authentication challenge by creating a Web Realm document associated with the Server document. By specifying the realm name to be the same as that used by WebSphere, you can affect a sort of simple SSO since most browsers will return the same userid and password for URLs within the same realm. Of course, this simple SSO scheme does not extend to the case where Domino and WebSphere are on different servers.

WebSphere can be told to use basic authentication for the Web components of any J2EE enterprise application. For an enterprise application configured to use basic authentication, WebSphere challenges the user for a name and password whenever an attempt is made to access a protected enterprise application Web resource.

Both Domino and WebSphere can be configured to perform user authentication via SSL V3 client certificates. However, there are some limitations to using client certificates on the Domino side especially for SSO. We'll discuss more about these limitations later in this chapter.

In addition to the user authentication functions, WebSphere provides for persistent or session-based authentication by means of a mechanism called Lightweight Third-Party Authentication (LTPA). It is this LTPA mechanism that was added to Domino and is used to provide SSO between Domino and WebSphere servers. Under LTPA, after a user is authenticated, the authenticating server generates a token, which is passed to the client as a cookie. On subsequent requests for protected resources within the same domain, the cookie is returned to the Domino or WebSphere server, and, if found to be valid, accepted as proof of the user's identity. Using LTPA, either Domino or WebSphere can perform the initial user authentication, and either can accept the generated LTPA token for authorization of subsequent requests.

Let's take a closer look at the LTPA cookie and token. The cookie consists of standard fields defined by HTTP set to specific values for LTPA and the cookie value itself set to the LTPA token. The cookie fields are set as shown in Table 12-1.

Table 12-1 LTPA Cookie Contents

Cookie Fields	LTPA Cookie Contents
Domain and Path	Set to the Internet domain name of servers recognizing this cookie, stored in plain text (unencrytped) format.
Cookie Expiration	Set to `transient`, meaning that the cookie is not stored on the browser system and is available only for as long as the browser is active.
Cookie Value	Set to the LTPA token.

The LTPA token consists of the contents listed in Table 12-2.

Table 12-2 LTPA Token Contents

User Data	User name and attributes stored as name/value pairs separated by a "$." One of the attributes is the `LDAP Realm`, which consists of the host name and port of the user directory used for user validation.
Expiration Date	The time and date the LTPA token will expire—that is, will no longer be recognized as valid by a server system.
Digital Signature	A digital signature of the User Data and the Expiration Date signed with the private key of the authenticating server (a private key generated specifically for LTPA).

The entire LTPA token is encrypted (using triple-DES encryption) using a key shared among all the servers participating in LTPA. Other servers can verify the LTPA token signature by using the LTPA public key of the signing server. The shared DES key and server public keys are kept in a LTPA key ring, which can be exported from WebSphere and imported to Domino and other WebSphere servers. (Note that Domino is capable of exporting an LTPA key ring, but this export is in a format that WebSphere cannot understand, so for SSO between WebSphere and Domino, the LTPA key ring must be generated and exported from WebSphere to Domino.)

The Role of the User Directory

An important aspect of both Domino and WebSphere authentication mechanisms (either basic or certificate-based) is that they must validate user names against a user directory of some sort. For SSO, naturally, the user name directory must be shared. Directory sharing is indeed possible with Domino and WebSphere since both products support the widely used Internet specification for directory access called the Lightweight Directory Access Protocol (LDAP).

A brief digression on LDAP and X.500 directory structure and naming may be helpful for our discussion of user directories. LDAP grew out of an effort to make the ITU/CCITT X.500 directory access protocols easier to implement on the Internet. It can be thought of as a simplified protocol for accessing X.500-based directory databases. The X.500 directory standard was intended to standardize and allow sharing of user (and other entity) information across organizations and countries. Its legacy is primarily the standards defining directory structure and naming. Briefly, with X.500, the directory information is organized as a tree with "entries" as nodes. Each entry in the directory tree consists of a set of attribute/value pairs and is assigned a unique name, referred to as a "distinguished name" (DN). A DN consists of a sequence of attribute value pairs. The pair sequence contains the DN of the parent entry and at least one pair for the entry itself. For example, an entry representing one of us might have a DN of: "cn=M.Laskey,ou=IBMGS, o=IBM,c=US," formed from the attributes common name (cn), organizational unit (ou), organization (o), and country (c). The set of attributes that may be included in an entry is defined by one or more "object classes." The set of attribute and object class definitions for a directory is specified by a "directory schema."

LDAP itself is an Internet (IETF-RFC) standard that defines a simplified query and update protocol to an X.500 structured directory. Because of its popularity and wide usage, it has overshadowed its X.500 parent. To be precise, we should write "LDAP accessible directory" since LDAP is a network protocol intended to simplify query and update access to an X.500 directory. However, LDAP has come to be synonymous with not only the access protocol, but the X.500 directory content as well. So we will use the term "LDAP directory" from here on to mean both.

Domino supports LDAP in two ways: 1) the Domino server supports access to its directory (the Domino public name and address book—names.nsf) via LDAP, and 2) the Domino address book can be configured to query an LDAP directory via the Directory Assistance database feature. This means that either the Domino address book database itself can be used as the shared user directory or that an independent LDAP directory can be established and used by Domino.

WebSphere refers to the user name directory as the "user registry" and supports LDAP access to the user registry for user authentication purposes. WebSphere makes the decision about what type of user directory to use easy: WAS requires an LDAP user registry type when the LTPA mechanism is configured. Therefore, an LDAP user registry is required for WAS/Domino SSO.

The Domino and WebSphere support for LDAP allows several user directory configurations for SSO. There is no best configuration. Which is best for you may depend on your experience, whether you're already using Domino or another LDAP directory (such as the IBM Directory Server or OpenLDAP), or perhaps cost. Some of the possible configurations include the following:

- Defining all users to a Domino directory and connecting WebSphere to the Domino directory via LDAP.
- Defining all users to a stand-alone LDAP directory and connecting both Domino and WebSphere to it.
- Defining Internet users to a stand-alone LDAP directory and internal (Notes-only) users to a Domino directory and connecting Domino and WebSphere to the LDAP directory.

User Naming for SSO

One of the most important items to consider when designing a Domino and WebSphere SSO configuration is how users will be named, or, more specifically, how the user entries in the user directory will be named. A user name is the primary data involved in authentication and authorization mechanisms. For authentication, a user name supplied by the user either via a login prompt (for basic authentication) or via a client certificate (for SSL client certificate authentication) is validated against the user registry. The following user name issues often arise when designing SSO applications:

- **Uniqueness.** Obviously, user names must uniquely represent clients. This becomes more difficult as the number of users increase.

- **Format.** Users may have to specify the user name in a certain format depending on which HTTP server is used.
- **Ease of Use.** Users want a simple name format to use and remember.
- **Legacy User Names.** Some existing user name formats may not fit those acceptable for SSO.

For basic authentication, the format of the user name that may be specified in the login prompt depends on who (Domino or WebSphere) is doing the authentication and the directory configuration being used. If the Domino HTTP server is handling the basic authentication, it will validate the user name and password against the Domino address book and any LDAP directories configured as extensions to the address book via a Directory Assistance database. If WAS is performing the basic authentication, it will validate against the user registry (which must be an LDAP directory for SSO).

Since both Domino and WebSphere perform user authentication against an LDAP directory, let's review how this is performed more closely. For basic authentication, there are two steps: first, an LDAP search request is issued against the LDAP directory using a "search filter," which is constructed from the name specified in the login prompt; then, if a unique user entry (DN) is returned by the search, an LDAP bind request is made with the DN of the user entry and the password from the login prompt. If the bind request is successful, the user is authenticated. For certificate-based authentication, since the user name in the certificate is validated via the SSL client authentication exchange, only the LDAP search is performed to validate that the user is known to the service, that is, exists in the user registry. We will see that the search filter used in the LDAP user name search can be configured in both Domino and WebSphere. This provides for a wide range of name formats, which can be supported in the login prompt or user certificate.

Domino has its own rules for how user names are authenticated against the address book. The following table summarizes the user name formats accepted by a Domino server using an address book and their corresponding LDAP directory attributes.

Table 12-3 Domino User Name Formats vs. LDAP Directory Attributes

Domino Name and Address Book Fields (Person Document)	LDAP Directory Entry Attributes
First user name (e.g., John Doe/Acme)	Common name (cn)
Alias (i.e., any Username other than the first)	N/A
Canonical name (e.g., cn=John Doe/o=Acme/c=US)	Distinguished name (DN)
Abbreviated name (e.g., John Doe/Acme/US)	Distinguished name (DN)
Short name/UserID (e.g., jdoe)	Userid (uid)
Last name	Surname (sn)
First name	Given name (givenName)

Note that although a wide range of user name formats makes it easier on the user, certain name formats could be ambiguous. The set of formats accepted by Domino can be restricted to just User name (cn), Canonical/Abbreviated name (DN) or Short name (uid) by changing the "Internet authentication" setting (in the Server document under the Security tab) to "Fewer name variations with higher security." In Domino R6, you can explicitly set the search filter used for the user name lookup for authentication if you use an LDAP directory via the Directory Assistance feature.

(Note: Apparently Domino performs slightly different searches on DIIOP requests vs. HTTP requests. For DIIOP session initiation, such as used when establishing a remote session via the Domino Java classes, only the User name variation is accepted.)

If WebSphere is authenticating, the name formats accepted are controlled by the "User filter" setting in the "Advanced LDAP Settings" page of the LDAP User Registry configuration. The default filter is

```
(&(uid=%v)(objectclass=inetOrgPerson))
```

which searches for the user name as a uid attribute value.

Since Short name/ uid is common to both Domino and WebSphere authentication and is simple to specify, it is a good candidate for the SSO name format. However, keep in mind that the UID values must be unique among all LDAP directory entries. A useful choice for the UID value would be a person's fully qualified Internet e-mail name or address. Internet e-mail names are familiar to users, flexible in format, and, most importantly, unique to each user.

For SSL client certificate-based authentication, once the client itself has been authenticated via the certificate, the user name in the certificate is compared to the directory entry name for authorization. Though this sounds simple enough in theory, several complications arise when using client certificates in practice. One complication is that the format of the user name in the certificate is dictated by the Certificate Authority (CA) that creates the certificate. The X.509 certificate standard requires that certificate user names be unique under a single CA. The certificate user name usually includes attributes not part of the user DN in the directory, such as locality and country code attributes. As we will see later, WebSphere provides a way to handle mismatches between certificate and directory user name formats via a special LDAP search filter for the certificate name lookup.

Domino's handling of certificate-based authentication is more stringent. Users that need to be authenticated by a Domino Web server via client certificates must exist in a Domino directory and not an LDAP extension. The client certificates themselves must be imported into Person documents via a Notes client or via the Domino Certificate Authority function. The client certificate name must match the Username value in the Person document. For WebSphere, only the certificate user name must be found in the user directory, and there is no requirement that the certificate itself be part of the directory entry.

Limitations/Problems with SSO

Although the implementation of SSO has been available in WebSphere and Domino for some time, it is an evolving effort. WebSphere and Domino functions are growing (e.g., Lotus Instant Messaging) and are being combined in more complex ways (e.g., WebSphere Portal Server), which sometimes result in problems with SSO function. In this section, we point out some of the limitations of the implementations and problem areas.

- **SSO Across Domains.** As mentioned previously, it is not possible to use SSO among servers that exist in different DNS domains. This is a limitation in the design of the cookie-based session sharing protocol implemented by Web browsers. Providing SSO across domains would require a more sophisticated "front-end security proxy" such as IBM's Tivoli Access Manager product.
- **The Multiple Identities Problem.** This problem occurs when the user registry established for SSO consists of both Domino and LDAP directories, and a single user is represented in both directories but has differing user names (DNs). Part of this problem is due to the WAS limitation of not supporting LDAP referrals on user name searches. (An LDAP referral is when the response to a user search query points to another directory.) Because of this limitation, the WAS user registry must consist of only a single LDAP directory. But since Domino requires users to be defined in the Domino directory for mail routing, we are forced to use the Domino directory extended by the LDAP directory and to have users defined in both directories. This problem also occurs in WebSphere Portal Server configurations, and we discuss it further in that context in the section on WPS later in this chapter.
- **LTPA Token Issues.** Because the LTPA token is used as the means for passing authenticated credentials, it must be generated in the same way by both WebSphere and Domino. There are sometimes problems where the encoding used for the name and text data is not the same, causing LTPA tokens to fail, especially when non-ASCII characters are part of the user names or text data. Also, the LTPA expiration times and keys must be carefully managed to avoid reuse or replay attacks involving the tokens. To avoid problems with tokens expiring too soon, the server's clocks and time zones must be synchronized across the domain in which SSO is to be used.

For SSO, since open standards are used, namely LDAP and LTPA, there are no major differences in SSO function among the different operating system platforms supported by WebSphere and Domino. However, there may be quirks in the implementations arising from operating system differences. For example, different systems may encode text data in different ways, causing problems with interpreting LTPA tokens.

Setting Up the SSO Environment Between Domino and WebSphere

Here we present the basic steps needed to set up the single sign-on environment for Domino R6 and WebSphere V5. Recall that for SSO between Domino and WebSphere, a common LDAP directory (user registry) is required. For this procedure, we discuss setting up the user registry either as a Domino public name and address book or as an LDAP directory.

1. **Create a user ID for WebSphere administration.**

   Before starting the configuration of the two servers, you must create an entry in the LDAP directory that will be used to access WAS administrative functions when WAS security is enabled. If using a Domino directory, you can create a Person document for the user "WebSphere Admin" with the User name field set to `WebSphere Admin`, the Short name/UID field set to `wasadmin`, and an Internet password. See Figure 12-1. Make sure you specify the User name field. If it is empty, WebSphere cannot look up this entry using LDAP even though you are specifying the short name. Note that the LDAP distinguished name for this entry will be "cn=WebSphere Admin, o=<domain>," where "<domain>" is set to the Notes domain of the Domino server.

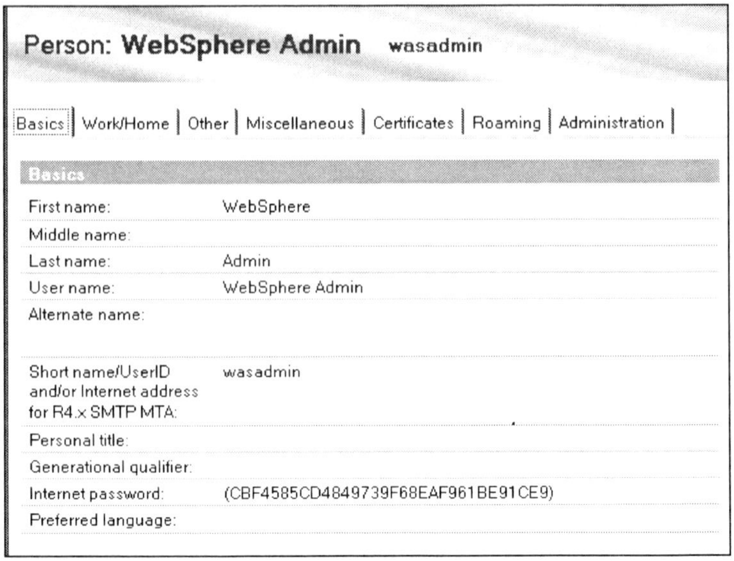

Figure 12-1 WebSphere administrator Person document in Domino.

   If using a non-Domino directory, you can add an entry to the directory as follows (expressed in the Lightweight Directory Interchange Format or LDIF):

```
# Entry for WAS Administration
dn: cn=wasadmin,o=MyOrg
cn: wasadmin
```

```
sn: wasadmin
uid: wasadmin
userPassword: secret
objectclass: top
objectclass: person
objectclass: inetorgperson
```

Note that the entry must have object class of `inetOrgPerson` since this object class defines the `uid` and `userPassword` attributes. This `inetOrgPerson` object class is also assumed by WAS for its default LDAP search filter. An LDAP search against the Domino directory shows that `inetOrgPerson` is also set by Domino:

```
CN=WebSphere Admin,O=MyNotesDomain
objectclass=dominoPerson
objectclass=inetOrgPerson
objectclass=organizationalPerson
objectclass=person
objectclass=top
givenname=WebSphere
sn=Admin
cn=WebSphere Admin
uid=wasdmin
userpassword=(12AFB84B93AFDF87FAFD8D1C78591B36)
```

2. **Ensure that the LDAP directory is accessible.**

   For a Domino LDAP directory, you can confirm that the Domino LDAP server task is running. For any LDAP directory, you can use the `ldapsearch` utility (which is supplied with Domino and usually with other directory software) to retrieve entries. For example, enter the following on a command line:

```
ldapsearch -h localhost "uid=wasadmin"
```

   (Note that for LDAP directories, you may need to specify the search base, such as "-b `o=MyOrg`." For LDAP directories that support SASL authentication, you may need to use the –x option to bypass SASL authentication where it has not been enabled.)

   We are now ready to configure SSO for WebSphere and then Domino. They must be configured in this order because the keys used to encrypt the LTPA token must be generated in WebSphere and then imported into Domino.

3. **Configure WebSphere Global Security for SSO.**

   Using a browser, access the WebSphere Web administrative console (the WebSphere server must be running).

First, we need to configure a user registry for WAS to use for user authentication. Since we are implementing SSO, we must configure an LDAP user registry. From the navigation frame on the left, select Security>User Registries>LDAP. In the Configuration form, specify the distinguished name (DN) of the directory entry that was created to represent the WAS administrator for Server User ID and specify the password, which was set in the directory entry for Server User Password. Note: once WAS global security is enabled, you will be required to supply this DN and password to access WAS administrative functions, including the Web console and server commands such as the stop server command.

For the Host field, you specify the host name for the LDAP directory. The Port field is set to the TCP/IP port on which the LDAP directory responds to requests (default is 389). Note: The LDAP host name and port form the LDAP realm name that is stored in the LTPA tokens generated by both WAS and Domino—the host name and port specified here must match the LDAP realm name specified in the Domino SSO configuration document (see step 4).

Specify the directory tree root name (that is, the root suffix you defined for the directory tree) as Base Distinguished Name. (If Domino is used for the directory, you can leave this field blank.) The remaining fields on this form can be left as is. When complete, the form should look similar to that shown in Figure 12-2. Select the **Apply** button at the bottom of the form to record the settings.

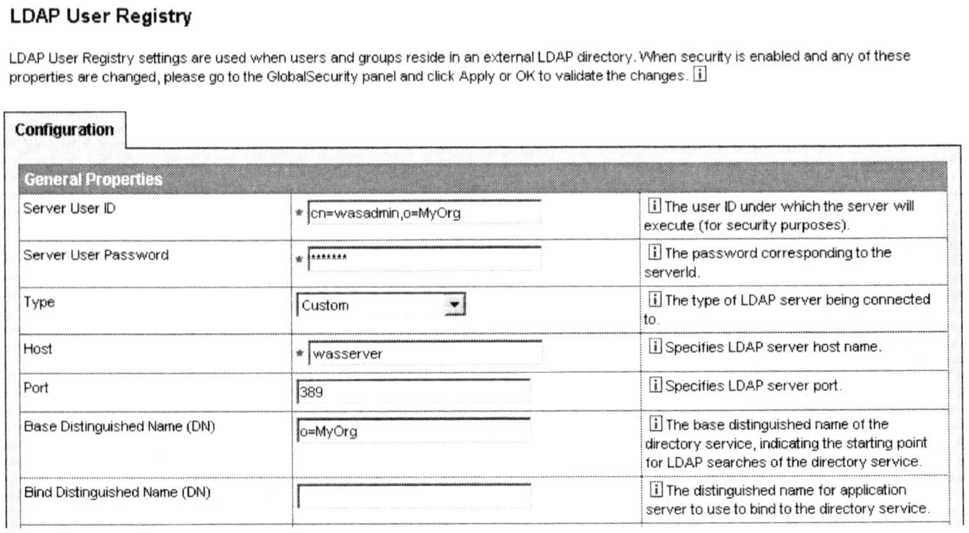

Figure 12-2 LDAP User Registry Settings for SSO.

Next, we need to configure the authentication mechanism for WAS to use. The only choice we have is the Lightweight Third-Party Authentication (LTPA) mechanism. Select **Security>Authentication Mechanisms>LTPA** in the console navigation frame. Specify a password in the field shown, and select the **Apply** button. This sets the password used for encrypting the LTPA server keys used to generate the LTPA tokens used for SSO. We will come back to this panel to export the keys after completing the security configuration.

At the bottom of this panel, click on the **Single Signon** link to display the SSO configuration panel. In this panel, check the **Enabled** check box and, optionally, specify a domain name for the set of servers to which SSO will apply. Select the **Apply** button.

Now we are ready to actually tell WAS to enable security using the user registry and authentication mechanism we have configured. From the navigation frame, select the **Security>Global Security** link. In the panel displayed, check the **Enable** checkbox and set the **Active Authentication Mechanism** and **Active User Registry** fields to `LTPA` and `LDAP`, respectively. Then select the **Apply** button. WAS verifies that the DN and password specified for the Server User ID exist and are valid in the directory specified for the user registry. If valid, you can then save this configuration and restart WAS to have the security configuration made active.

Note that enabling security will force the use of SSL for the WAS administrative console login. If not explicitly configured, WAS uses a self-signed server certificate shipped with WAS for the SSL server authentication. For more information about SSL and how to configure it differently, see the WebSphere Info Center.

Once WAS is restarted, the Web administrative console page requires login (over HTTPS) with the Server User ID and password as shown in Figure 12-3. You can use either the fully qualified DN or just the value of the uid attribute as the login User ID.

The final step in configuring WAS for SSO is to export the LTPA server keys so that they can be used by the Domino server. To do this, log in to the WAS Web administrative console and select **Security>Authentication Mechanisms>LTPA** in the console navigation frame. Specify a file name in the Key File Name field; then select the **Export Keys** button. If a path is not included in the specified file name, the file is created in the WAS install path. This file contains the LTPA server keys that you will need to import to Domino.

4. **Enabling SSO in Domino.**

   Support for SSO in Domino is also referred to as "multi-server session-based authentication" in the Lotus help documentation. As we discussed earlier, it is a type of HTTP session-based authentication provided by the Domino Web server function. Here we present in detail the tasks involved in configuring your Domino servers for SSO.

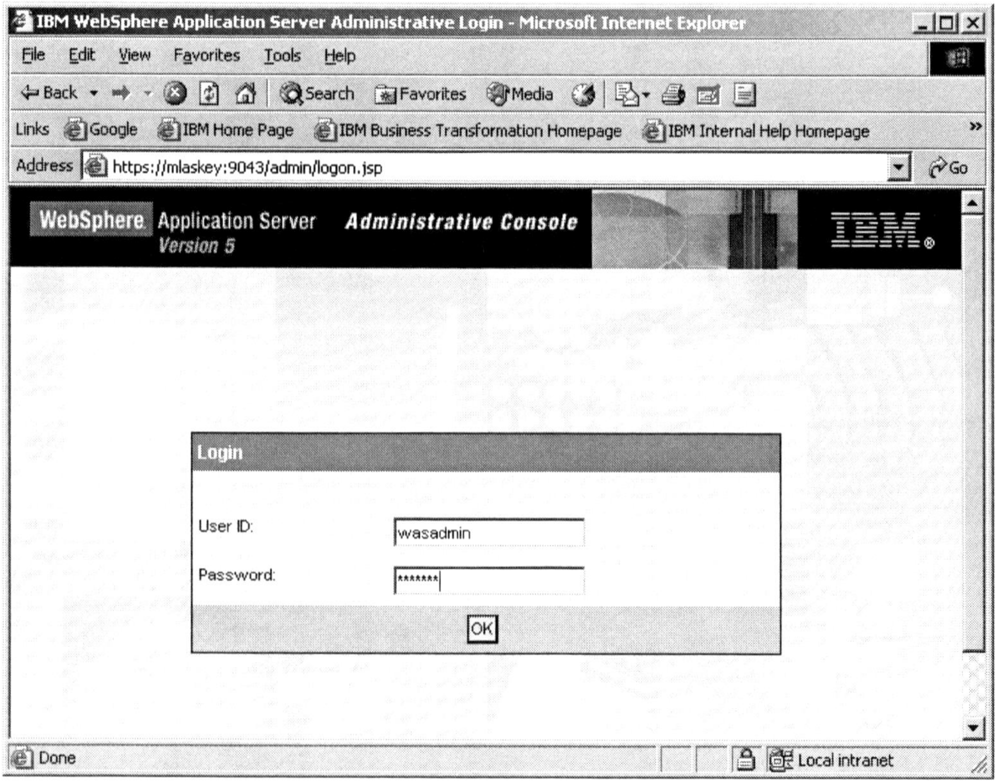

Figure 12-3 WAS administrative console login.

There are two required tasks for configuring SSO in Domino:

1. Configure Domino Web SSO. This must be done once for a Domino domain.
2. Configure Domino HTTP session-based authentication. This must be done for every Domino server in the domain that will support SSO.

If you are using a non-Domino LDAP directory, there is a third task: configuring the Domino directory to reference the external LDAP directory.

4.1. Configuring Domino Web SSO.

To configure Domino Web SSO, you must use either the Domino Notes or Administration client accessing a running Domino server because the configuration requires importing the LTPA keys, and this is not supported via a Web browser or via local access to the address book database. Also, you must ensure that the location document of your Domino client is pointing at a server in the domain for which you wish to configure SSO. This is necessary because the public keys of the participating Domino servers need to be accessed and used to encrypt the LTPA key imported from WebSphere.

Open the Domino server's address book database (names.nsf) and display the Servers view. Select the **Web...>Create Web SSO Configuration** action. The form shown in Figure 12-4 is displayed. The Configuration Name field is preset to LtpaToken, which you can leave as is. If any of your Domino servers are at release 5, you must use this value for the name. The value for the Organization field should be left blank—this is because we are creating the SSO configuration stand-alone and not as part of an Internet Site configuration. Set the DNS Domain field to the Internet domain name of the set of servers among which you want to use SSO. Choose the Domino servers to apply this SSO configuration to in the Domino Server Names field (group names are not permitted).

Next, from the action bar at the top of the SSO Configuration form, select the **Keys...>Import WebSphere LTPA Keys** action, and then specify the file path and password of the LTPA keys file exported from WAS. If the import is successful, you will see a dialog message to this effect. You will also see that the LDAP Realm field is set by the import. Ensure that host name and port set in this field match the hostname and port of the LDAP directory (user registry) server you configured for WAS. Also ensure that there is a '\' (backslash) character preceding the colon between the hostname and port number. Save the SSO Configuration document.

Note that the DNS Domain field is required for SSO. Also note that once SSO is configured, a domain name will be required as part of a URL issued to a Domino or WebSphere server. The reason for this is that the browser (HTTP client) uses the domain name portion of a host address as a criteria for sharing cookies among a group of servers—it only shares cookies among servers with a common domain name. Domino or WebSphere servers will not generate LTPA cookies for URLs containing just IP addresses or host names, so SSO will not work if the fully qualified URL host addresses are not used. However, this does not mean that you must set up a Domain Name System (DNS) server. For a test network, you can simply add fully qualified host names to your server's host name definition file (for example, the /etc/hosts file on Unix/Linux systems or the WINNT\system32\drivers\etc\hosts file on Windows 2000 systems).

4.2. Configure Domino HTTP Session-Based Authentication.

For each Domino server that is to use SSO, you need to configure it for HTTP session-based authentication. (If you are using the Domino R6 Internet Site feature, you can do this configuration via a single Internet Site document. We will show only the per server approach.)

Edit the Server document for the Domino server and move to the Internet Protocols/ Domino Web Engine notebook page. In the Session Authentication field, choose `Multiple Servers (SSO)`, then choose the `LtpaToken` configuration for the Web SSO Configuration field (see Figure 12-5).

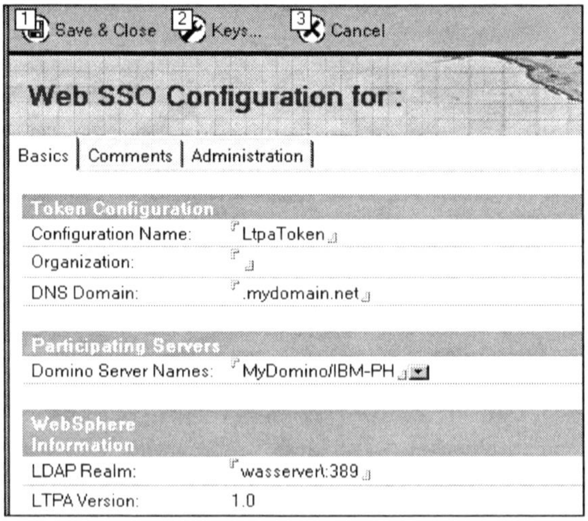

Figure 12-4 Domino Web SSO configuration.

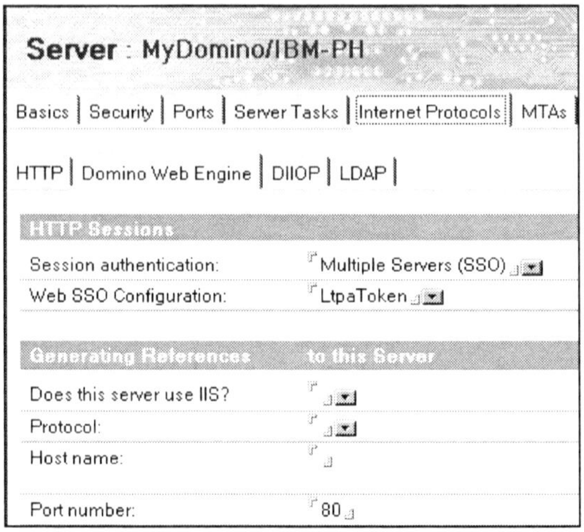

Figure 12-5 Server HTTP session-based authentication.

Note that the Web SSO Configuration field is a keyword list of the Web SSO Configuration document(s) you configured above. If the Organization field is *not* blank in the Web SSO Configuration document, you will *not* see it in this list. You should only specify the Organization field if using an Internet Site configuration.

Next, verify that at least basic (name and password) authentication is configured for the Domino server's HTTP port. Open the Ports>Internet Ports>Web notebook page in the Server document and check that at least the Name & password field is set to `Yes` under Authentication options for the HTTP port (80) and, if you are using SSL, the HTTPS port (443).

4.3. Configuring Domino Directory Assistance.

This task applies only if you are using a separate, non-Domino LDAP directory for the shared user registry between Domino and WebSphere. If so, you need to configure the Domino directory to make use of the LDAP directory. This is done by creating a Directory Assistance database and pointing the Domino server documents in the Domino directory to this database.

Create the Directory Assistance (DA) database from the da50.ntf database template. In this DA database, create a Directory Assistance document. On the Basics notebook page, select **LDAP** for Domain type. Choose any names for the Domain name and Company name fields, but make sure that the domain name specified here is different from your Domino domain name as it represents resources external to the Domino directory. Select at least the **Notes Clients & Internet Authentication / Authorization** option for the Make this domain available to setting. If you want to be able to specify LDAP group names in Domino ACLs, set Group Authorization to Yes. Finally, enable use of this DA document (see Figure 12-6).

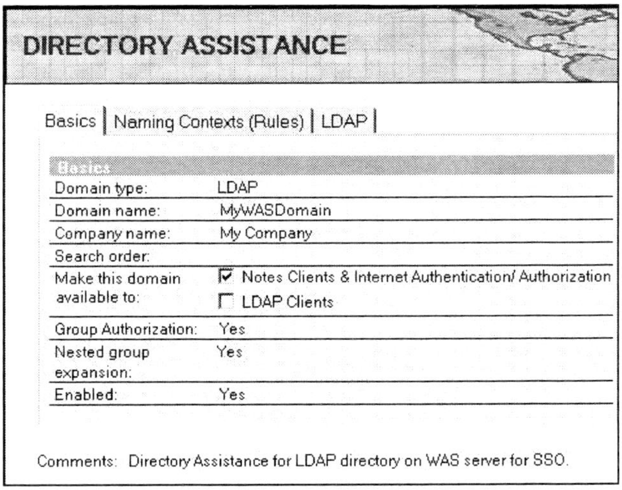

Figure 12-6 Directory Assistance basics.

On the Naming Contexts (Rules) notebook page, ensure that the default rule is enabled and is set to allow authentication of Internet users based on credentials (name and password) found in the LDAP directory (see Figure 12-7). Leaving the rule with all asterisks for the name components results in the rule being applied to any LDAP (X.500) distinguished name.

On the LDAP notebook page, you configure the LDAP directory server access. You will need to set at least the Host name and Base DN for search fields. You may need to set the Authentication credential fields if you have configured your LDAP directory to require authenticated user access for searching.

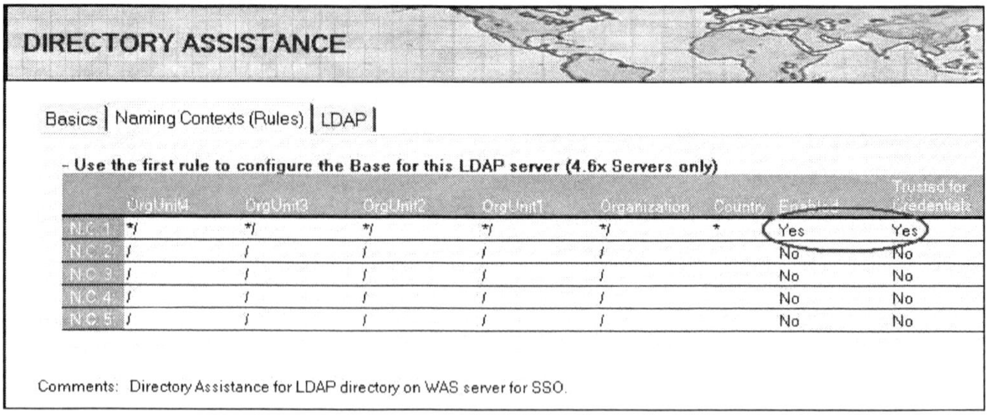

Figure 12-7 Directory Assistance naming rules.

The Type of search filter to use field tells Domino how to lookup users in the LDAP directory when doing authentication, authorization, or mail routing. You can leave it set to **Standard LDAP** if your LDAP directory entries are based on the `inetOrgPerson` object class (that is, contain the Common Name (cn), userid (uid), or mail attributes) or `GroupOfUniqueNames` class for groups. The search filter used for authentication in this case is as follows:

```
(|(cn=%*)(uid=%*)(mail=%*))
```

(Refer to the IETF RFCs 2251 and 2254 for a description of LDAP search filter syntax.) This filter says "look for a user entry with a common name or userid or e-mail ID equal to the entire value entered in the login dialog." You can customize the search filters if you want to do the user name lookup based on different attributes or portions of the login name, for example, if you want to use e-mail addresses for the login names. Keep in mind that your search filters must result in a unique user name being returned by the lookup. (Note: the search filter shown above is different from the default search filter published in the Domino Administrator Help documentation, but it was determined based on tracing the LDAP requests from Domino to the LDAP directory. You should either explicitly set the search filters or trace the LDAP requests to be sure of what your configuration will use.)

After creating and saving the Directory Assistance document, edit the Domino Server document(s) and specify the name of the DA database you created in the Directory Assistance database name field on the Basics notebook page.

After saving these changes, the Domino server is set up to use the common LDAP directory (user registry) for user authentication and to use LTPA token cookies for SSO between protected Domino and WAS resources.

5. **Setting up WebSphere Application Security.**

Here we give the basic steps involved with protecting the components of a J2EE application running under WAS V5. We'll focus on how to set up a J2EE application for authentication and authorization of its Web users using the SSO mechanism, LTPA. Just to be clear, our focus here is on the "network" security aspects of J2EE applications versus the security aspects of how the application is running on the server operating system itself.

As discussed in the introduction to this chapter, the J2EE network security model is role-based, where authorization to access application resources is granted to defined roles. The roles are mapped to users or groups in the user registry. A major benefit of this approach is that the component security function can be designed in terms of logical roles and then mapped to the user name space at deployment time, without having to modify component code.

A J2EE application must be constructed and deployed as an Enterprise Application in order to apply security functions. The security related configuration settings are specified in the deployment descriptors at each component level, application, EJB, Web, etc. Much of a J2EE application's security can be specified in the deployment descriptors rather than having to write code. Access to any J2EE resource, servlets, JSPs, or EJBs can be specified as part of application assembly. We'll look at how to configure security function using WSAD, although the same could be done using the WAS Application Assembly Tool.

5.1. **Define security roles.**

You define the set of security roles that are used by the application in the application deployment descriptor. Here you also define the mappings between security roles and the user registry. In WSAD, this is done by opening the Application Deployment Descriptor to the Security page and adding roles. For each role, you can assign it to a set of users or groups in the user registry (LDAP directory). You also can assign a role to the special groups Everyone or All authenticated users, which grant access as their names imply. Figure 12-8 shows the role `BookAuthor` assigned to the group `cn=Group1, o=MyOrg` in the user registry.

5.2. **Configure Web application security.**

You configure security for Web components in the Web deployment descriptor. There are three sets of items that must be configured: the set of security roles, the security constraints for the Web component resources, and the type of login authentication. The security roles for the Web resources are usually the same as defined for the application, although this is not necessary. (If a different set of roles exist for a Web component, you can map these to the application roles via "security role references.")

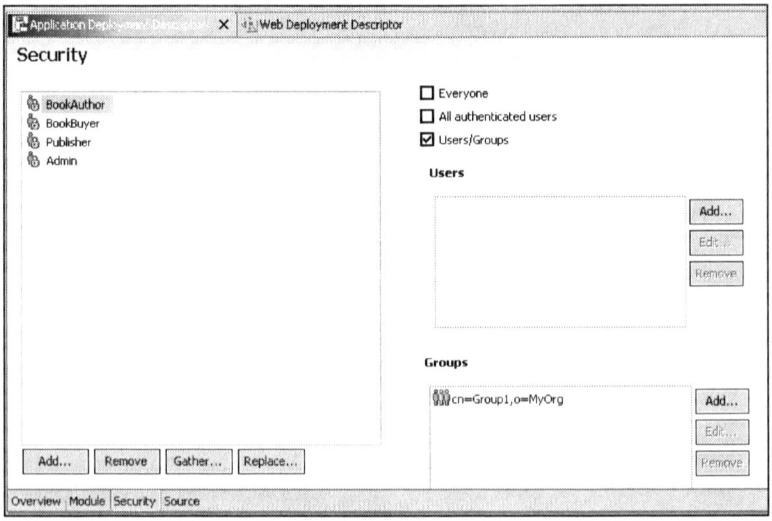

Figure 12-8 Application security role definition.

The most important configuration items are the security constraints. The constraints govern how the application server decides that a Web resource needs to be protected, and, if so, which roles are allowed access and whether authentication is required. A constraint defines which roles are authorized to access which Web resources. The Web resources in a security constraint are specified as "Web resource collections" in terms of URL patterns and HTTP methods. When presented with a Web resource request URL, the WAS server goes through the set of security constraints and compares the URL to the Web resource collection definitions. If the URL matches a specified pattern and the request type matches one of the specified HTTP methods, the server ensures that the user is authenticated and belongs to one of the constraint roles.

To specify security roles and constraints in WSAD, open the Web Deployment Descriptor to the Security page. There are two tabs at the top of this page for the roles and constraints. The Roles page shows the set of roles defined to the application. The Constraints page allows the addition of constraints and the specification of Web resource collections and authorized roles for each constraint. Figure 12-9 shows an example of the Constraints page.

We have found that specifying the URL patterns for the Web resource collections can be tricky. The patterns apply to the portion of the URL after the context root. A single forward slash "/" indicates to apply the constraint as the default for the Web application. You can use an asterisk at the beginning or end of a pattern (but not both) to match any character. It may take some trial and error to get the pattern definition correct for the set of resources you want to protect. We also found that if you change the constraint definitions, you must restart WAS for the change take effect; you cannot just restart the Web application. Fortunately, you can test the constraints within WSAD via a WebSphere Test Environment (WTE) server. (It is even possible to configure the WTE server for LTPA and LDAP directory usage—just enable the adminstrative console application under the WTE configuration page and access it to configure global security. It is also possible to turn on WAS tracing for the security functions. This is helpful to see how WAS is performing the URL pattern matching for the Web resource collections.)

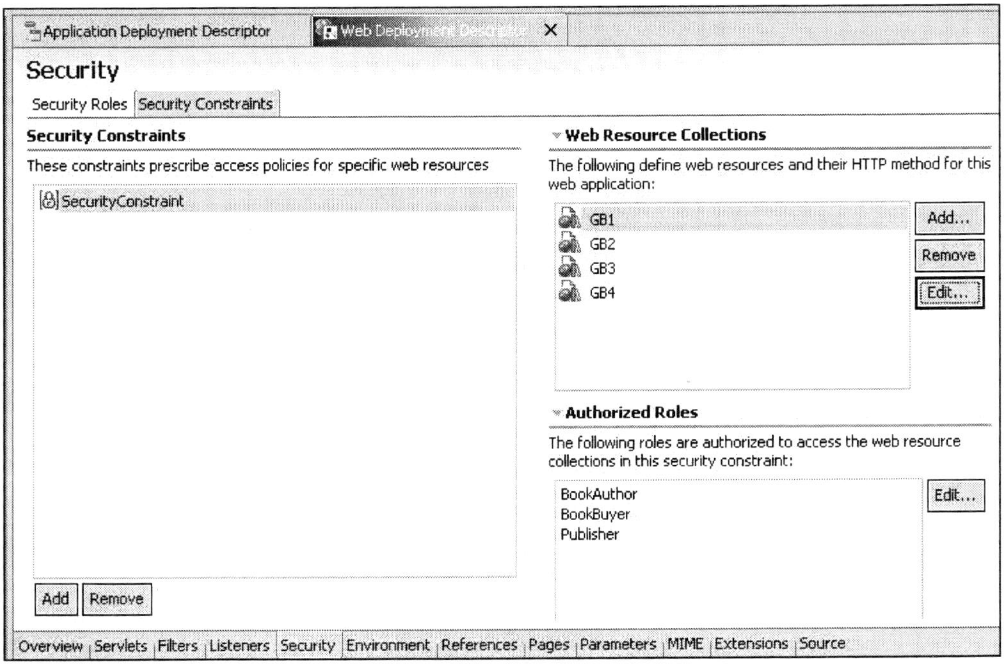

Figure 12-9 Web application security constraints.

The remaining configuration item for Web application security defines how WAS should perform user authentication. In WSAD, this configuration is done on the Pages page of the Web Deployment Descriptor notebook. The choices for login authentication types are `Basic`, `Form`, and `Client-Cert`. `Basic` and `Form` refer to HTTP basic authentication, where `Basic` indicates that the login is done via browser prompt and `Form` via custom HTML login forms. `Client-Cert` indicates that the authentication is done using SSL client certificate-based authentication.

6. **Configuring Domino application security.**

   As mentioned in the first part of this chapter, Domino uses a resource-based model for providing access control to its resources. All Domino databases contain an ACL that specifies which users can do what to the database content. Since Domino applications consist of sets of Domino databases and the views, forms, documents, and agents within these databases, securing a Domino application consists of configuring the database ACLs.

The Domino server will require user authentication when access is requested to a database whose ACL restricts access to specific users—that is, where the ACL has Default and Anonymous entries set to "No Access." To permit access for a user to a database, you add the name of that user or a group containing that user to the ACL along with the user/group access level. (It is much easier to manage access control using group names in ACLs.) Users requiring access to Domino and WebSphere resources via SSO must be defined in the shared user registry. Any groups to be used must also be defined in the shared user registry. Then these user or groups are added to ACLs using their distinguished names (DNs), which are unique in the user registry.

Since the user registry for SSO between Domino and WebSphere must be an LDAP (X.500) directory, the DNs consist of a set of attribute/value pairs such as, "cn=A. User,o=MyOrg,c=CC" or "cn=CustomerGroup,o=MyOrg,c=CC." DNs can be added to Domino ACLs by specifying the attribute/value format, except that "/"s must be used for the separators instead of commas ("cn=A. User/o=MyOrg/c=CC").

Note that Domino also provides server security, which controls access to the Domino server itself. If server security is used, the users and groups permitted for application access must also be permitted server access. This permission is configured in the Domino server document in the Domino directory (which is not required to be part of the user registry).

Testing Your SSO Configuration

To test an SSO configuration, create a simple J2EE application with a protected servlet and a Domino database with an ACL granting access only to certain users in the LDAP directory. Deploy the J2EE application to a WAS server (or to a WTE server) and map the security role to the same set of users permitted to the Domino database. Via a browser, attempt to access the servlet and authenticate with one of the user names in the LDAP directory. Then access the Domino database URL. You should not be prompted to re-authenticate on the database request. After closing the browser (to clear the cached name and password and LTPA token), try the access in the reverse order, Domino database and then servlet. This time you should not be prompted to re-authenticate for the servlet request.

Using SSO with SSL Client Certificate Authentication

You can specify the SSL client certificate authentication is to be used for authentication of a Web application by modifying the deployment descriptor for the Web module login configuration via the Application Assembly Tool or WSAD. Select the **CLIENT-CERT** authentication type and `Integral` or `Confidential` for the **User Authorization**. (The `Integral` and `Confidential` settings force the application to be accessed over SSL.)

Naturally, you must configure your HTTP servers for SSL and client certificate authentication, usually over port 443, the standard HTTPS port. For Apache-based servers (such as the IBM HTTP Server), this is done via the httpd.conf configuration file. For the Domino HTTP server, it is done via the Server document in the Domino directory under the Ports> Internet Ports>Web page. When configuring the authentication options, you should set client certificates to be required rather than optional, which will force the client certificate to be sent in response to the initial resource request. This is necessary so that the certificate can then be passed to WAS via the WAS plug-in. To require client authentication for Domino HTTPS, specify "yes" for client certificate authentication and "no" for both the Name & password and Anonymous authentication options.

The HTTPS port configuration will require that a server key pair and certificate be set up on the server side for the SSL server authentication handshake. You will need to provide the server keys and certificate in a format acceptable to the HTTP server. (For Domino HTTP server this format is the "keyring" or .kyr format.) Also a set of trusted signer certificates that can validate the client certificates will need to be specified on the server side. Usually, these trusted certificates consist of the Certificate Authority (CA) signer certificates that are obtained from the CAs used to sign the client certificates. For an application made available on the open Internet, you can require your clients to use commercial CAs for certificate signing. Keep in mind, however, that Domino provides its own CA function, which you can make available to your clients and also use for signing server certificates.

Once the SSL port is configured on the HTTP server, ensure that WAS will accept this port under the virtual host name defined for the HTTP server. You may need to add the SSL port to the Virtual Host definition for the application server.

Finally, there is one more area of server-side configuration required for WebSphere SSL client certificate authentication. This additional configuration is required to establish SSL between the HTTP server WAS plug-in and the WAS application server.

On the WAS side, ensure that an SSL port is configured for the application server. The default WAS SSL port is 9443. You should specify a different server private key and certificate and trusted signer certificate files for this port other than the default self-signed server certificate that is provided by IBM. To do this, you need to first create a new SSL "repertoire" that represents a set of SSL key files. Using the WAS Web administrative interface, open the SSL page under the Security branch and create a new SSL Repertoire. On the SSL Repertoire configuration page, you can specify the server private key/public certificate file and password and the trusted signer certificates file (these may be the same file if one file contains all). These files must be in either Java Key Store (JKS) or PKCS#12 format. The WAS help documentation describes how to create these files containing self-signed certificates using the IKeyMan utility supplied with WAS. However, self-signed certificates are generally not as secure as CA signed certificates. If a commercial CA cannot be used, then using the Domino-provided CA function to sign the server certificate is the next best thing. Once the SSL repertoire is created, it can be associated with the 9443 port, via the Servers>server<n>>WebContainer>HTTPTransports configuration page.

On the WAS plug-in side, you must edit the plug-in configuration XML file to tell the plug-in to use the SSL port configured above to connect to WAS for SSL requests. You must also specify a trusted certificates file so that the plug-in can authenticate the WAS server certificate on the SSL handshake. As is often the case with SSL, this trusted certificates file must be in a format different from that accepted for the WAS SSL port. Here the file must be in "CMS Database" format or ".kdb" format. A file in this format can be created by using the IKeyMan utility supplied with the IBM HTTP Server. If the WAS server certificate is CA-signed, then this trusted certificates file need only contain the CA signer certificate. If the certificate file is password protected, then you must create a "stash" file to specify the password to the plug-in. Here is a section of the plug-in configuration file showing the specification of the SSL port and certificate and stash files:

```
<Server Name="default_server">
  <!-- The transport defines the hostname and ports that the
  ➥Web server plugin will use to communicate with the
  ➥application server. -->
  <Transport Hostname="washost" Port="9080" Protocol="http"/>
  <Transport Hostname="washost" Port="9443" Protocol="https">
     <Property Name="Keyring" value="/plugin/config/
     ➥trustedcerts.kdb"/>
     <Property Name="Stashfile" value="/plugin/config/
     ➥trustedcerts.sth"/>
  </Transport>
</Server>
```

If you encounter problems with the plug-in after configuring for SSL, ensure that you have the latest fixes for the plug-in library files from the IBM WebSphere product Web site. There have been several fixes released for the plug-in, some specific to the Domino HTTP server plug-in.

With the HTTP server set for SSL and client certificate authentication and SSL enabled between the HTTP plug-in and WAS, you are ready to test and use client certificate authentication for your Web application components within an SSO domain. Well, almost. The client certificates need to be installed into the user browsers (or clients). (The Domino CA function handles this very nicely as part of the certificate signing process.) In addition, you should ensure that the client certificate user names can be looked up in the WAS user registry. As mentioned earlier, WAS attempts to match the certificate user DN with a user DN in the LDAP directory, or you can specify a search filter for WAS to use for the certificate user name look up.

Now we are finally there. With all of this set up, the initial authentication for a protected Web resource is made using the client certificate, and all other protected accesses are authenticated via the LTPA token.

However, as mentioned previously, there is a problem with client certificate authentication to Domino resources. An initial client certificate authentication to a Domino resource won't work if the certificate user is defined only in a non-Domino LDAP directory. This is because Domino requires that the user certificate be stored in the Domino directory for certificate authentication. If client certificate authentication is used with SSO without storing the certificates in Domino, then the initial authentication must be to a WAS resource.

Using SSO with Domino Java Classes

As discussed in Chapter 9, "Accessing J2EE Elements from Domino," it's often useful to invoke Domino functions from J2EE components using Domino Java classes, especially from servlets or JSPs. Here we discuss some points about using SSO in this context. Let's consider a servlet, which needs to present data from a Domino database, and we want the access to the database to be made under the same user authenticated for the servlet.

The Domino Java classes require a user name and credentials to be passed to the `NotesFactory.createSession()` method in order to establish the session under that user. Most commonly, the user name and its Internet password are passed as text strings to `create-Session()`. If basic authentication is being used to authenticate the servlet, we could extract the user name and password from the HTTP header. But if SSO is configured, there is an easier way. Fortunately, the `createSession()` method also admits an LTPA token for the user name and credential data. We can simply find the `LTPAToken` cookie in the HTTP header and pass its value, the LTPA token itself, to `createSession()`. Using the LTPA token for the Domino session also allows the servlet to use either basic, form-based, or client certificate authentication interchangably.

Unfortunately, things are not quite so simple. If the servlet request itself triggers the authentication, the LTPA cookie is not passed to this initial servlet request and thus is not available to be passed to `createSession()`. There are a couple of ways to work around this problem. One way is to detect the missing LTPA cookie and force the client to reissue the servlet request, this time with the LTPA cookie as part of the request, via the `HTTPServletRequest.sendRedirect()` method. Another way is to extract the user name and password from the authentication header and use these to establish the Domino session instead. (But this second technique works only if basic authentication is used). The following listing shows the code required to both obtain the LTPA cookie and extract the user name and password from an authentication header if necessary:

```
public void doGet( HttpServletRequest theReq,
➥HttpServletResponse theResp )
   throws ServletException, IOException
{
   // See if LTPA token exists, if so we'll use it to
   ➥identify/authenticate
   // the user to Domino.
   //
   String aLTPAToken = null;  String aUser = null;  String
   ➥aPass = null;
```

```
javax.Servlet.http.Cookie[] aReqCookies =
➥theReq.getCookies();
if ( aReqCookies != null )
{
   for ( int i = 0;  i < aReqCookies.length;  i++ )
   {
      javax.Servlet.http.Cookie aCookie = aReqCookies[i];
      if ( aCookie.getName().equalsIgnoreCase( "LTPAToken" )
)
         aLTPAToken = aCookie.getValue();
   }
}

// If the LTPA token is not found, try and get the
➥userid/password.
//
if ( aLTPAToken == null )
{
   // Get the Authentication header if it exists, then the
   ➥userid/pw.
   if ( theReq.getAuthType().equalsIgnoreCase( "Basic" ) )
   {
      String aAuthStr = theReq.getHeader( "Authorization" );
      // Auth header is "Basic <base-64 encoded
      ➥'userid:pw'>"
      if ( aAuthStr != null )
      {
         sun.misc.BASE64Decoder aDecoder = new
         ➥sun.misc.BASE64Decoder();
         String aUserPw =
           new String( aDecoder.decodeBuffer(
           ➥aAuthStr.substring(6) ) );
         int aSep = aUserPw.indexOf( ':' );
         if ( aSep > 0 )
         {
            aUser = aUserPw.substring( 0, aSep );
            aPass = aUserPw.substring( aSep+1 );
         }
      }
   }
```

```
        }

        try {
            // Set up Domino session based on the LTPA or user/pw
            ➥info above.
            //
            Session aSession = null;
            if ( aLTPAToken == null )
            {
                aSession = NotesFactory.createSession( aHost, aUser,
                ➥aPass );
            }
            else
            {
                aSession = NotesFactory.createSession( aHost,
                ➥aLTPAToken );
            }

            // ...
        }
        catch ( Exception e )
        {
            System.out.println( "Create session error: " + e );
        }
    }
```

Troubleshooting Your SSO Configuration

If you do have a problem with SSO between Domino and WebSphere, check the following:

- Ensure that the Web SSO Configuration document in the Domino Directory has been replicated to all servers participating in the SSO relationship. You should also do this if you have modified this document.
- Ensure that you are using the same LDAP directory for both Domino and WebSphere and that the directory is accessible from all servers in your SSO domain. The LDAP utility ldapsearch provided by Domino comes in handy here.
- Ensure that fully qualified host addresses are used in URLs when accessing the protected resources from a browser. The LTPA cookies are only generated for requests that specify a fully qualified host address.

- Ensure that the LDAP Realm Name specified in the Domino Web SSO configuration document is the same hostname and port specified as the LDAP user registry in WAS.
- Ensure that the user registry (LDAP directory) contents contain the attributes that will satisfy the search filters used for authentication.
- Ensure that the access control information associated with the protected resources is specified correctly. For example, distinguished names in Domino database ACLs should be specified as "cn=SomeName/ou=Dept/o=MyOrg," with "/" separators and not commas.
- Ensure that the time zones are specified correctly for the servers participating in SSO.
- Re-import the LTPA keys to Domino on any change to the WAS security configuration. (The LTPA keys may be regenerated by WAS on a configuration change.)
- Restart the Domino server after any change to the Web SSO Configuration document.

The following trace logs are useful for determining problems with an SSO configuration:

- The WAS SystemOut.log and SystemErr.log files under $WASROOT/logs/servern.
- A trace log file generated by enabling the following WAS traces: `SASRes=all=enabled:com.ibm.ws.security.*=all=enabled`. This trace is useful in determining problems with security constraint specifications.
- The HTTP WAS plug-in log file (specified in the plug-in XML configuration file).
- The Domino Web Server Log (if Domino HTTP server is being used).
- The HTTP server log file (if another HTTP server is being used).
- A trace log for the LDAP directory server.

Supported Configurations: Domino and WebSphere

At the time of the writing of this book, the most recent version of Domino generally available was R6.5. The first version for which SSO can be enabled was R5.0.5. On the WebSphere side, things are slightly more complicated. The first version of WebSphere Application Server (Advanced Edition) for which SSO can be enabled is version 3.5, with PTF (fixpack) number 1, sometimes expressed as version 3.5.1. The versions of WAS and Domino described in this chapter are WAS 5.0.1 and Domino 6.02 CF2.

Configuring SSO for WebSphere Portal Server

We now present material concerning SSO and the WebSphere Portal Server (WPS). The purpose of this section is to discuss the issues related to using Lotus Domino and Lotus Collaborative products within the WebSphere Portal environment. It will discuss several scenarios related to having the Lotus Collaborative products work cohesively within the WebSphere Portal framework as well as the configuration of Domino Directory Assistance in this environment. The focus here is on using the native facilities in WebSphere Portal and Domino to provide an SSO experience.

This section is targeted at Domino and WebSphere Portal administrators or IT architects who will integrate WebSphere Portal into an existing Domino environment using the native SSO capabilities of these products. We do not discuss implementing SSO with other products such as Tivoli Access Manager.

As it relates to WebSphere Portal, SSO provides the ability to log into the WebSphere Portal Server and provide access with those user credentials to the Domino environment, such as Domino, Lotus Instant Messaging (IM), QuickPlace, and other Domino-based tools. Without an SSO relationship between WebSphere Portal and the Domino environment, users would need to log in to the Domino server each time they accessed a portlet containing information from a Domino-based application or service. Additionally, there are WebSphere Portal API's and services, such as People Awareness, that do not provide a facility for login. Even through these services do not provide distinct login facilities, they do require authentication through SSO in order to operate.

WebSphere Portal provides a Credentials Vault capability. The Credentials Vault passes the username and password to the back-end application through the HTTP Basic Authentication Header. For the Domino server to accept credentials passed through this header, the server must be configured for Single-Server SSO mode, since, in Multi-Server mode, the Domino server will only accept credentials passed through the LTPA mechanism. With the Credentials Vault approach, the user authenticates once, and WPS forms a set of credentials for that user. The credentials are then stored in an encrypted database table and passed to the back-end application each time the user accesses a portlet requiring authentication. For details on configuring the Credentials Vault, see the WebSphere Portal InfoCenter.

The option that most organizations work toward is to provide an automatic method of using the current credentials from WebSphere Portal to authenticate with back-end applications.

How Does It Work?

As we have discussed before, SSO function is based on the ability to share user authentication credentials between the application servers. For WebSphere and Lotus Domino, this credential sharing is done through a mechanism called Lightweight Third Party Authentication (LTPA). Both WebSphere (and thus WebSphere Portal and any other WAS-based products) and Domino use LTPA. The LTPA mechanism is based on the use of an HTTP cookie. An HTTP cookie is simply a set of data generated by a server and given to a browser to return on subsequent requests to URLs within the same Internet (DNS) domain. For LTPA, an encrypted session cookie is set in the user's browser and contains information that a WebSphere or Domino application server can decrypt and use to determine that the user is authenticated within the DNS domain that the cookie covers.

We note that even without using the LTPA mechanism it is still possible to provide SSO behavior from WebSphere Portal. This is because there are two different methods that are used to authenticate the user into the Domino environment from WebSphere Portal. To understand how these methods work it is important to understand the points of integration from WebSphere Portal into Domino.

WebSphere Portal to Domino Server

Portlets integrate WebSphere Portal and backend applications. Portlets act as a sort of proxy when the user requests information from an external application, and that information is displayed in the WebSphere Portal page by the portlet.

The Notes mail portlets demonstrate this behavior. The WebSphere Portal makes the request for data to the Domino server in the user's name. WebSphere Portal requests information from the Domino Server using the current user's credentials, which are collected from the session information. The portlet then formats the request and sends it to the Domino server. When the data is returned, it is formatted and integrated into the WebSphere Portal page and sent to the user's browser for viewing.

When you log into the WebSphere Portal, you use an LDAP attribute that is typically a unique abbreviation of the user's name, employee serial number, or some other unique identifier. After the WebSphere Portal receives validation of the user name and password or existing LTPA token from the LDAP server, it gathers other information that includes the full Distinguished Name (DN) of the user. This information is then cached on the WebSphere Portal until the user logs off of the server or his session expires. Once this information is gathered, the portlet builds an HTTP, IIOP, or LDAP request and sends it to the Domino server. When the Domino server receives the request, it gets the user credentials and uses the Domino authentication and authorization facilities to see if the user is able to obtain the information. If authentication and authorization are granted, the information is retrieved and handed back to WebSphere Portal in XML format. If the user does not have authorization to the requested information, an error is returned to WebSphere Portal.

There is another special case that is used by many of the Domino portlets. When in the configuration mode of the Domino portlets, the list of available databases is retrieved via the DIIOP interface of the Domino server. The only difference between IIOP and HTTP connections is the protocol used to talk to the Domino server; the LTPA information that is sent to the Domino server is the same as with an HTTP request.

For this level of SSO to work, it is necessary for the DN and associated password to match between Domino and WebSphere Portal. This is a side effect of the level of SSO provided in WebSphere Portal to Domino server integration. The user's WebSphere Portal username and password are being passed along.

Browser to Domino Server

The second set of transactions involving SSO occurs between the user's browser and the Domino server. These transactions occur when the user opens a Lotus Notes document from the WebSphere Portal interface, or when any Domino application is opened in an iFrame portlet. If you take a look at the links in the Notes mail portlet or Notes view portlet, you will notice that the URLs point directly to the Domino server and bypass WebSphere Portal.

Since the WebSphere Portal is not involved in the transactions, the user's identity is passed via the LTPA cookie that was created when the user authenticated to the WebSphere Portal. In other words, the standard LTPA mechanism is used for SSO between the portlet and Domino. Recall that the LTPA cookie contains the user's DN that was retrieved from the LDAP directory at login. The encrypted LTPA cookie information is passed along with the HTTP request to the Domino server.

For the LTPA SSO to work, the LTPA cookie that is created by WebSphere must have been generated with an encryption key that is valid on the Domino server. As discussed in previous sections of this chapter, the LTPA key can be exported from WAS via the Administrative application and imported into Domino, providing a common key between the two environments. While Domino servers can consume the key created by WebSphere, Domino does not have the capability to export its own LTPA keys for use by WebSphere. Thus if LTPA were being used for other Domino-based applications such as Lotus IM or QuickPlace, a new set of LTPA keys generated by WPS must be imported into Domino in order for these applications to be used under WPS with SSO.

Options for User Directory Sharing

The following sections discuss the options for establishing a shared user directory between WPS and Domino. Recall that a shared user directory or registry is one of the requirements for providing SSO. These options reflect the most common environments found at customer sites. If you have an existing environment, you can focus on the section that most closely matches your environment. If you are creating a new environment, you can read each section and select the scenario that best fits your requirements.

The scenarios fall into two categories:

- One user directory shared by WPS and Domino
- Separate user directories shared via LDAP

One Directory Serving Both WPS and Domino

This scenario represents environments where a single LDAP directory is in place for the WebSphere Portal environment. If Domino exists in the environment, then the Domino Server provides the LDAP services, since it is one of the supported LDAP servers. If Domino will only be used for the support of QuickPlace and Lotus IM (which are Domino-based applications) within WebSphere Portal, it can be configured to use a non-Domino LDAP directory. WebSphere Portal must be configured to authenticate users against the LDAP directory. Recall that WAS servers (hence WPS servers) can only be configured to authenticate against a single LDAP directory.

It is possible to use the Domino directory as the LDAP directory for an entire enterprise. The Domino server provides support for LDAP access to its directory and thus can be used for authentication from WebSphere Portal and other applications. If Domino will be used as the enterprise LDAP, the environment is considered a single directory environment.

Separate Domino and WPS Directories

In these scenarios, typical of environments that have existing Domino infrastructures, there are separate directories for Domino and WebSphere Portal. Typically, the Domino directory contains information on each of the Domino users, while the LDAP directory contains a super-set of this population. The LDAP directory must contain a user record for every user that will access WebSphere Portal. In the case where a given user is also a Domino user, there is certain information that must be reconciled between the two records. This is typically done either manually or with tools such as IBM Directory Integrator (IDI). We will discuss how to manually reconcile the LDAP and Domino directory entries later in this chapter.

For this approach to work, the LTPA mechanism must be established between the WebSphere Portal and Domino environments. Configuring WPS for LTPA is similar to what is required for a standard WAS server. See the earlier sections in this chapter and information found in the Portal InfoCenter and WebSphere Portal Readme file for details.

Using a Domino Directory

WebSphere Portal can be configured to use the Domino directory for authentication. This is the simplest environment since the user credentials and passwords are derived from the same source and thus are the same when accessed either via LDAP or from within Domino. IBM recommends that you use this directory configuration if you will make use of the Lotus Collaborative Components (e.g., the Notes mail, Notes view portlets).

Apart from configuring Domino to provide LDAP access to the Domino directory, you will have to ensure that the directory includes the necessary groups and users for WPS use and that it is configured to allow anonymous access to certain fields in the Person and Server documents. For Domino 6.x versions, the configuration will also include adding a new attribute to the Domino LDAP schema database. Consult the current WebSphere Portal InfoCenter and Readme file for more information on configuring Domino directory for WPS use.

Lotus IM and QuickPlace servers can be configured to use the Domino Directory either natively or via LDAP.

Using a Non-Domino Directory

WebSphere Portal supports IBM Directory Server, Microsoft Active Directory, iPlanet, and Novell eDirectory. Because LDAP is used for directory access, other directory software, such as OpenLDAP, may be used as well (although not officially supported). In this scenario, Domino must be configured to use the separate LDAP directory for at least user authentication. In the same way as for WAS directory sharing, this can be done via the Domino Directory Assistance feature.

In order for mail to route inside of the Domino environment there must be directory records containing mail information such as mail file and mail server for each user. This e-mail related information can be specified in a Person document in the Domino directory or in entries using an appropriate schema in an LDAP directory.

Domino and Directory Assistance

The Domino server has a mechanism for linking to an external LDAP directory, called Directory Assistance (DA). This mechanism allows the Domino server to search one or more LDAP directories after searching the Domino directory for user authentication. DA is enabled by creating a Directory Assistance database from the provided template, entering the name of the database in the Server document, and then filling out a document in the DA database.

Group and user management can be done entirely from the LDAP directory. References to users' names in Domino database ACLs, group membership, and other contexts can be specified as the user's LDAP DN. The only difference is that the commas used in the LDAP DN must be replaced with the "/" character when entered in Domino. For example, "cn=Elizabeth Somebody,ou=users,o=YourCo," must be entered as "cn=Elizabeth Somebody/ou=users/o= yourco.com."

Both the QuickPlace and Lotus IM environments should be configured to authenticate against the same directory as the WebSphere Portal. With QuickPlace, the LDAP directory is accessed directly, without the need for Directory Assistance. Early versions of Lotus IM relied on Directory Assistance to get information from LDAP, and it was recommended that Lotus IM should be configured to use the native Domino directory. However, the most recent version of Lotus IM can obtain user information directly from an LDAP directory.

The use of Directory Assistance works in environments where the same users will not be listed in both the LDAP and Domino directories. If Domino is only being implemented to support QuickPlace and Lotus IM, this type of configuration can be used. In more complex environments with existing Domino infrastructures, where more Domino function will be used, it is necessary to configure multiple directories.

The Multiple Identities Problem

Although it is the natural choice for customers with existing directory infrastructures, the scenario in which separate directories are used for WebSphere Portal and Domino leads to a situation we refer to as the Multiple Identities problem. This problem often surfaces when WPS is built on top of an existing Domino installation and has caused consternation among the early adopters of WPS and Lotus collaboration components. Let's look at the nature of this problem more closely in order to understand how to configure the various parts to resolve the associated issues.

We will begin our discussion of the Multiple Identities problem by defining an example environment that parallels what is found in a typical environment. Our example company is called YourCo. YourCo has a master LDAP directory with users existing in branches for each city under the OU for users. For our example employee, Elizabeth Somebody, we have the following information in her LDAP directory entry, which we refer to as her LDAP identity.

Table 12-4 Example LDAP Directory Entry

Field	Value
Dn	Elizabeth Somebody,ou=Boston,ou=users,dc=yourco,dc=com
Uid	esomebody

In the existing Domino environment, Elizabeth's Person document contains the following values, which we will refer to as her Domino identity.

Table 12-5 Example Domino Directory Entry (Person Document)

Field	Value
First Name	Elizabeth
Last Name	Somebody
User Name	Elizabeth Somebody/Boston/YourCo
Elizabeth Somebody	Short Name Somebody

Let's say Elizabeth uses Lotus IM within a portal. Elizabeth logs into the WebSphere Portal using her LDAP uid, esomebody. As part of the authentication process, the WAS retrieves her full DN from the LDAP directory and uses it to build an LTPA cookie. When she accesses a WebSphere Portal page as an authenticated user, the Lotus Collaborative Component (LCC) of the WebSphere Portal collects her user information (common name, LDAP uid, and DN) from WebSphere Portal. LCC then attempts to initiate a Lotus IM Links connection with the Lotus IM server. In establishing this link, the common name is supplied to Lotus IM to initialize and provide people awareness for Elizabeth.

When Elizabeth accesses a WebSphere Portal page that contains a Notes portlet, her DN is passed to the Domino server over HTTP. The Domino server then searches the Domino Directory for the DN, where it will not be found (at this point, the LTPA cookie is not used).

The seemingly obvious solution is to enable Directory Assistance on the Domino to include the LDAP server in the authentication path. When this is done, the search for the DN will include the LDAP directory, where a match will be found and the user authenticated. The problem is that Elizabeth is authenticated as "cn=Elizabeth Somebody,ou=Boston,ou=users,dc=yourco, dc=com," and Domino does not know that this is the same user as "Elizabeth Somebody/Boston/YourCo." Everything that Elizabeth does in the Domino environment will be done as her LDAP identity, not her Domino identity. For example, when Elizabeth sends mail, the From field will contain her LDAP DN. She will be able to address and send mail, but if a recipient replies to the mail they will receive a "name not found in directory" error because there is no Domino Person document with the exact LDAP DN.

Furthermore, when Elizabeth creates documents in Domino databases, the author metadata will contain her LDAP DN. If a view is sorted on the Authors field, documents she creates through WebSphere Portal will sort in a different location than documents created through the Notes Client or than those created when accessing the Domino application directly from a Web browser.

Elizabeth will not have access to databases where her Domino identity is listed in the ACL, either explicitly or through group membership. The same applies to documents where there are restrictions through either reader or editor fields.

Because of these limitations, we are forced to abandon the use of DA as a solution to this problem. What is really needed is a method to map the LDAP identity to the Domino identity. Fortunately the Domino server provides this for us.

The User Name field of the person document is a multi-valued field and can contain as many variations of the user's name as needed (although, as a best practice, each of these entries should be unique in the directory). When a match is found against one of the entries in this field, Domino maps the identity to the first entry in the field for the purposes of authorization. This is normally the hierarchical version of the user's name. To take advantage of this capability and overcome the issues discussed here, we need to turn off DA and then add the user's LDAP DN (replacing commas with "/") after the second line of the User Name field.

Now when the Domino server receives the full DN, it searches the Domino directory. This initial search still fails. If DA were enabled on the server, DA would now search the LDAP directory, and you would end up with the same problems you had earlier. Instead, it examines all of the entries of the User Name field and finds a match of the full LDAP DN (with the commas translated to /). The user's LDAP DN is then associated with the first entry, and the user is authenticated as the Domino identity. All of their actions in the Domino environment are now carried out with their Domino Identity and those in the WebSphere Portal environment done with their LDAP identity.

This same solution works when the LTPA cookie is used. Since the LTPA cookie is created with the user's LDAP DN, the Domino server will find that name in the User Name field and associate it with his or her Domino identity.

Configuring to Avoid the Multiple Identities Problem

The first thing to do is configure the Domino environment to establish the LTPA relationship between the WebSphere Portal and Domino servers. Directory Assistance is turned off by default. However, if you are using DA to cascade multiple Domino address books, it can be left enabled—just do not include the WPS LDAP directory in the Directory Assistance search path.

In each user's Person document you will need to add at least one entry to the User Name field. The first line of this field should already contain the full hierarchical Domino user name. The second entry should be the common name of the user (cn) (First Name, Last Name). These entries are set up by Domino when a user is registered and should be left as is. Additional entries, such as other forms of the common name, maiden name, and nicknames, may exist and are ok. After that you should include the user's LDAP DN in Domino format ("/" replacing commas).

In the preceding example, the user's short name in LDAP matched the one in the Domino directory. If this is not the case, then the LDAP uid must also appear in the User Name field of the Person document. The same applies to other LDAP attributes, such as common name (cn). If the common name of the LDAP entry is different than the second line of the User Name field, the common name must be added to the field to allow people awareness to work.

The following table illustrates the different values needed in the Domino User Name field.

Table 12-6 Example Directory Entries for WPS

LDAP Entry	Old Domino Entries	New Domino Entries
cn=Elizabeth Somebody,ou=Boston, o=YourCo uid=euser	User Name = Elizabeth Somebody/Boston/YourCo Elizabeth Somebody Beth Somebody Shortname = esomebody	User Name = Elizabeth Somebody/ Boston/YourCo Elizabeth Somebody Beth Somebody cn=Elizabeth Somebody/ ou=Boston/o=YourCo Short Name = esomebody
cn=Elizabeth Somebody,ou=Boston, o=YourCo uid=y32483	User Name = Elizabeth Somebody/Boston/YourCo Elizabeth Somebody Beth Somebody Shortname = eSomebody	User Name = Elizabeth Somebody/ Boston/YourCo Elizabeth Somebody Beth Somebody cn=Elizabeth Somebody/ ou=Boston/o=YourCo y32483 Short Name = esomebody
cn=Beth Somebody, ou=Boston,o=YourCo uid=euser	User Name = Elizabeth Somebody/Boston/YourCo Elizabeth Somebody Shortname = esomebody	User Name = Elizabeth Somebody/ Boston/YourCo Elizabeth Somebody Beth Somebody cn=Beth Somebody/ ou=Boston/o=YourCo Short Name = esomebody

Since WebSphere Portal may pass user name and password for some of the authentications, you also need to synchronize the Internet password between the Domino and LDAP directories. This is absolutely critical for full Domino integration with the WebSphere Portal. Remember, there are still transactions that occur between the Domino Server and WebSphere Portal using the credentials used to log on to WebSphere Portal.

These requirements will require ongoing maintenance. This can either be done manually or with tools such as IBM Directory Integrator.

Lotus IM Configuration

The Lotus IM server should be configured to authenticate against the native Domino Directory to allow people awareness. With version 3.0, the Lotus IM documentation recommends that an LDAP directory be used. However, in a multiple directory environment configuring Lotus IM to use the Native Domino directory provides for more flexible People Awareness.

The specific example is the use of the NotesView portlet. If People Awareness is turned on for a column that contains the common name of a user, either directory (LDAP or Native Domino) can be used. If the column contains the full canonical name of the user, Lotus IM will not be able to determine online status if Lotus IM is using the LDAP directory.

Unlike QuickPlace, which makes LDAP calls to the directory server, Lotus IM uses Directory Assistance when authenticating against LDAP. Since Directory Assistance is necessary for Lotus IM to operate, DA must be configured on the Lotus IM. However, this is the only server where it should be configured. Although Lotus IM awareness will work, you will have the problems discussed earlier when authenticating to Domino applications running on the Lotus IM server.

QuickPlace Configuration

When using QuickPlace 2.08, you should configure the server to use Domino as the directory. When configuring the underlying Domino server for SSO, you must use a special version of the domcfg.nsf file. This version is available from `www.lotus.com/support`. This file contains an updated default domain login document that performs some additional processing to maintain the LTPA cookie. If you use the default domcfg.nsf that is provided with Domino, the LTPA cookie will be overwritten when you access a QuickPlace, and when you return to WebSphere Portal you will need to log in again.

QuickPlace 3.0 should be configured to access the same LDAP directory as the WebSphere Portal. If you are accessing an existing QuickPlace environment that has migrated from an earlier version, this may be a problem. You may need to use the `changemember` command to update the user names in the QuickPlaces to their LDAP version. This will allow users to maintain their current QuickPlace capabilities in the environment.

If you have a requirement that you must use either the Native Domino directory or Domino Directory through LDAP, then you will need to obtain a patch that supports the appropriate name mapping.

SSO for WebSphere Portal and Domino Observations

SSO is rarely simple to implement and must be well thought out. The methods outlined in this section will allow you to configure your WebSphere Portal and Domino environments to leverage their respective strengths. As WebSphere Portal and Domino evolve, other solutions will be available. In more complex environments, it may be desirable to consider the version of the WebSphere Portal Server that includes Tivoli Access Manager.

CHAPTER 13

Clustering

"WebSphere Application Server Network Deployment (WAS ND) V5 provides clustering support that lets you set up multiple instances of the WAS server on one or more systems to increase application server performance and availability."

—Timothy Eden, *e-Pro Magazine*, December 2003

"Domino cluster technology lets a group of Domino servers work together as if they were a single Domino server although the servers have different names and can be used separately even when they're clustered. Domino clustering can operate over a WAN and facilitates transparent failover from one server farm to another."

—John Lamb, *e-Pro Magazine*, November 2002

Both Lotus Domino and WebSphere offer what each calls "clustering." Clustering in Domino, however, is quite different than clustering in WebSphere—even though the basic concept is the same (i.e., to have a group of servers appear as if they are one server). Clustering is a basic way for both Domino and WebSphere to provide high availability (e.g., one server in a cluster can crash, but the cluster is still available and fully functional) and high scalability (e.g., just double the number of servers in the cluster and you've doubled the capacity of the cluster). In this chapter, we discuss how to implement clustering for both Domino and WebSphere. Chapter 14, "Performance and High-Availability Issues," covers the more general topic of building high-performance, high-availability Web systems that can include clustering and many other techniques, such as the use of IP sprayers.

WebSphere and Domino Clustering

If you have a system that includes both WebSphere and Domino, you can use clusters for each, and they can serve somewhat different purposes. This is discussed in the sections that follow.

Cluster Examples

Examples of both Domino clusters and WebSphere clusters are given in the following sections. Two "real world" examples of Domino clustering across the wide area network to provide very high availability and some aspects of disaster recovery are discussed in the "Domino Clustering" section. A "real world" example of using WAS V5 clustering for an IBM high-availability Web conferencing offering is discussed in the "WebSphere Clustering" section.

Domino Clustering

The use of Domino cluster technology can have a significant impact on the performance and reliability of your Domino system. This section gives "lessons learned" and recommendations on the use of Domino clusters across the wide area network based on IBM's extensive experience with the product gained during many large outsourcing projects. IBM Global Services has implemented Domino clustering for many large commercial customers at IBM server farms as part of an outsourcing agreement with these customers, including several large banks where security, reliability, and scalability of the Domino system have been of paramount importance. Domino cluster technology has regularly been used between server farms, across the wide area network, to provide very high-system availability and many aspects of disaster recovery.

Domino clustering works well over a wide area network, and this section emphasizes the availability improvement benefits and also disaster recovery aspects of clustered Domino servers over the WAN. Although Domino clustering is based on "real-time" replication in order to always keep two Domino servers "in sync," scheduled replication, where the requests can be queued up, is usually set up on an hourly basis to solve the problem of the connection between the two clustered Domino servers going down (e.g., the WAN). Thus Domino servers that are cluster replicated over the WAN can tolerate even long disruptions in the connection between the clustered servers, since scheduled replication ("regular" replication) will quickly get the two servers in sync again when the connection is repaired. In our experience, Domino cluster replication over the WAN works very well, and a design where both Domino servers are normally used (i.e.,"active/active") has many advantages.

Domino cluster technology allows a group of Domino servers to work together to appear to the user as if they were a single Domino server. The cluster is used to provide high-availability of data and services, to balance workload, and to significantly increase the number of users your system can support (scalability). Servers in a Domino cluster use cluster replication (real-time replication) to keep all servers in the cluster the same. If one server in a cluster goes down, users are transparently switched to another server in the cluster (failover). Cluster workload balancing ensures that heavily used servers are able to pass requests to other cluster servers and that work is evenly distributed across the cluster. Additional servers can easily be added to a cluster to allow the Domino service, with workload balancing, to "scale" to a very large number of users.

Domino clustering can be used successfully over a wide area network and allows transparent failover from one server farm to the other. Having two widely separated server farms (in different time zones) in addition to providing high availability also provides some aspects of disaster recovery. A "disaster" at one server farm would still leave a fully functional Domino system at the other server farm that would be available immediately and transparently through Domino cluster failover. Domino clustered severs across the wide area network have allowed us to easily migrate Domino applications from the customer to IBM or from one IBM server farm to another server farm without any down-time.

Operating System (OS) clustering also can be used to cluster Domino servers. OS clustering provides high availability but not load balancing. Two popular examples of OS clustering are AIX HACMP and Microsoft OS Clustering (MSCS). OS clustering supports hardware failover for multiple servers, or nodes, in a cluster sharing the same common disk device. This cluster capability differs significantly from the application clustering provided by Domino. HACMP provides failover protection so that if two servers running Domino are set up for failover, and one server becomes unavailable, the second server will come up, use the shared disk subsystem, restart the respective applications, and continue with the processes. It will use the same disk where the Domino data resides and the same IP address that the Domino server is using. Therefore, the failover function is transparent to clients. Unlike Domino clustering, HACMP provides only failover protection. Also if the shared disk device OS clustering is using goes down, the system is down. Thus it is very important to have a mirrored (RAID 1) shared disk for OS clustering.

Domino clustering provides failover and load balancing for Notes clients, but it does not provide Web (HTTP) access. Most Domino servers used for Web applications need HTTP failover, so OS clustering is required. Domino does have the ICM (Internet Cluster Manager) server task that does allow for HTTP failover, but ICM has limitations and is often not the best solution for Domino Web server failover.

Implementing Domino R5 Clustered Servers

Domino clustering was first made available in Release 4.0 as "Notes for Public Networks" working with telecommunication and Internet service providers. This was the technology developed for service providers such as AT&T Network Notes. The release for general customer use was with Domino R4.5, where Domino clustering was part of the Domino Advanced Services license. Clustering ships with the Domino Enterprise Server license in Domino R5. This section will give summary steps for setting up Domino clusters for Domino R5. It is not difficult to set up Domino clustering. Often the biggest problem is deciding which servers and which databases need cluster replication. Often the decision is essentially to make all clustered servers "clones" of each other so that they back each other up completely. Making all servers in the cluster look the same makes it easy to plan which application databases should be clustered—since they all will be.

Requirements

- All servers in the cluster must be running the Enterprise Server license.
- Each server can only be a member of one cluster. The maximum number of servers supported within a single cluster is six.
- All servers in the cluster must support TCP/IP and be on the same Notes name network.
- All servers must be in the same Domino domain and use the same Domino directory.
- Every server in the cluster must have a hierarchical server ID (i.e., any flat IDs must be converted to hierarchical IDs).
- Notes clients must be running Notes release 4.5 or higher to exploit cluster failover. Also the Notes clients must be running TCP/IP.

Creating New Clusters

1. Open the Domino Administrator on an existing Domino server.
2. Click the **Configuration** tab.
3. Expand the Server section and click **All Server Documents**.
4. Select the server(s) to add to the cluster from the results pane.
5. Click **Add to Cluster**.
6. Select **Yes** when prompted to continue.
7. Select **Choose New Cluster** rather than selecting an existing cluster to add the servers to.
8. Enter the name for the new cluster (recommendation—keep the name simple by avoiding spaces, etc.).
9. Select **Yes** to add the servers to the cluster immediately or **No** to submit a request using the Administration Process. Selecting **Yes** is the fastest method and is the recommended option. If the server is part of the cluster, it is immediately replicated to the other cluster servers. If the server is not part of the cluster, replicate the changes to one of the servers added to the cluster.

Of course, you must have the correct access to the Domino servers to create or modify clusters. The minimum access required is Author access and Delete Documents access along with Server-Modifier and ServerCreator roles in the Domino directory and at least Author access in the Administration Requests database to create a cluster. As a result of creating a new cluster, all of the required Domino cluster activities will start, such as the Cluster Administrator, the Cluster Manager, the Cluster Database Directory Manager, and the Cluster Replicator. The Cluster Administrator adds the Cluster Database Directory Manager tasks (CLDBDIR) and Cluster Replicator task (CLREPL) to the ServerTasks setting in the NOTES.INI. The Cluster Administrator also starts the Cluster Database Directory Manager (CLDBDIR) and the Cluster Replicator

(CLREPL). The Cluster Database Directory Manager modifies the Cluster Database Directory (CLDBDIR.NSF) to include documents for each database on the server. All Domino databases on the server will be listed in CLDBDIR.NSF and will be given a status of "enabled." The Domino administrator should "disable" or remove those databases that are not to be cluster replicated. Also the administrator needs to make replicas of any database that needs to be on other servers. Of course, every database that is on the server where you originally created the cluster does not have to be cluster replicated. IBM, for example, has one Domino cluster implementation where 30,000 large databases used for an EDI application are spread over five Domino servers. Five servers are needed to handle the load for the 30,000 databases. However, a typical database exists on only two of the five clustered servers. Domino clustering technology (i.e., the Cluster Manager and Cluster Database Directory Manager) keeps track of which database is where and uses that information for failover and load balancing.

New servers can be added to or removed from the Cluster using the Domino Administrator. For example, to add servers to a cluster: Open the Domino Administrator on an existing Domino server, click the **Configuration** tab, expand the Server section and click **All Server Documents**, select the server to add to the cluster from the results pane, click **Add to Cluster**, select **Yes** when prompted to continue, select the name of the cluster you want to add the serve to, and then click **OK** and select **Yes** to add the servers to the cluster immediately (recommended). To remove servers from a cluster: Open the Domino Administrator on an existing Domino server, click the **Configuration** tab, expand Cluster and click **Clusters**, select the document representing the server to be removed from the cluster in the results pane, click **Remove from Cluster**, and select **Yes** to remove the server from the cluster immediately (recommended).

Implementing OS Clustering for Domino Servers

OS clustering for Domino servers can be implemented with different operating systems. This section will give a summary on implementation steps for using OS clustering with the Microsoft NT/2000 system. This OS clustering is called Microsoft Cluster Server (MSCS). MSCS supports hardware failover for two NT/2000 servers in a cluster sharing the same common disk device. MSCS can be implemented in either an Active/Passive or Active/Active configuration. In an Active/Passive configuration, only one node provides services to the clients at a time. The passive node is reserved for failover if the active node fails. The Active/Active configuration allows both nodes to provide services to clients. With the Active/Active configuration you have to install two partitioned servers on each node. For simplicity, this section will give steps for the Active/Passive MSCS configuration. MSCS clusters were supported starting with Domino release 4.6.2; however, these steps will assume Domino R5.

Requirements

- All Domino servers and clients must be configured to use TCP/IP.
- The Domino program files must be installed on a local, non-shared drive for each node in the MSCS cluster.

- The path for the Domino program files must be the same for both nodes in the cluster (e.g., c:\lotus\domino\).
- The Domino data directory must be installed in a shared drive. Shared drives are drives that both servers can access. Typical shared drives are based on NAS (Network Attached Storage) or SAN (Storage Area Network) technology.
- Each Domino server must be assigned a static TCP/IP address, distinct from the address of the cluster nodes. Thus there are at least two TCP/IP addresses active on the Windows 2000 server that will run Domino. These are the server's public IP address and the virtual server's IP address, created as a cluster resource and reserved for Domino server usage. The virtual server's IP address is in the NOTES.INI (the NOTES.INI is put in the shared disk drive in the domino\data directory).
- The server ID cannot have a password, or the Domino server restart or failover will not complete because the Domino server will wait for a password prior to initializing and processing client requests. When the server ID is created, set the password length to zero.
- Prior to completing the MSCS configuration, the following groups/resources must be created: Domino Server Resource Group(s), IP Address Resource(s), Network Name Resources(s), and Assign the physical disk resource to the Domino resource group, Domino Server Resources. Use the Microsoft Cluster Administrator program, which installs automatically on both nodes in a Windows 2000 cluster. To start the administrator, select **Start>Programs>Administrative Tools>Cluster Administration**. Create the resource group for the Domino server by selecting **File>New>Group**.

Installing Domino on the First and Second Nodes

1. Use the MSCS Cluster Administrator to move the group(s) containing the shared drives to be used for the Domino data directory to the first node.
2. Launch the Domino Server installation program and proceed through the license terms, etc.
3. Specify the destination folder for the program directory as a private, non-shared drive.
4. Select **Customize** when selecting the type of Domino server and enable Domino as an NT service.
5. Specify the destination folder for the data directory as one of the shared drives of the cluster.
6. Select the program folders to which to add the new installation.
7. Complete the installation program.
8. Move the NOTES.INI file from the Domino program directory to the Domino data directory of the server.

9. Follow the preceding steps to install Domino on the second node. Specify the same directory for the program files and data directories.

Configuring the Domino Servers

1. Open the Domino Administrator on an existing Domino server.
2. Click the **Configuration** tab.
3. Register the Domino server by clicking **Registration>Server** in the Tools pane. The server name must be the same as the network name resource defied in MSCS. Enter the password length of zero.
4. On the second cluster node, start the Domino server, which will launch the configuration program.
5. Complete the Domino configuration.
6. Exit and restart Domino. This launches the server.

Now the Domino servers are ready to run under MSCS. From the Cluster Administrator, right-click the Domino server group, and select **Bring Online**. You can stop Domino by using the Cluster Administrator. Right-click the Domino server group and select **Take Offline**. When the Domino server is running with MSCS, the Notes clients must be configured with TCP/IP.

You can also choose to implement MSCS and Domino clusters on the same system. This solution can be very attractive, for example, if there are several geographical sites in the Notes network and when high availability is required at each site. Figure 13-1 shows an example of a system with both MSCS clusters and Domino clusters. In this example there are two sites connected over a wide area network. There are four Domino servers in the network with two MSCS clusters and two Domino clusters. At Site 1, the two Domino servers have been configured as Active/Active MSCS clusters. Likewise, at Site 2, the two Domino servers have been configured as Active/Active MSCS clusters. The Domino clusters combine servers across the wide area network. Server 1 at Site 1 and Server 3 at Site 2 form one Domino cluster while Server 2 and Server 4 form the second Domino cluster. In this setup, there is no cluster replication traffic between the local nodes, and Microsoft Cluster service provides fault tolerance for the servers on one site. Domino cluster replication is used between the two sites over a high-speed WAN link. When the servers are configured in this way, even if both servers on Site 2 go down, client requests will be forwarded to the replica databases at the server on Site 1 over the high-speed WAN link. This example indicates that with the benefits of both OS and Domino cluster technology, the two different clustering technologies together can be complementary and provide high availability.

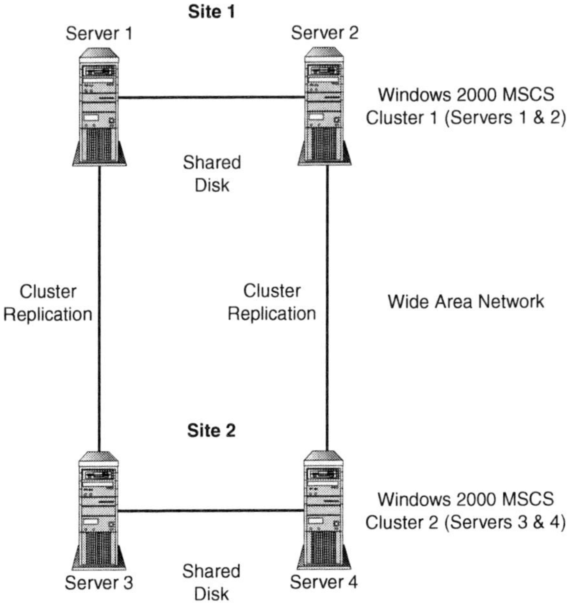

Figure 13-1 Domino clustering combined with MSCS clusters.

Living with Domino Clusters

As mentioned earlier, each Domino cluster consists of up to six Domino servers, within the same Notes domain. A domain may contain multiple clusters, but an individual server may only be a member of one cluster. Replication within the cluster is supported only with the TCP/IP protocol, so TCP/IP must be one of the protocols available to your Domino system if you use clusters. TCP/IP is also recommended for your Notes client connections because it will provide better failover performance than other protocols. When a Notes client connects to a server using a specific protocol, the client must use the same protocol when it fails over to another server in the cluster. If you're also using Domino as a Web server, then TCP/IP is required for all Web client access, and you might as well just use that protocol. Domino clusters are controlled by the Cluster Manager, the Cluster Database Directory (CLDBIR), the Cluster Database Directory Manager, and the Cluster Replicator. OS clusters (e.g., MSCS) are often limited to only two servers, and in the Active/Passive mode, only failover is available—not load balancing.

Database Compaction with Clustered Servers

With R4 of Domino, the server needed to be taken down to compact names.nsf and log.nsf. However, with Domino R5 at the IBM server farms, we regularly compact all databases (including names.nsf) while the server is up. We do bring down some of our R5 NT Domino servers to compact because of the length of time involved. Because of the size of the databases, keeping the server up took all day for the compaction process. We now bring the Domino server down (one of two clustered servers) and run a batch file that runs four compact processes simultaneously. The process still takes eight hours to complete. We take the Domino servers at one server farm down on Saturday morning for compaction and the Domino servers at the other server farm on Sunday morning so that there is no system down-time. With OS clustering of Domino servers, there is only one copy of each database (on the shared disk system), so you do not have the luxury of bringing one of the clustered pair of servers down for compaction while keeping the Domino system available. Bringing one server down at a time for compaction of databases is a definite plus for Domino clusters.

Planning, Installation, and Setup

Planning, Installation, and Setup for clusters in your Domino system requires knowledge of such things as database size, network bandwidth, server capacity, the number and location of expected users, database usage, and volume of data. For best results, use the Administration server to create new clusters. You must have at least Author access and Delete Documents access in the Domino Directory and at least Author access in the Administration Requests database to create a cluster. Some of the planning factors to consider are

- **Database size.** For larger databases, you may want to create fewer replicas.
- **Bandwidth.** If you have minimum network bandwidth, you'll need to minimize replication between databases.
- **Server capacity.** If the Domino servers are running on high-end servers, they will better handle additional replicas.
- **Number and location of expected users.** For larger audiences, they will probably experience greater performance if the usage is dispersed among multiple replicas.
- **Database usage.** If you expect a high usage/utilization for a database, additional replicas will probably improve performance. However, usually creating more than three replicas of a database is overkill.
- **Volume of data.** If the number of documents and the size of the data contained in the documents are expected to be large, creating multiple replicas can decrease performance. You will have to factor the server performance and LAN configuration to determine if bandwidth is a concern.

Load Balancing

Load balancing is available to provide administrators the capability to define threshold values to redirect users in order to distribute the load so that no servers become overloaded. When a server reaches its availability threshold (the Cluster Manager marks the database BUSY) or if the server reaches its maximum users limit, subsequent requests to the database will be redirected to another database replica within the cluster. The Cluster Manager uses the Cluster Manager Database Directory to determine the location and availability of the database replicas prior to redirection. If there are no other database replicas within the cluster or all the other servers have a BUSY status, the original database is opened. The availability threshold does not affect replication; therefore, replication will occur even if the database is currently in a BUSY state. Again, since OS clusters (e.g., MSCS) are often limited to only two servers and are set up in Active/Passive mode, so there is no load balancing available.

There are two basic ways for automated load balancing based on server workload:

- **Setting the server availability threshold parameter.** By setting the server availability threshold, you decide when your server becomes BUSY. When the server enters the BUSY state, it load balances new users' requests to another server in the cluster. The NOTES.INI parameter `Server_Availability_Threshold` can be set to a value in the range of 0 to 100. Each server in a cluster periodically determines its own workload, based on the average response time of client requests recently process by the server. The value of 100 indicates a lightly loaded server (fast response times), and 0 indicates that the server is heavily loaded (slow response times). You can find out the current value computed for a server with the SHOW CLUSTER Domino console command. It is recommended that if you want to use this type of load balancing, start off with a conservative number such as 80. This means that when the server availability threshold drops below 80, additional users will be load balanced to another server. If users report unfavorable response time on a server with a threshold setting of 80, change the setting to 85 to help improve the load balancing.
- **Setting the maximum number of users parameter.** Limiting the number of user connections that a server will support is another way to distribute the workload around the cluster. This is done by setting the NOTES.INI parameter: `Server_MaxUsers`. For example, you could set `Server_MaxUsers=100`, and then the 101st user would be load balanced over to another server in the cluster. A setting of 0 signifies an unlimited number of active users.

With Domino clustered servers, end users can also determine load balancing. For example, as an end user you can have on your Lotus Notes desktop icons for the same database from two or more servers. If you are dissatisfied with database performance on one server, try the same database on another server. Another type of "manual" load balancing would be to have everyone in one group use Server A as their primary access to a database, while everyone in a second group would use Server B as their primary access to a database. This type of manual load balancing, although far from ideal, is still much better than having everyone accessing a database on Server A and using Server B only during automatic cluster failover (i.e., where Server A and Server B are essentially in Active/Passive mode).

Also, Domino Administrators can manually set a server as BUSY or a database as "out of service," which is another type of manual load balancing. To set a server as BUSY, the administrator could go to the Domino console and type set config `Server_MaxUsers=n` where n is a number less than the current number of active users on the server. The Cluster Manager monitors this setting, and if the number of active users reaches the maximum number, the server state changes to the MaxUusers state. A server in the MaxUsers state will not allow any additional user sessions to be established on the server. Users attempting to access a server that has reached its maximum number of active users will failover to a replica on another server in the cluster. To set a database as "out of service," the administrator opens the Domino Administrator on an existing Domino server, clicks the Files tab, expands the Directory to locate the data database and select the database in the results pane, selects Cluster in the Tools pane or right-mouse clicks the database and selects Cluster, and selects the "out of service" status from the Manage Clusters dialog box. Marking a database "out of service" will force users to failover to another replica database. This might be done because you are temporarily performing maintenance on the database, or it might be done to reduce some of the load on the server.

Using Scheduled Replication to Backup Cluster Replication

Since cluster replication is event-driven and database updates are stored in memory, there is a small chance that some database updates may be lost in the event of a network or hardware failure. In order to solve this problem, Lotus recommends that you configure scheduled replication within clusters at least once a day as a backup to cluster replication. At IBM server farms we typically configure scheduled replication on a one-hour basis for clustered databases. This has proven to work quite well for our high-availability servers and databases that are clustered across the wide area network between server farms. Since there is no cluster replication with OS clustered Domino servers, scheduled replication to backup cluster replication is not an issue.

Using Domino Clusters Across the WAN for Disaster Recovery

Domino servers that are clustered across the wide area network present an opportunity to use these systems for disaster recovery. This section gives two "real world" examples of using cluster replication across the wide area network for disaster recovery.

Example 1—PC Banking High-Availability Domino System

This PC Banking case study gives a highly successful example of Domino clustering across the wide area network for a very high-volume, high-availability system. The PC Banking system involves 16 Domino servers at two of IBM's server farms called Universal Server Farms (USFs). The USFs for this case are at Schaumburg, Illinois, and West Orange, New Jersey. Domino was chosen for this banking application both for security reasons and for Domino clustering, which has allowed us almost unlimited scalability. Each server farm contains eight Domino servers, which are clustered across the two USFs for high availability. The bank's 800,000 PC Banking customers use a Web browser with access through the Internet to the bank's 20 Web servers (running Sun Solaris) called Terminal Controllers (TCs). This system provides high availability and security through the use of Domino.

The PC Banking system has identical configuration at each of the two server farms. Load balancing between server farms is directed by TC Web servers. Server capacity is such that either server farm will be able to handle 100 percent of the load in case of lost access to one of the farms. Cluster replication between the farms is used to keep the databases on the servers at each farm in sync. This architecture for PC Banking at the farms is shown in Figure 13-2.

The Customer TCs use the Notes API via IBM's commercial Virtual Private Network (VPN) to access the Domino servers at IBM to send requests to the bank's 450 Customer Service Representatives (CSRs) using the Domino workflow application. The CSRs respond to customer requests (e.g., new checkbook, transfer between accounts, bill payments, etc.) by accessing the clustered Domino servers with a Notes client. Load balancing for this clustered system is manual (i.e., the CSRs have icons for workflow databases for both clustered servers on their Notes workspace and can choose either server). Domino clustering allows transparent failover from one server farm to the other. Having two widely separated server farms (in different time zones) in addition to providing high availability also provides some aspects of disaster recovery. A "disaster" at one server farm would still leave a fully functional Domino system at the other server farm that would be available immediately and transparently through Domino cluster failover. The benefits of having clustered servers across the wide area network cannot be overemphasized.

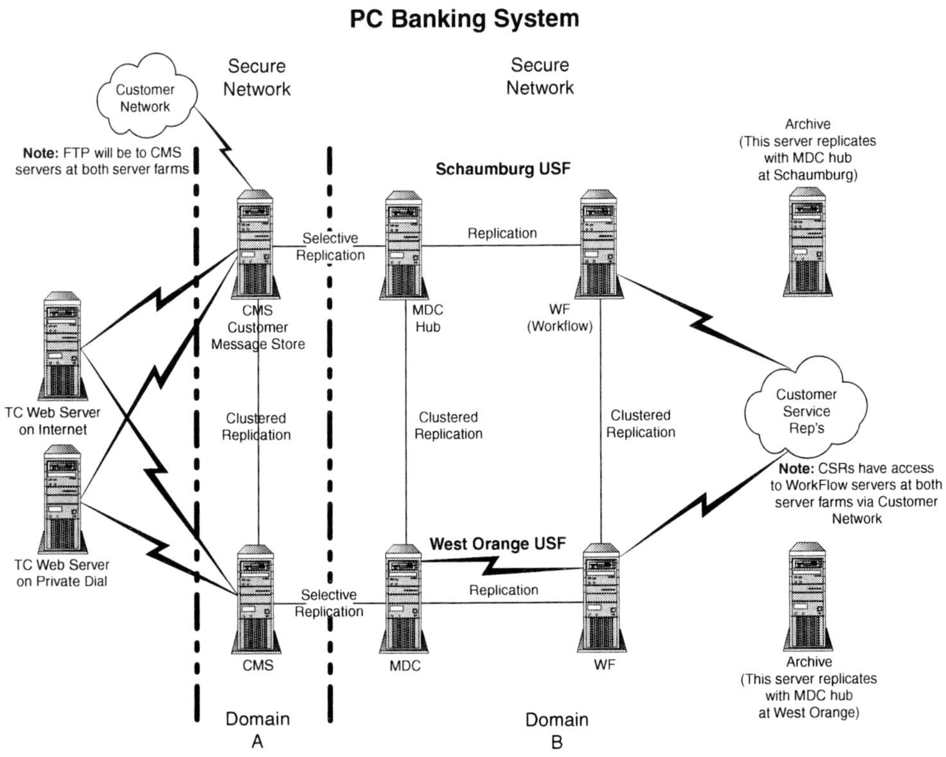

Figure 13-2 PC Banking architecture with Domino clustering over a wide area network between IBM server farms.

The only significant concern that IBM has found for servers clustered across the wide area network is network bandwidth. For the PC Banking example, discussed above, network bandwidth between the IBM server farms was not a big concern since the documents involved in cluster replication were text messages and small in byte count even though document changes are frequent and of high volume. T1 bandwidth (1.544Mbps) has proven to be more than adequate to support all cluster replication between the two sites (four Domino servers at each site). If the databases being cluster replicated have changes that tend to include a large number of bytes (e.g., graphics files), then network bandwidth requirements across the wide area network must be carefully determined.

Example 2—Domino 6 Mail Servers Clustered over the Wide Area Network

This example of using Domino clusters across the wide area network involved migration to AS/400 Domino Mail Servers as part of a consolidation from NT Domino servers.

Working from a "Statement of Work," the combined IBM/Customer team made several decisions on how best to quickly proceed with migration of all 15,000 mail users in the United States to two large iSeries (AS/400) servers at Secaucus, New Jersey, and West Des Moines, Iowa. The important decisions included the need to leverage the performance of the new servers as much as possible (by using cluster replication across the WAN—with both servers sharing the load and using Domino 6) and at the same time minimizing any migration impact on end-users (by keeping the same Domino server names and IP names (not numeric IP addresses) on the iSeries servers and not changing the Notes level software on any of the client workstations). The customer Project Manager for this endeavor created a detailed Project Plan giving dates for the migration of each customer location to the iSeries servers. Included in the plan were fallback procedures. For example, if we found during testing with 30 to 40 IT personnel that Domino 6 was not stable or reliable enough to continue the project with that release of Domino then the fallback would be to use Domino R5.0.11 and continue the migration. Likewise, the recommendations that follow state that the migration of locations to the iSeries should not require any network changes. However, the Project Plan includes contingency plans, in case network changes are required to make the migration for some sites to the iSeries a complete success.

The following recommendations were used to configure the iSeries Domino servers to begin the migration. Details on the reasons for these recommendations/decisions are given in the next major section of this chapter. Where appropriate, recommended implementation details are also given.

- Use Domino Clustering across the WAN for iSeries Servers and load balance based on geography (active/active configuration).
    - Make use of iSeries Domino server configuration information based on experiences with Credit Card Company and Canadian Insurance Company.
    - Maintain Domino server names and IP names when consolidating servers to iSeries in order to make migration transparent to end users.

- Use Domino 6 on iSeries boxes for all U.S. Mail Servers to significantly improve Network and Server performance.
  - Use Domino 6 Streaming Replication and Network Compression over the WAN between iSeries Servers.
  - *Do not use the two major iSeries servers for Domino applications* since all partitions on the iSeries will be Domino 6, and the iSeries servers should not be LPARed. Application servers should be consolidated with the same level Domino as currently used (mostly R5).
- Use separate NIC for Cluster Replication and start implementation of mail on iSeries with current network infrastructure (i.e., *no migration dependency on network enhancements*).
- Encourage replication of user mail files to each workstation to greatly improve end-user experience and reduce both network bandwidth requirements and server CPU requirements.
- Use BRMS backup and restore information from experiences as input to customer backup and restore strategy.
- *Do not plan on upgrading or reconfiguring Notes client software during the mail server consolidation.* By maintaining all Domino server names and IP names when consolidating on the iSeries servers, there will be no need for any changes to the Notes client software. The only exception will be for 30 or 40 users in the customer IT group who will be involved in pilot testing before any actual consolidation.
  - Many Notes/Domino 6 improvements for the end-user will have to wait until client software is updated. However, just upgrading the server Domino software in the initial consolidation is the best procedure to greatly reduce consolidation risk and to greatly speed the initial consolidation effort for all U.S. mail on the iSeries servers.

Network Bandwidth Required for Cluster Replication Over the WAN for Domino Mail Servers

IBM's "rule of thumb" for supporting clustered Domino mail servers across the WAN is that one T1 will handle 1000 registered users. Thus if there are 15,000 registered users on the two AS/400 (iSeries) mail servers, the IBM rule of thumb would be that 15 T1s would be needed to support the cluster replication traffic (about one-half the bandwidth of the 45Mbps T3/DS3 between the two iSeries servers). Cluster replication traffic would be required no matter how the 15,000 registered users are load balanced across the two servers (e.g., 7500 on each or 15,000 on one server and none on the other) since the databases on the two servers have to stay in sync; however, the cluster replication traffic would be more balanced (i.e., balanced traffic in each direction) if the servers are load balanced. Use of Domino 6 with "Streaming Replication" and "Network

Compression" should further reduce network requirements for cluster replication by up to 50 percent. By far the best way to verify network bandwidth requirements for cluster replication is to measure and monitor actual traffic and then project future requirements based on user growth.

Disaster Recovery Aspects of These Domino Clusters Over the Wide Area Network

Domino servers that are clustered across the wide area network present an opportunity to use these systems for disaster recovery. The preceding iSeries discussion is a "real world" example of using cluster replication across the wide area network for disaster recovery.

This example of using Domino clusters across the wide area network actually goes back to 2001. At that time the project involved migration to AS/400 Domino Mail Servers as part of a consolidation from NT Domino servers. At the time of our study, most mail and application servers were on NT R4.6.5 platforms with five-year-old hardware. There was one AS/400 test server at One World Trade Center in New York City. IBM Rochester, Minnesota (the "home of the AS/400") was involved in the sizing of the AS/400s required for the mail users. An AS/400 workload estimating tool was used for this effort. The latest Gold release of Domino (R5.0.8 or later) would be used for the AS/400. We told the customer that for one of our IBM Global Services outsourcing contracts with a major commercial bank, we found significant Domino NT server performance improvements when moving from Domino R4.6.4a to R5.0.7. For this customer, just upgrading the release of Domino reduced CPU utilization by half and significantly improved several performance aspects of the bank's applications. We therefore told this customer that moving from R4.6.5 to R5 would significantly improve their Domino server performance whether the upgrade is on the NT or AS/400 platform. The strategy for deploying AS/400 servers was to have consolidated AS/400 Domino mail servers in New York City and Des Moines, Iowa, and these servers would be connected with redundant T3 (45 Mbps) bandwidth.

The architecture for this design is given in Figure 13-3. This design allows for failover, load balancing, and disaster recovery for all the customer's Lotus Notes mail users in the United States.

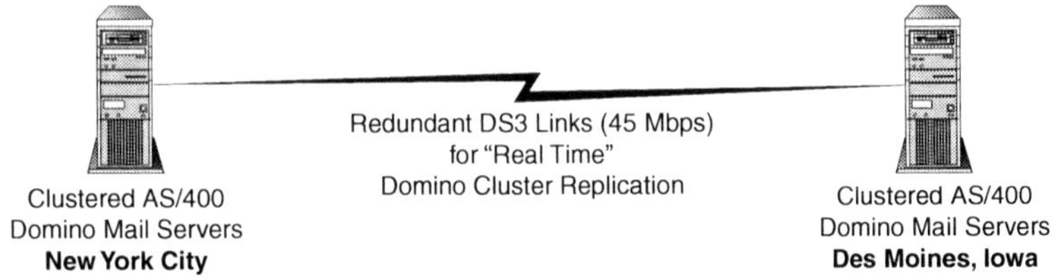

Figure 13-3 Domino Mail System clustered over a wide area network for disaster recovery.

> **NOTE**
> The tragedy at the World Trade Center happened the day after IBM gave a presentation to the customer on our recommendations. Thus, part of our work with the customer was spending over a week helping them restore the Domino servers destroyed in the World Trade Center disaster. Backup tapes had been stored off-site, and those were used to help restore the Domino servers. Clustered Domino servers over the wide area network, although implemented too late for the WTC disaster, will provide immediate disaster recovery in the future. The project described in this section, now implemented with Domino 6, has this disaster recovery system in place.

Conclusions

The benefits of clustered Domino servers can be realized either with Domino clustering or OS clustering. However, only Domino clustering allows two clustered servers to be separated by a wide area network, hence Domino clustering is necessary in the design of Domino systems for immediate disaster recovery. OS clustering will not work with that scenario because with OS clustering, two servers share a hard disk array and thus must be fairly close together. IBM has used OS clustering of Domino servers in cases where HTTP failover is required as part of an overall Web server system, and this has worked well. With bandwidth becoming less expensive, disaster recovery designs with Domino clustered servers over the wide area network are becoming more and more feasible. In addition, Domino 6 will produce significant reductions in network bandwidth requirements through data compression and data streaming techniques.

WebSphere Clustering

WebSphere Application Server Network Deployment (WAS ND) V5 provides clustering support that lets you set up multiple instances of the WAS server on one or more systems to increase application server performance and availability. WebSphere Application Server (WAS) clustering increases application availability by allowing an application to continue running without interruption even when one of the WAS instances, or an entire server in a multi-server configuration, fails. Setting up WAS V5 clustering is fairly straightforward, but this section will give you some helpful hints on how end up with the most effective solution.

WAS Cell and Cluster Overview

First, we'll review some V5 concepts, keeping in mind the WAS clustering architecture has changed significantly since V4. A WAS-managed application server is simply an instance of WAS running on its own JVM in its own operating system process. In this section we will use the simpler term, WAS instance. With WAS ND, you can run multiple WAS instances and manage them together as a single cell. (Don't be confused by this term. A WAS cell isn't an elemental artifact; rather, it's the most encompassing level of a WAS clustered topology.) Each cell has a single master configuration repository that contains information for all WAS instances that are part of the cell. The scope of a cell is also referred to as an administrative domain.

There is one Deployment Manager process running for the cell, which provides a central point of administrative control for all the cell's WAS instances. The Deployment Manager maintains the master configuration repository and propagates this information throughout the cell, as explained next. (With the WAS base product, each WAS instance is independently responsible for managing its own configuration information.)

A node comprises one or more WAS instances from the same cell running on the same server (i.e., physical computer). A server can have more than one node. Each node has its own local configuration repository, which contains information for all WAS instances that are part of the node. There is also a node agent process running for each node, which runs on the same server as the node and provides a local point of administrative and process control for all the node's WAS instances. A cell's deployment manager communicates with all its nodes' respective node agents to propagate and synchronize configuration information across the cell. However, even when there is temporarily no communication between the deployment manager and a node agent, the node agent can continue to coordinate execution of the node's WAS instances using the local configuration repository.

A cluster is a set of WAS instances within a cell that have the same applications deployed on them. A WAS instance that's a member of a cluster is sometimes referred to as a clone. A cluster's WAS instances may all be in the same node (or in several nodes all on the same machine); this topology is known as vertical clustering. If each WAS instance in a cluster is from a node on a different machine, the topology is called horizontal clustering. WAS clustering also supports a mix of these two topologies; for example, a cluster might include two WAS instances from one node and a third instance from another node on a different machine.

Vertical clustering alone provides protection against operating system or WAS software failures that disable a single WAS instance in the cluster. Horizontal clustering provides additional protection against system hardware failures that don't bring down all the servers in a cluster. Of course, horizontal clustering requires multiple machines and may cost more as a result. Both vertical and horizontal clustering can be good strategies to increase performance.

There are other potential failure points to consider when you design your complete topology. For example, you may run one or more Web (i.e., HTTP) servers on different systems than you use for WAS. You have to consider similar issues for these Web servers, such as whether to run multiple copies of them on one or more systems, as well as what type of network redundancy you should implement so the Web servers can still communicate with WAS instances even when a network connection fails.

Setting Up WAS Clustering

Before we look at installation and configuration details, it may help to take a high-level view of the steps required to get an application running on a WAS cluster. For example, in order to set up horizontal cluster on two computers, Server1 and Server2, here are the basic steps:

1. Install and configure prerequisite hardware, software, and network connections.
2. Install the IBM HTTP Server (IHS) on Server1. This can be done as part of Step 3, when you install WAS on Server1. (For a production environment, however, you would normally install IHS on one or more machines other than the ones used to run WAS.)
3. Install the WAS base product on Server1. During installation, name the newly created node Node1.
4. Install the WAS base product on Server2. During installation, name the newly created node Node2.
5. Install the Deployment Manager (from the WAS ND installation CD) on Server1 (or Server2, but only on one or the other). During installation, name the new cell Cell.
6. Install current WAS base product and WAS ND fixpacks on Server1 and Server2.
7. At the command line on Server1, execute an `addNode` command to add Node1 as a node in Cell.
8. At the command line on Server2, execute an `addNode` command to add Node2 as a node in Cell.
9. Use the WAS Administrative Console to define a new cluster, named ClusterH, within Cell.
10. When you create ClusterH, specify two new WAS instances to create and add to ClusterH: WAS1, which you specify as a member of Node1; and WAS2, which you specify as a member of Node2.
11. Use the WAS Administrative Console to configure virtual host(s) (e.g., `default_host`) to accept HTTP requests on the ports used by WAS1 and WAS2.
12. Use the WAS Administrative Console to generate a new Web server plug-in configuration file and copy this to the proper directory for IHS.
13. Deploy your enterprise application(s) on ClusterH.

You need to license the WAS ND product (or WAS Enterprise, which contains WAS ND features) to implement clustering. The WAS ND package includes the CD for the WAS base product (which you use for Steps 2, 3, and 4 above) and a CD for the "ND" enhancements, including the deployment manager (which you use for Step 5). IBM Getting Started manuals and the WAS V5.0 System Management and Configuration Redbook provide detailed guides to the installation process, so those resources should be used in addition to this section which provides some advice in several areas based on experience in setting up a cluster.

For a Windows 2000/NT environment, be sure to decide before the installation whether you want IHS and/or WAS to run as a service and which Windows 2000/NT user(s) you want the service(s) to run under. For WAS, be sure the user has the following rights:

- Logon as a service
- Act as part of the operating system

To modify user rights, go to the Windows Control Panel and select Administrative Tools>Local Security Policy>Local Policies>User Rights Assignment. For a Solaris environment, you must create the mqm and mqbrkrs users before installing WAS. These users are necessary in order to get the Deployment Manager to work properly with the node agent.

It's typical for a high-availability installation to run IHS or another Web server on a different system(s) than WAS, so your first step in a production environment is usually to install the Web servers, if necessary, and then install the WAS plug-in on these system(s). If you're installing IHS from the WAS base product CD, you can install the WAS plug-in at the same time. Be sure you have your Web server running properly before you move on to install WAS.

WAS Installation

Repeat the basic WAS installation on each of the systems you want to participate in the cluster. The Getting Started manual suggests not installing the WAS samples, JMS, or other services you don't intend to use. Especially if you're going to be using multiple systems, you can get a significant boost from omitting these features. However, you may want to initially install the samples so you can use them to verify a successful installation on each machine. Once you have your entire cell up and running, and you have installed and verified your own clustered applications, you may want to remove the samples.

During a WAS installation, the installation wizard prompts for node and host names, supplying the machine name as the defaults. It is recommended that you follow a consistent convention for node names (e.g., Node1, Node2), rather than using machine names (recall that there can be more than one node on a machine, and node names must be unique within a cell). The host name is the name by which the physical machine is known on the network. Other WAS nodes use the host name to connect to this node. This host name can be a fixed IP address or the same as the machine name, or it can follow some other naming convention, as long as it resolves to a network node (e.g., network card) on the machine where you're installing WAS. Fixed IP addresses are the least flexible, but they have the advantage of not requiring a DNS service to be available to make connections among WAS nodes. Whichever approach you take, be sure all systems on your network can successfully find and connect to all the WAS instance host names.

We suggest you select the option to run WAS as a Windows service, which (1) allows the WAS instance to be automatically started and (2) provides Windows support for process recovery.

Deployment Manager Installation

At this point in the setup, you have application servers installed on multiple machines, but there is no coordination among them. The next step is to install the deployment manager—which will provide the coordination—on a single machine. The deployment manager can run on the same machine as a node, or it can be on its own machine. Using a separate machine may improve performance, but it will definitely improve application availability.

The WAS ND Getting Started manual provides detailed deployment manager installation steps, so we'll just offer a few pointers. You can select other features, including Web services gateway, UDDI registry, and embedded messaging client, at the same time you install the deployment manager, but these aren't necessary for clustering. (If you've already installed embedded messaging as part of the WAS base product installation, be sure not to select it for installation with the deployment manager.) As with WAS, it is recommended that you select the option to run the deployment manager as a Windows service.

The installation prompts you for node, host, and cell names. I suggest you use a descriptive name, such as CellManager, which incorporates the cell name (e.g., Cell). The host name is the name by which the physical machine on which the deployment manager runs is known on the network. If you install the deployment manager on the same machine as a node, you can use the same host name for both the node and the deployment manager. The cell name is any arbitrary name for the cell that will be managed by the deployment manager.

Note that when you install the deployment manager on a system that already has the WAS base product installed (as from the previous step), you actually end up installing a second Administrative Console enterprise application, which may cause a warning about port conflicts for the administrative ports. You can ignore the warning because a later step—adding the node to the cell— will automatically delete the Administrative Console enterprise application for the node, leaving only the Administrative Console enterprise application for the deployment manager.

Fixpack Installation

After you install the WAS base product and the deployment manager, you should download and install the fixpacks in the following order: Fixpack 1, Installer Update, and Fixpack 2. If you're installing on a Windows 2000/NT system, be sure all services such as the MQ service are stopped before you do the Fixpack 1 install.

Adding Nodes to the Cell

Once the deployment manager is installed, you execute an addNode command on each machine to add that machine's node to the cell. (If you have multiple nodes on a machine, you execute one addNode command for each node you want to be part of the cell.) The WAS ND Getting Started manual describes this step (see Figure 13-4). Just be sure the deployment manager is running before you attempt to add nodes. Add each node in turn, making sure you can successfully start the node manager (i.e., with a startNode command) for a newly added node before you attempt to add another node.

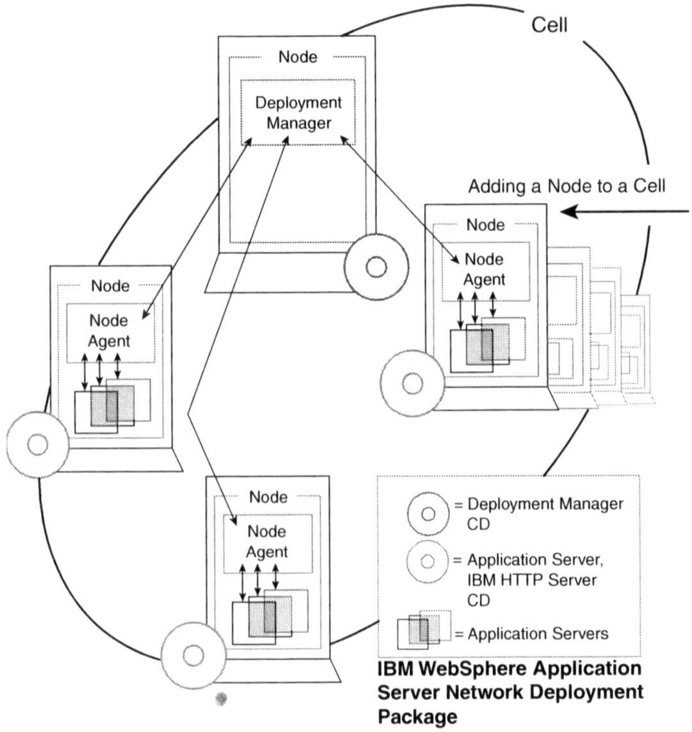

Figure 13-4 Adding a node to a cell.

Once you've added nodes to a cell, you can use the Administrative Console Web interface to manage all the nodes' configurations. With a cell, the Administrative Console provides an interface to the deployment manager, which in turn maintains the master configuration repository. This approach makes it convenient to centrally administer all your WAS servers.

Creating a Cluster

To create a cluster, run the Administrative Console, select Server>Clusters, and click the **New** button. On the initial wizard page, you need to assign a cluster name (e.g., ClusterH), which can be any name you choose. There are three important options to consider:

Prefer local enabled. This option indicates an EJB request should preferentially be routed to an EJB container on the same system. I suggest you leave this unchecked so requests can be routed to any server.

Create replication domain for this cluster. This option indicates memory-to-memory replication for sharing persistent session data across servers. Select this option so you have support for WAS V5's new method of providing session data failover.

Do not include an existing server in this cluster. Select this option so you can create new WAS instances to be members of the cluster. Creating new members simplifies some configuration steps. When you click **Next**, you'll see the page where you can create new WAS instances. On this page, enter a name for the first new WAS instance (e.g., WAS1) and identify which node (e.g., Node1) to create the instance in. We suggest you select the following options on this page:

- **Unique Http Ports**—to avoid port conflicts among the WAS instances
- **Create Replication Entry in this Server**—to use memory-to-memory replication for sharing persistent session data with this server
- **Default application server template**—to use the WAS-supplied configuration defaults

When you click **Apply**, the new WAS instance is created. You can create the rest of the WAS instances in your cluster by entering a new name and selecting a target node, leaving the rest of the options the same, and then clicking **Apply**. After you've added all the new WAS instances to the cluster, you can click **Finish** to complete the creation of the cluster.

Adding Ports to Virtual Hosts

When you set up a cluster as described, each WAS instance listens for HTTP requests on a different port. It's common when you install a Web application to associate a virtual host (such as `default_host`) with the Web module. This virtual host should be configured to accept HTTP requests from any port used by a member of the cluster on which you plan to deploy the application. So, for example, if a cluster has two members that listen to ports 9081 and 9082 respectively, you should be sure the virtual host associated with any Web application that runs on the cluster also listens to these two ports. You can add a port by selecting Virtual Hosts><host>>Host Aliases in the Administrative Console and clicking New. Enter "*" for the Host Name, and supply the appropriate port number.

Generating a New Web Server Plug-In File

After you've completed the installation and configuration steps, be sure you save your configuration changes in the Administrative Console. Select the "Synchronize changes with Nodes" option when you do the save so the configuration is propagated to all the nodes. Next, you should select Environment>Update Web Server Plugin in the Administrative Console to create a new Web server plug-in file. Then copy the Plugin-cfg.xml file to the system(s) on which your Web server(s) are running.

Deploying Enterprise Applications

The only thing you need to do differently when installing an application to a cluster rather than to a standalone instance of WAS, is to select the target cluster from the list box on the Map modules to application servers page of the Administrative Console's application installation wizard. In WAS V5, deployment manager and node agents handle propagation of application artifacts, as well as application configuration.

Session Management and Persistence Considerations

WAS supports the Java servlet session mechanism to preserve user state within a Web application. Within a cluster, WAS provides two ways to share a particular session's data among WAS instances in the cluster so that if a WAS instance running an application fails, the application can continue running on another WAS instance with the same session state. This capability requires session persistence to be enabled, using either a database as the sharing mechanism (as was available in WAS V4) or using the new memory-to-memory feature. If you selected the Create Replication Domain for this cluster option when you created a new cluster (as suggested above), your clustered applications will have session persistence, and thus session failover, support.

Be sure to read the "Configuring session management" chapter in *WAS V5.0 System Management and Configuration* for various options and considerations with session failover support. Although WAS provides the underlying support, you may need to make some application changes (e.g., to make session data serializable) to have full application failover capability. With WAS ND, setting up a cluster for basic application failover is fairly simple and provides substantial increase in application availability.

High Availability for Enterprise WebSphere Components

Figure 13-5 shows a general design for high availability for all WebSphere components. This type of high-availability WebSphere design is sometimes referred to as the "Gold Standard" since all components are redundant and highly scalable. The next section describes a high availability WebSphere/Domino system for a very high-availability, high-volume Lotus Web Conferencing system at an IBM server farm that closely follows this "Gold Standard" design.

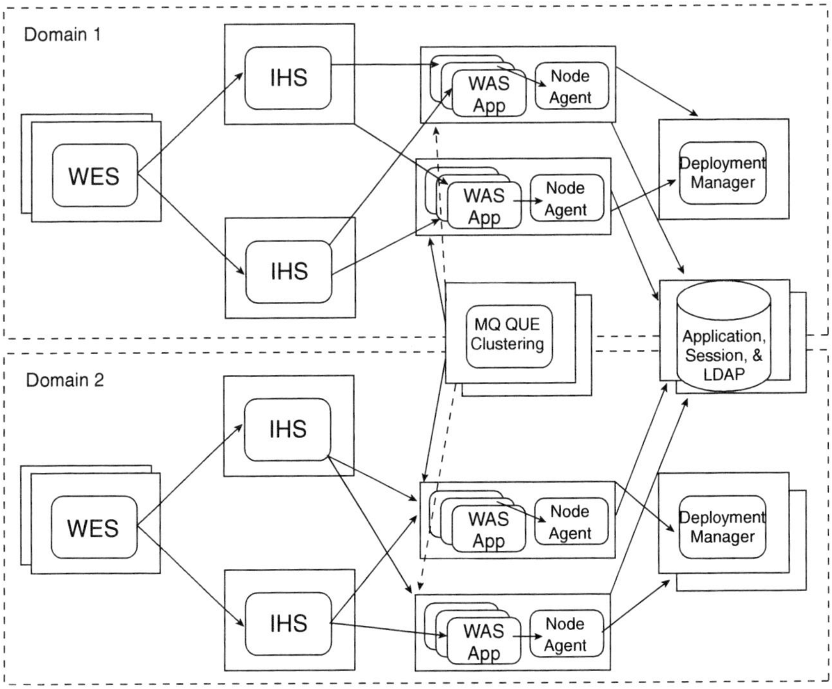

Figure 13-5 High availability for all WebSphere components.

Example of WAS Clusters for a High-Availability Web Conferencing System

This "real world" example is based on IBM's On Demand Workplace Web Conferencing offering. IBM's On Demand Web Conferencing system is a hosted offering designed to allow customers to securely share Web conferencing resources. ODW Web Conferencing is built around the Lotus Web Conferencing 3.1 product (a.k.a Lotus Sametime 3.1) including the Enterprise Meeting Server (EMS) option. The diagram shown in Figure 13-6 shows the Web conferencing logical application components at an IBM server farm. The 23 servers, including WebSphere Application Servers, Domino 6 servers, WebSphere MQ Servers, LDAP servers, SMTP servers, and DB2 servers, all provide for failover and load balancing. The Web conferencing Enterprise Meeting Server (EMS) machines rely on WAS 4 clusters. The WebSphere Everyplace Subscription Manager (WESM) boxes rely on WAS 5 clusters for high availability. The other servers have high availability provided by other techniques as described in the next section.

Failover and Load Balancing for Web Conferencing Servers

In addition to installation of the application code, the ODW Application Installation Guide describes the configuration and code needed for failover and load balancing:

- AIX Highly Available Cluster Management Program (HACMP) Active/Active configuration for the WebSphere MQ servers.

- HACMP Active/Passive configuration for the DB2 servers.
- LDAP Master/Peer configuration.
- WAS clustering for the EMS and WESM servers.
- HACMP Active/Active configuration for the SMTP servers.

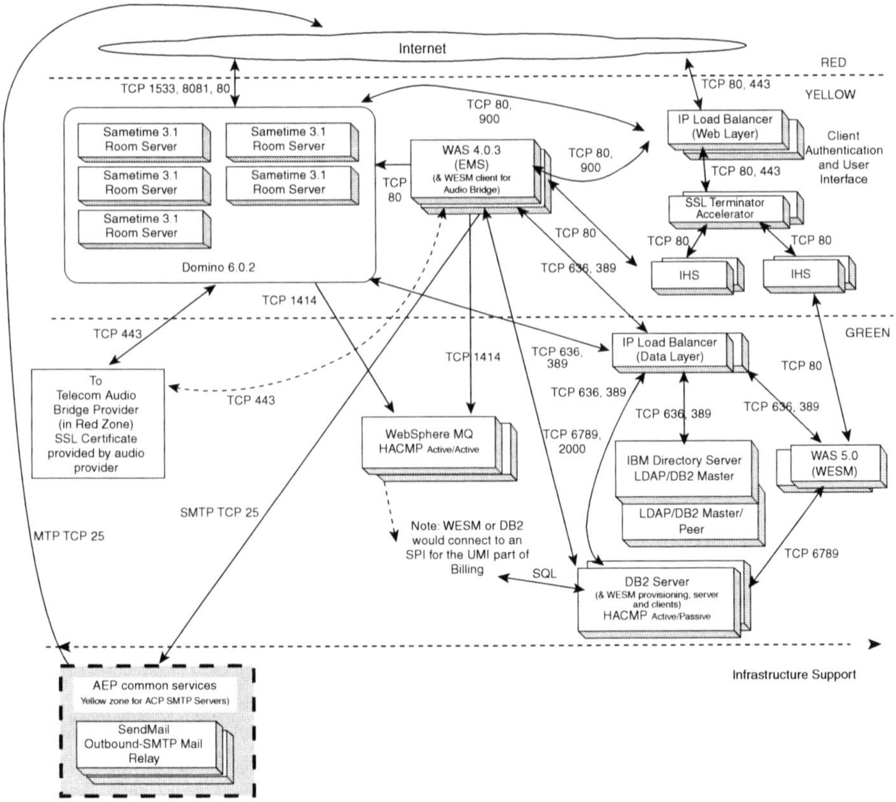

Figure 13-6 Server components in IBM Web Conferencing System.

Brief Description of Each Web Conferencing Server Type

This section briefly describes each of the servers given in Figure 13-6.

Sametime 3.1 Room Servers. Sametime 3.1 on Domino 6.02 servers are where the actual Web conferences are conducted. The Room Servers will be shared by multiple customer organizations. Usage data is contained in log files that have no user access.

WAS 4.0.3 (EMS). The EMS (Enterprise Meeting Server) server runs on WebSphere Application Server (WAS). EMS is an option with the Lotus Web Conferencing 3.1

(Sametime 3.1) product. EMS is used for meeting management and to load balance meetings across the Sametime 3.1 Room Servers. EMS utilizes the LDAP directory and the DB2 data store.

IHS. The IBM HTTP Server systems provide the HTTP user interface. Multiple servers are used for redundancy and horizontal scalability. The EMS and WESM systems each require their own set of IHS servers since EMS and WESM require different levels of WebSphere Application Server.

IBM Directory Server (LDAP/DB2). We will utilize the IBM Directory Server as an LDAP credential store. Multiple servers will be used for redundancy and future horizontal scaling. This server stores information about customer user IDs, passwords, etc. The Directory will be implemented using the IBM Directory Server product (formerly known as Secureway). There will be replicas of the directory for horizontal scalability. The directory is used for user credentials (authentication).

WAS 5.0 (WESM). The WESM (WebSphere Everyplace Subscription Manager) server runs on WebSphere Application Server (WAS). We will utilize WESM to enroll customers, authenticate users, hook to the billing system, etc. WESM utilizes the LDAP directory and the DB2 data store.

DB2 Server. Meeting information and log data are stored in the DB2 system. The DB2 server is utilized by both WESM and EMS (with different instances of DB2). The LDAP server will include its own local copy of DB2. The DB2 server also houses the WESM provisioning server and "out of the box" WESM provisioning clients.

WebSphere MQ. Messages from the room regarding changes in meeting state (active, scheduled, or finished) and room server availability are passed to EMS via the MQ server. The MQ Server passes health messages regarding the state of a room server (e.g., 20 meetings with 200 users, memory utilization 50 percent, absence of message indicates servers down, and EMS begins failover process). The WebSphere MQ servers use HACMP in active/active mode for failover and some load balancing.

SMTP Outbound Mail. The SMTP outbound mail relay servers will initially (Phase 1 of the project) be part of the ODW CORE system. However, for later phases of ODW Web Conferencing, we will use SMTP mail (inbound and outbound) via Application Enablement Program (AEP) common services. ODW Web Conferencing will also make use of other AEP common services as they become available (e.g., LDAP services).

CHAPTER 14

Performance and High-Availability Issues

"Failure is not an option."

—Gene Krantz, Apollo 13 Flight Director, at LotuSphere '99, in Orlando, Florida

Almost every company needs to have high performance and high availability for their Web servers. However, there are many different aspects of high availability. At one time, the only avenue for achieving high availability in the UNIX realm was through fault-tolerant technology. *Fault tolerance* relies on specialized hardware to detect a hardware fault and instantaneously switch to a redundant hardware component—whether the failed component is a processor, memory board, power supply, I/O subsystem, or storage subsystem. Although this cut-over is apparently seamless and offers nonstop service, a high premium is paid in both hardware cost and performance because the redundant components do no processing. More importantly, the fault-tolerant model does not address software failures, by far the most common reason for down-time. *High availability* views availability not as a series of replicated physical components, but rather as a set of system-wide, shared resources that cooperate to guarantee essential services. High availability combines software with industry-standard hardware to minimize down-time by quickly restoring essential services when a system, component, or application fails. While not instantaneous, services are restored rapidly, often in less than a minute. The difference between fault tolerance and high availability, then, is this: A fault-tolerant environment has no service interruption, while a highly available environment has a minimal service interruption. Many sites are willing to absorb a small amount of down-time with high availability rather than pay the much higher cost of providing fault tolerance. Additionally, in most highly available configurations, the backup processors are available for use during normal operation. This chapter discusses the different ways to achieve high availability for your Domino and WebSphere systems.

WebSphere and Domino Performance Issues

In this chapter, we discuss performance issues for both Domino and WebSphere and also discuss performance issues of a combined Domino and WebSphere system. We start off discussing some Domino performance issues, cover WebSphere performance issues, and then discuss issues when using a combined WebSphere/Domino system.

> **NOTE**
> Throughout the discussions, examples of "real world" performance issues for both Domino and WebSphere are discussed.

Performance Aspects of Domino on Multiprocessor Servers

What are some of the performance aspects of using Domino on multiprocessor servers? Are there advantages to running Domino on a 4-way server with 200MHz processors rather than a single processor server with an 800MHz processor? What about a single 1,000MHz processor vs. four 200MHz processors? With Domino R5, it's possible to run multiple MAIL.BOX files, multiple indexers, multiple cluster replicators, etc. This section gives numerous recommendations on optimizing Domino for multiprocessor servers. Often the Lotus guideline for the number of concurrent tasks, such as indexers, is based on the number of processors in your server (e.g., use multiple indexers up to the number of processors in your server). So, perhaps that's a significant advantage of a multiprocessor server. IBM Global Services has implemented Domino systems on many types of multiprocessor servers and has looked into these issues. A second important aspect of multiprocessor Domino servers is partitioning. Under R4 of Domino, partitioning a multiprocessor mail server usually allowed administrators to register many more users without sacrificing performance. With R5 of Domino, the performance and capacity improvement using partitions has changed to the extent that it is often better not to use partitions for Domino mail servers.

Introduction

To boost Domino R5's performance, you may have considered running it on multiprocessor servers. But doing so can prompt a lot of questions. Should you run Domino on a four-way server with 200MHz processors or a one-way 800MHz processor server? Or what about a single 1,000MHz processor versus four 200MHz processors? With Domino R5, you can run multiple tasks, such as multiple Mail.BOX files, Update tasks, and cluster replicators to improve your server's performance. But what are your best choices?

Another important consideration is whether to partition multiprocessor Domino servers. With Domino R4, partitioning a multiprocessor mail server typically lets administrators register numerous users without sacrificing performance. But with Domino R5, performance and capacity have improved to the extent that it's often better to avoid using partitions for Domino mail servers. IBM Global Services has implemented Domino systems on many types of multiprocessor servers and has examined these issues. In this section, we share the lessons we've learned and provide recommendations for the use of Domino on multiprocessor servers.

Overview of Multiprocessor Servers

Many large Domino servers (including UNIX, NT, and the AS/400) have multiple processors that use Symmetric Multiprocessing (SMP)—a computer architecture in which tasks are distributed among two or more processors. Other computer architectures, such as Massively Parallel Processing (MPP), are also used for multiprocessor servers—specifically, large servers that may have hundreds of processors (e.g., IBM's chess-playing supercomputer "Deep Blue").

A few years ago, high-performance multiprocessor machines carried a price tag of $100,000 or more. The multiprocessor market consisted of proprietary architectures that demanded a higher cost due to the scale of economics. In 1995, Intel helped to change this scenario by bringing high-performance computing to the mainstream with its Pentium Pro processor. The Pentium Pro was a high-performance processor with built-in support for multiprocessing. This support, coupled with the low cost of components, enabled computer manufacturers to build high-performance multiprocessor machines inexpensively. Today, a four-processor machine costs less than $12,000.

SMP is the primary parallel architecture employed in these inexpensive machines. SMP architecture is a tightly coupled multiprocessor system where processors share one copy of the operating system (OS) and resources, which often include a common bus, memory, and an I/O system. However, the machine is handicapped unless it has an OS that can take advantage of its multiprocessors.

In the past, manufacturers of multiprocessor machines were responsible for the design and implementation of a machine's OS. Frequently, machines could operate only under the OS that the manufacturer provided. However, typical multiprocessor Domino servers are built from common PC components and have open architectures. This makes possible an environment where any software developer can design and implement an OS for these machines.

Two mainstream examples of such operating systems are UNIX and Microsoft's Windows NT/2000. These operating systems exploit the power of multiprocessors by incorporating multitasking and multithreading architectures. Although the OS implementations of these architectures are different, the results for Domino servers are similar. Windows NT/2000 and UNIX/Linux are two of the most popular Domino servers. The AS/400 (iSeries), another popular Domino server, also uses SMP architecture and can contain up to 24 processors.

Lotus Guidelines

Lotus's guidelines for the number of allowable concurrent tasks, such as Update (a default task that continually checks and updates folders and views), is often based on the number of processors in your server. For example, you can use up to the same number of multiple Update tasks (also called indexers or view-indexers) in your server as you have processors. You also can use *one less* Update task.

Updating is a CPU-intensive process, so you might presume that it's useful to leave one processor free to handle other tasks while multiple indexers employ the other processors. And actually, as we'll see, Domino on most multiprocessor systems does work that way. With Domino, a single task is run on one processor. Because the Domino Update task is threaded, it can take advantage of running multiple processes without having multiple CPUs. In addition, tests conducted at Lotus's Enterprise Center of Competence (ECoC) indicate that you can run R5 with a single partition (on a four-way server) and achieve the same (or slightly better) capacity and performance of a multipartition configuration.

With Domino R4 on early versions of multiprocessor NT hardware, many technical people believed that a mail server with more than two processors was a waste because the server would effectively use only two of the processors.

We've captured several graphs using the NT Performance Monitor on a customer's production mail server (Figure 14-1). The Perfmon screen captures showed an even distribution of CPU utilization over the four processors. When we captured the Perfmon screen, the average overall CPU utilization was a moderate 26 percent for the mail server. Of course, the same even distribution of processing is also true for Domino R5. But does that even distribution hold true when one CPU-intensive Domino server task (e.g., updaters, replicators, or cluster replicators that are sized based on the number of processors in the server) is running? The next few sections discuss our findings for these tasks.

Multiple Update Tasks

Update tasks read a request from the queue, remove it from the queue, and then perform the indexing functions on the view. A single Update task works on one database that's pulled from the queue. If a second request enters the queue, the next Update task removes the request from the queue and starts working on it. If both requests are for the same database, the two tasks will work in tandem on that database. But more likely, the two tasks will work on different databases.

Multiple Update tasks can update different view indexes simultaneously within the same database. However, they cannot update the same view or run on the same full-text index (because it's one index) at once.

Multiple Update tasks are helpful from a response-time standpoint when they work on different databases because two Update tasks, for example, can empty a queue twice as fast. Additionally, multiple Update tasks that continually update views reduce users' wait time for an indexer to complete and display the view. At IBM server farms, we use three Update tasks per each four-way NT server running Domino to improve performance for users accessing large databases.

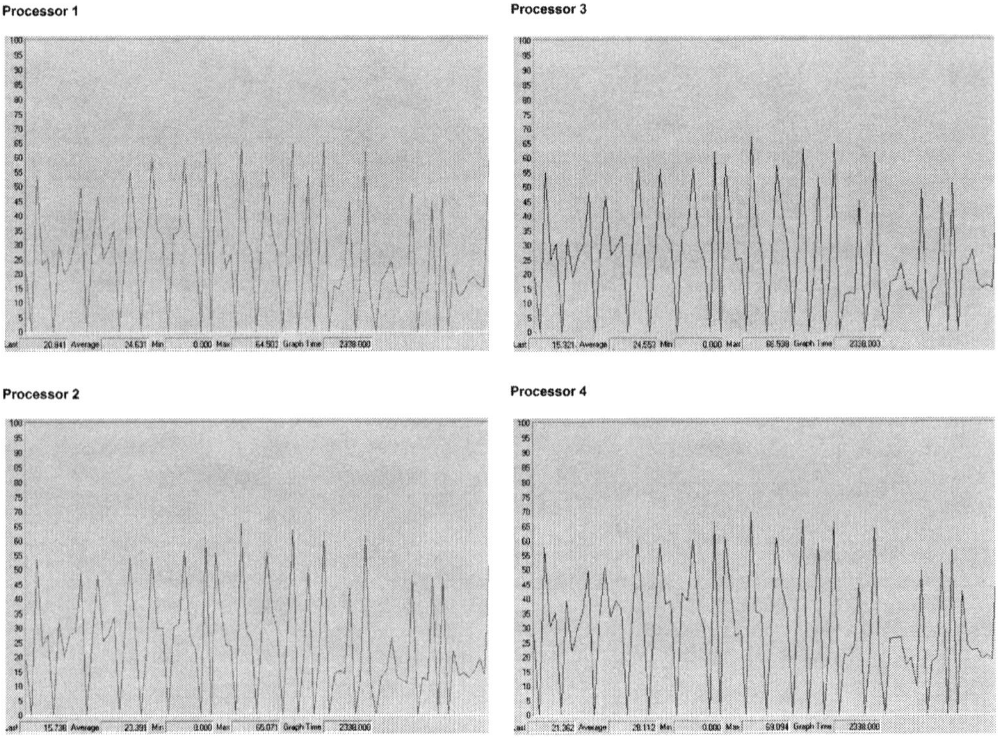

Figure 14-1 Perfmon display of CPU utilization for four-processor R4.6 mail server.

Beginning with R4, you can run multiple Update tasks on a server. To enable multiple Update tasks, you must perform one of the two following steps:

1. Add the parameter "`Updaters = # of desired Update tasks to run on the server`" to the Notes.INI file.
2. Manually load the Update task multiple times at the server console by typing **`LOAD UPDATE`**.

Multiple Update Tasks Test Results

We ran our tests at IBM's server farms using a test system that duplicated a customer's production system. The test system used a four-processor Netfinity 7000 server running Domino R5.0.7 server code. Here's part of the test scenario we used:

Test 1 (using one Update task)

1. Obtain a copy of the large workflow database (about 750MB) from the production system that must be indexed before you can open views.

2. Obtain a copy of the second production database that must be indexed before you can open views.
3. Have one Update task available.
4. Start the capture of CPU utilization for each processor.
5. From the Domino server command line, type **`UPDATE`** to update the second database.
6. From a user workstation, open a view on the workflow database (that must be indexed) and time how long it takes to open.
7. Save CPU utilization results for each of the four processors.

Test 2 (using one Update task with the same test steps as Test 1 but without the Domino server `UPDATE` command)

1. Restore the two databases to their original state (i.e., copy the two databases from saved copies outside of Domino\Data).
2. Start the capture of CPU utilization for each processor.
3. From a user workstation, open a view on the workflow database (that must be indexed) and time how long it takes to open.
4. Save CPU utilization results for each of the four processors.

After we ran these two tests, we tried the same tests again with two Update tasks defined. Our test results indicate that multiple Update tasks can have a significant impact on the consistency of performance when many users employ numerous large databases. Our tests with NT servers showed that the processing for each Update task uses only one processor in a multiprocessor server. So the concept of running three Update tasks on a four-way NT server and leaving one processor free for other tasks is valid. For some of our test scenarios, running three Update tasks proved optimal for our large workflow application.

Multiple Mail.BOX Files

Support for multiple Mail.BOX databases, new in R5, can improve performance for Domino mail systems. IBM recommends configuring multiple Mail.BOX databases on both dedicated R5 SMTP servers (that handle SMTP/MTA function for remaining R4 servers) and R5 mail servers to solve bottleneck problems. We usually recommend that SMTP/MTA and mail servers have three or four Mail.BOX databases because these servers are typically four-way NT platforms.

Before support was available for multiple Mail.BOX databases, a corrupted Mail.BOX could prevent mail flow on a server. But by moving to multiple Mail.BOX databases, you can enable mail flow to continue even if one Mail.BOX database is corrupted. Although the use of multiple Mail.BOX databases is a handy feature for R5 Domino mail servers, you still must determine how many Mail.BOX databases to define. Lotus usually recommends two Mail.BOX databases and suggests that more is a waste. But is that true?

To test the multiple Mail.BOX feature, we first set up a multiprocessor server with one Mail.BOX and ran a test script to drive the mail system to measure delays and monitor CPU and I/O utilization. Next, we set up a multiprocessor server with two Mail.BOX databases and repeated the prior procedure. Finally, we set up a multiprocessor server with three Mail.BOX databases and repeated the procedure used in the previous two test scenarios. According to our results, test scripts that drive heavy mail flow and define three or four Mail.BOX databases can be optimal for some mail systems.

Multiple Scheduled Replicators and Cluster Replicators

Domino supports both multiple scheduled replicators and multiple cluster replicators. Of course, you must determine how many of these replicators to run to optimize performance on your Domino system. But what criteria can you use to determine the optimal number of replicators to use?

> **Scheduled Replication.** You can load two instances of the schedule replicator task per processor (but not more than eight). Each replicator can process one server at a time and queue up to five requests. To specify that a server loads multiple replicators whenever it starts, add the `Replicators=n` variable to the Notes.INI file, where n is the number of replicators the server should run. Each replicator occupies about 3MB of RAM when idle, so adding replicators even when they're not in use carries a memory cost. Replication in R5 is much more efficient than in R3 and somewhat more efficient than in R4. Domino 6 provides significant performance improvements in both scheduled replication and cluster replication via data compression and data streaming techniques.
>
> **Cluster Replication.** The Cluster Replicator task synchronizes database replicas within the cluster and ensures that clients see up-to-date information. When you make changes to a database in a cluster, the Cluster Replicator pushes the change to all other replicas in the cluster. Therefore, the Cluster Replicator is event-driven rather than schedule-driven, as in traditional replication. One Cluster Replicator is automatically run for each server in the cluster. However, you can run more than one Cluster Replicator to improve performance. Multiple cluster replicators work in tandem, sharing the replication workload. To enable multiple cluster replicators, append an additional CLREPL task to the Server Tasks setting in the Notes.INI file or type **`load CLREPL`** at the server console.

Our test results indicated that implementing both multiple scheduled replicators and multiple cluster replicators in a production system can improve performance. We usually run two scheduled replicators and two cluster replicators for Domino servers at our server farms.

Monitoring and Performance Analysis

All multiprocessor Domino server platforms, such as NT, UNIX , and the AS/400, have tools to monitor each processor's performance.

The NT server includes the Performance Monitor and Task Manager, which lets you monitor your system's performance. The Performance Monitor has more capabilities than the Task Manager does. It can monitor how efficiently your computer is running, determine whether any components are causing system bottlenecks, plan for future growth, and set alert conditions.

You use Task Manager—a new utility for NT 4.0— to display an overview of system CPU and memory usage. Task Manager also provides monitoring capabilities for the Performance Monitor's more commonly tracked options. Task Manager is easy to access and requires no configuration. You can use it immediately to display graphs of CPU utilization for each of your Domino server's processors. With Task Manager, you can quickly determine the impact on each processor when a Domino task, such as an indexer or replicator, kicks off.

You use the UNIX command `vmstat` to monitor CPU utilization on each processor. For example, `vmstat 3.5` gives you a running status every three seconds, five times. Using `vmstat` also provides information about paging, disks, and memory use. Other UNIX commands are:

- `sar`—provides similar data as `vmstat` for UNIX V environments.
- `ps`—provides a list of current processes and their use of resources.
- `iostat`—provides status and usage of I/O resources (e.g., disks).
- `lsps -a`—provides statistics about the use of paging space.

Using Operations Navigator (OpsNav) on the AS/400 and iSeries lets you monitor and manage your server. OpsNav also provides Management Central functions, which you can use to observe a wide range of performance metrics, including CPU utilization, disk arm utilization, memory usage, and LAN utilization. Management Central's realtime monitoring capability lets you drill down to tasks and resources that may be causing or experiencing high-resource utilization and determine the contributing factors. It also provides data archiving, replay, and exporting capabilities for studying performance trends.

For more comprehensive performance monitoring of the server, the IBM Performance Tools for iSeries includes data collection and reporting capabilities for all system resources and tasks, along with an advisor function that analyzes data and provides appropriate configuration and tuning recommendations.

Using Domino Partitions on Multiprocessor Servers

Now let's look at examples of using Domino partitions on different platforms such as NT, UNIX, and AS/400 servers. These examples reflect the experiences of IBM and Lotus as well as R5 test results.

At its test lab in Cambridge, Massachusetts, Lotus tested various aspects of performance with and without partitions on four-processor UNIX servers (RS/6000) and four-processor NT servers (Compaq Proliant 6500) running Domino. Lotus conducted the tests using Domino R4.5 and R5.0.2 server code.

The first test system consisted of an RS/6000 SP 332MHz four-way SMP thin node with 3GB of RAM, two 4.5GB SCSI disks for rootvg and paging, and forty-eight 9.1GB SSA disks for Notes data and logs. The server contained one partition with 3,500 registered users and two other partitions with 1,750 registered users each.

As Figure 14-2 shows, the probe response time for both partition scenarios is flat to 2,000 users but increases at 2,500 users. Notice that, at 1,500 users, CPU utilization for one partition is 80 percent and for two partitions is 95 percent. As you can see, using one partition works as well as using two partitions. Domino R5's capacity was 2,000 or more users, and for R4.5, it was about 650 users. Each server showed a fast response time up to capacity.

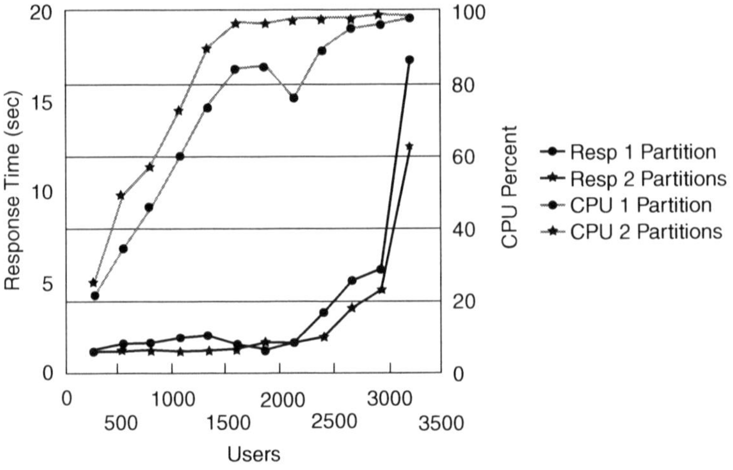

Figure 14-2 Probe response time with CPU utilization for 1 partition and 2 partitions.

The second test system consisted of a Compaq Proliant 6500 four-processor NT 4.0 machine with 1 GB of RAM and 90 GB of hard drive. Lotus conducted the test on a large customer application to determine the performance of R4 against R5. The results indicate that Domino R5 showed some performance improvement and had a much higher capacity than R4, especially when R5 ran on SMP systems.

IBM's guidelines for partitioning Domino servers have changed from R4 to R5. For Domino R4.5 mail servers, IBM recommends using one partition for every two processors on UNIX or NT platforms. This arrangement allows more mail users per physical box. With R5, IBM usually doesn't partition two-processor UNIX or four-processor NT platforms. IBM's tests show that partitioning doesn't improve the number of mail users allowed per physical box. But at IBM server farms, we partition the OS/400 subsystem architecture in AS/400 and iSeries Domino boxes to isolate each partition from the others (see "Partitioning on the AS/400" later in this chapter).

Other Considerations

With Domino R5, a four-way 200MHz server performs similarly to a one-way 800MHz server. But because SMP architecture carries some overhead, a single processor server should always perform somewhat better than a multiprocessor server. For servers greater than four-way, performance is a function of SMP architecture. Of course, availability may be a factor if losing CPU is a concern (it usually isn't). Therefore, one advantage of using a four-way NT server running Domino is that you get four times the processing power of a single processor server with the same processor speed. Conversely, if you need to replace your four-way 200MHz Domino server with an equivalent server, a one-way 800MHz server will do the job.

Another aspect of SMP processing is how it affects response time. Figure 14-3 shows the results of response time tests conducted in an IBM lab using three iSeries servers: a two-way 255MHz 170-2409, a one-way 450MHz 270-2423, and a one-way 540MHz 270-2423. These tests implemented a Web shopping application employed by 100 Web shopping users. As Figure 14-3 shows, using two 255MHz processors won't provide a faster response time than using one 450MHz processor will. The reason is that every SMP architecture requires some overhead, although the amount of overhead varies by architecture (see "NotesBench Performance of Domino on Multiprocessor Servers," for more information about various SMP architecture overhead).

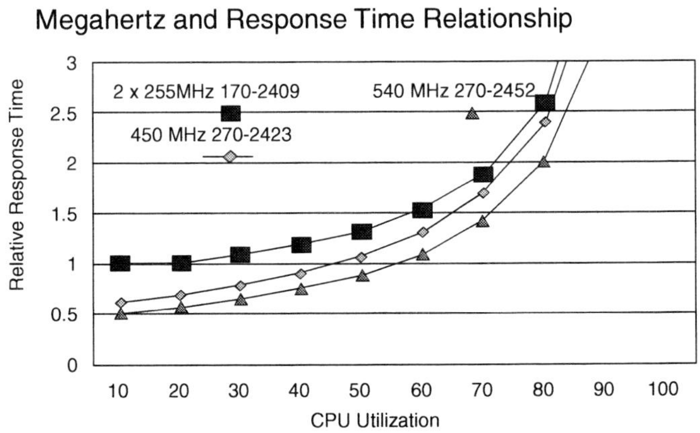

Figure 14-3 Response time for single and multiprocessor servers.

The tests were made in an IBM lab using a Web shopping application involving 100 Web shopping users and three iSeries servers: 2×255MHz 170-2409, a 450MHz 270-2423, and a 540MHz 270-2423.

You'll also want to consider limiting the risk of having all your mail users on one large Domino server without partitions should that server fail. For example, suppose your company has 4,000 Domino mail users, and you want to limit the number of those affected by the failure of one Domino mail server to 1,000 users. To accomplish this, you'd want to partition a large four-way

Domino server into four partitions rather than run a large Domino server with no partitions. This logic assumes that if one partition fails, it doesn't bring down the other partitions.

A final point concerns Lotus's recommendation. Using a guideline based on the number of processors a server has to determine the maximum number of multiple tasks to run does make sense. However, the guideline should be also be based on the overall power of your server, not on the number of processors it has. We could argue that for most relatively new servers, a four-way server is typically about four times as powerful as a one-way server. But since processor power continues to increase rapidly (according to Moore's law, it doubles every 18 months) and because many slightly older servers are still in use, that argument is open to a lot of exceptions.

What's Next for Domino on Multiprocessor Servers?

Since its start with Notes R4.0, Domino on SMP servers has matured into a technologically solid and reliable product. Some server SMP architectures (e.g., iSeries) have little SMP penalty for Domino servers greater than four-way. In fact, NotesBench data indicates little penalty up to a 24-way iSeries Domino server. Multiprocessor UNIX Domino servers are also viable. This is not the case, however, for Intel-based Domino servers greater than four-way. But because Intel-based platforms are steadily becoming more powerful, eight-way (and higher) Domino servers based on Intel platforms will likely approach the effectiveness of current iSeries and UNIX servers.

Notes/Domino 6 (ND6) has many performance improvements related to the topics I've discussed in this chapter . With bigger, more powerful Domino servers to come, the need will be even greater for Domino network performance improvements. ND6 provides significant reductions in network bandwidth requirements via data compression and data streaming techniques. Lotus tests indicate that ND6 could decrease network traffic up to a 50 percent. This reduction will improve performance of all Domino's network aspects, including scheduled replication and cluster replication, and will make Domino clustered replication across the WAN even more feasible—which is critical to Domino's continued success for the more powerful Domino servers of the future.

NotesBench Performance of Domino on Multiprocessor Servers

One way to determine whether a four-way server with 200MHz processors is equivalent to Domino on a one-way 800MHz processor server is to look at NotesBench performance numbers. Because Symmetric Multiprocessing (SMP) architecture contains some overhead, a single 800MHz processor server will generally perform better than a four-way server with 200MHz processors. Also not all SMP architectures exhibit the same efficiency.

Table 14-1 contains information from the NotesBench Web site (`http://www.notesbench.org`) that shows how Domino servers with various SMP architectures differ. Notice that the last column displays the number of Notes users per MHz, which is useful for comparing the efficiency of SMP systems. The number of users per MHz for the iSeries doesn't seem to vary much for the 24-way, eight-way, two-way, and one-way systems. Although the pSeries section contains only two entries for UNIX, these systems typically have efficient SMP architectures for numerous processors (more than four). The Intel boxes, however, differ significantly in the efficiency of Domino on multiprocessor servers. The single-processor Intel box supports twice as many users per MHz as the eight-way server does. As a colleague mentioned, this is one reason we don't see Domino solutions being rolled out on eight-way and larger SMP NT/2000 servers.

Extrapolating from Table 14-1, it's clear that a one-way 800MHz processor Intel box should make a significantly more powerful Domino server than a four-way server with 200MHz processors will. But for iSeries and UNIX servers, it appears that a one-way 800MHz processor is only slightly more powerful than a four-way box with 200MHz processors. The single-processor server should always be somewhat more powerful than the equivalent multiprocessor MHz SMP platform, but the amount of extra power depends on the SMP architecture used.

Partitioning on the AS/400

IBM's commercial server farm runs a 12-way AS/400 with 2.5 terabytes of hard drive for a large commercial customer. Domino R5 runs on the AS/400 and contains 12 partitions—10 are mail partitions with 1,600 users each, and two are application partitions with 2.5 terabytes of hard drive shared by all partitions in the pool.

The customer wanted to limit the number of users to about 1,500 per partition. Although Domino works well with fewer large partitions, increasing the number of partitions (and, hence, the number of Domino servers) can reduce the risk of impact should one of the Domino servers fail.

The administration cost of using 12 partitions on Domino is as follows:

- Administration cost is *not* reduced for monitoring.
- Hard drive pooling allows significant administration savings because you don't have to balance a load based on hard drive use.
- You upgrade or apply fixes to Domino only once for all 12 partitions.
- You upgrade or apply fixes to the operating system only once for all 12 partitions.

Through its experiences at its commercial server farms, IBM also realized these aspects of partitioning:

- The AS/400 proved robust with partitions; it allows Domino to automatically restart and lets programmers with NT skills work with the AS/400 administration interface.
- NT experience shows that if one Domino partition fails, it takes down the whole server. This rarely happens with the AS/400 (iSeries) because its subsystem architecture isolates each subsystem, and in this case, it isolates one partitioned Domino server from the others.
- Windows 2000 proved more robust than NT.
- Unlike the AS/400, you can't pool hard drive space on NT or UNIX.

Table 14-1 NotesBench Results for Multiprocessor Servers

Server	Benchmark	Benchmark CPU Utilization	Number of Processors	Speed of Processors (MHz)	Total MHz Used for Benchmark	Benchmark Users per MHz
iSeries						
iSeries 400 Model 840	100500	99.3%	24	600	14,299 MHz used	7.03 users per MHz
iSeries 400 Model 840	75000	93%	24	500	11,160	6.72 users per MHz
iSeries internal Benchmark for 830 8way**	22900	70%	8	540	3,024	7.57 users per MHz
iSeries internal Benchmark for 820 4way**	11810	70%	4	600	1,680	7.03 users per MHz
iSeries internal Benchmark for 820 2way**	6660	70%	2	600	840	7.86 users per MHz
iSeries internal Benchmark for 820 1way**	3110	70%	1	600	420	7.4 users per MHz
INTEL						
Compaq ProLiant 8500	14000	96%	8	550	4,224	3.3 users per MHz
Compaq ProLiant 8500	10000	95%	4	550	2,090	4.78 users per MHz
IBM Netfinity 5600	8600	89%	2	933	1661	5.18 users per MHz
Compaq ProLiant DL360	5800	94%	1	933	877	6.61 users per MHz
pSeries						
IBM RS/6000 Enterprise Server S80	57600	78%	24	450	8,424	6.84 users per MHz
IBM RS/6000 Enterprise Server M80	28032	99%	8	500	3,960	7.08 users per MHz

Implementing High Availability for Your WebSphere Application Servers

What are some of the ways to implement high availability for your WebSphere Application Servers (WASs)? IBM Global Services has implemented high-availability WAS servers in different ways for customers at our commercial server farms. In some cases, IBM uses AIX Clustering with HACMP or Microsoft OS Clustering (MSCS). In other cases, IP sprayers (e.g., Network Dispatchers, Cisco Local Directors, Cisco Distributed Directors, etc.) provide failover and load balancing capability. There is also the possibility of using a single server that has very high availability with redundancy built in (e.g., fault-tolerant systems such as IBM z Series mainframes or Tandem computers). This article gives "lessons learned" and recommendations on the use of different techniques for implementing high-availability WebSphere servers.

IBM's WebSphere Web Application server (WAS) product provides a powerful base for building sophisticated Web (Internet or Intranet) applications. For most Web servers, high availability is of paramount importance. Availability is not an all-or-nothing proposition. Availability can be thought of as a continuum. Reliable hardware and software provide the base level of availability. Advanced features such as RAID devices provide an enhanced level of availability. High-availability software provides near continuous access to data and applications (e.g., OS clustering where a hot standby server might take a minute or so to come online in a failover situation). Fault-tolerant systems ensure the constant availability of the entire system, but at a higher cost. WAS V4 "server groups" and WAS V5 "clusters" can be used with high-availability techniques to provide load balancing. So a good starting point for high availability would be to use WebSphere's built-in failover capability using the Work Load Management (WLM) facility.

OS clustering can be used along with the failover capability in WebSphere to enhance the overall system failover. OS clustering provides high availability but often not load balancing. The primary goal of high-availability clustering software is to minimize, or ideally, eliminate, the need to take your resources out of service during maintenance and reconfiguration activities. Two popular examples of OS clustering are AIX HACMP and Microsoft OS Clustering (MSCS). OS clustering supports hardware failover for multiple servers, or nodes, in a cluster sharing the same common disk device (thus, OS clustering is also referred to as hardware clustering). OS clustering often provides failover protection so that if two servers running WAS are set up for failover, and one server becomes unavailable, the second server will come up, use the shared disk subsystem, restart the respective applications, and continue with the processes. It will use the same disk where the WAS data resides and the same IP address that the WebSphere server is using. Therefore, the failover function is transparent to clients. However, if the shared disk device OS clustering is using goes down, the system is down. Thus, it is very important to have a mirrored (RAID 1) shared disk for OS clustering. Often OS clustering provides only failover protection—not load balancing—but newer versions of products such as HACMP will also provide some load balancing. Often IP sprayers such as Network Dispatcher are used in conjunction with OS clustering to provide very high availability and load balancing.

High-availability systems are an excellent solution for applications that can withstand a short interruption should a failure occur but which must be restored quickly. Some industries have applications so time-critical that they cannot withstand even a few seconds of down-time. Many other industries, however, can withstand small periods of time when their database is unavailable. For those industries, basic OS clustering can provide the necessary continuity of service without total redundancy.

Use of AIX HACMP for Server Failover

IBM's tool for building UNIX-based, mission-critical computing platforms is the HACMP for AIX software. The HACMP for AIX software ensures that critical resources are available for processing. HACMP for AIX has two major components: high availability (HA) and cluster multi-processing (CMP).

Until recently, the only avenue for achieving high availability in the UNIX realm was through fault-tolerant technology. *Fault tolerance* relies on specialized hardware to detect a hardware fault and instantaneously switch to a redundant hardware component—whether the failed component is a processor, memory board, power supply, I/O subsystem, or storage subsystem. Although this cutover is apparently seamless and offers non-stop service, a high premium is paid in both hardware cost and performance because the redundant components do no processing. More importantly, the fault-tolerant model does not address software failures, by far the most common reason for down-time. *High availability* views availability not as a series of replicated physical components, but rather as a set of system-wide, shared resources that cooperate to guarantee essential services. High availability combines software with industry-standard hardware to minimize down-time by quickly restoring essential services when a system, component, or application fails. While not instantaneous, services are restored rapidly, often in less than a minute. The difference between fault tolerance and high availability, then, is this: A fault-tolerant environment has no service interruption, while a highly available environment has a minimal service interruption. Many sites are willing to absorb a small amount of down-time with high availability rather than pay the much higher cost of providing fault tolerance. Additionally, in most highly available configurations, the backup processors are available for use during normal operation.

"Cluster multi-processing" is a group of loosely coupled machines networked together, sharing disk resources. In a cluster, multiple server machines cooperate to provide a set of services or resources to clients. Clustering two or more servers to back up critical applications is a cost-effective high-availability option. You can use more of your site's computing power while ensuring that critical applications resume operations after a minimal interruption caused by a hardware or software failure. Cluster multi-processing also provides a gradual, scalable growth path. It is easy to add a processor to the cluster to share the growing workload. You can also upgrade one or more of the processors in the cluster to a more powerful model. If you were using a fault-tolerant strategy, you must add *two* processors, one as a redundant backup that does no processing during normal operations.

The latest version of HACMP software (AIX HACMP version 5.1) optimizes availability by allowing for the "dynamic reconfiguration" of running clusters. Most routine cluster maintenance tasks, such as adding or removing a node or changing the priority of nodes participating in a resource group, can be applied to an active cluster without stopping and restarting cluster services. HACMP software allows you to combine these components into a wide range of cluster configurations on RS/6000 systems, providing you with flexibility in building a cluster that meets your processing requirements.

Nodes form the core of an HACMP cluster. A node is a processor that runs both AIX and the HACMP for AIX software. The HACMP for AIX software supports pSeries (or RS/6000) uniprocessor and symmetric multiprocessor (SMP) systems and the Scalable POWERParallel processor (SP) systems as cluster nodes. To the HACMP for AIX software, an SMP system looks just like a uniprocessor. SMP systems provide a cost-effective way to increase cluster throughput. Each node in the cluster can be a large SMP machine, extending an HACMP cluster far beyond the limits of a single system and allowing thousands of clients to connect to a single database.

In an HACMP cluster, each node is identified by a unique name. A node may own a set of resources—disks, volume groups, file systems, networks, network addresses, and applications. Typically, a node runs a server or "back end" application that accesses data on the shared external disks. HACMP software supports from two to eight nodes in a cluster, depending on the disk technology used for the shared external disks. HACMP/ES clusters can have up to 32 nodes in a cluster.

Use of Microsoft's OS Clustering (MSCS)

Microsoft's OS Clustering (MSCS) supports hardware failover for multiple servers, or nodes, in a cluster sharing the same common disk device. MSCS provides failover protection so that if two servers running WebSphere are set up for failover, and one server becomes unavailable, the second server will come up, use the shared disk subsystem, restart the respective applications, and continue with the processes. It will use the same disk where the WAS data resides and the same IP address that the WebSphere server is using. Therefore, the failover function is transparent to clients. MSCS provides only failover protection. Also, if the shared disk device OS clustering is using goes down, the system is down. Thus, it is very important to have a mirrored (RAID 1) shared disk for OS clustering.

Microsoft Cluster Server ("MSCS") is clustering software that ships with Microsoft Windows NT Server—Enterprise Edition. MSCS supports clusters of two, specially linked servers running Microsoft NT Server—Enterprise Edition. The primary function of MSCS occurs when one server in a cluster fails or is taken offline. With MSCS, the other server in the cluster takes over the failed server's operations. Clients using server resources experience little or no interruption of their work because the resource functions move from one server to the other.

MSCS is comprised of two main components: clustering software and the Cluster Administrator. The clustering software enables the two servers of a cluster to exchange specific types of messages that trigger the transfer of resources at the appropriate times. The clustering software has two primary components: the Cluster Service and the Resource Monitor. The Cluster Service runs on each cluster server. It controls cluster activity, communication between cluster servers, and failure operations. The Resource Monitor handles communication between the Cluster Service and the application resources. The Cluster Administrator is a graphical application that is used to manage a cluster. It runs on any version of NT (server, workstation) that has Service Pack 3 or later installed. In MSCS, a cluster is a configuration of two nodes, each of which is an independent computer system. Together, these independent servers create a "server cluster." The cluster appears to users as a single server. For MSCS, both nodes must be running NT Server (or Windows 2000) Enterprise Edition. The network applications, data files, and other tools you install on the nodes are the cluster resources, which provide services to network clients. A resource is hosted on only one node at any time.

What About the Role of IP Sprayers Such as IBM's Network Dispatcher?

Network Dispatcher is a TCP connection router that supports failover and load sharing across several TCP servers. Network Dispatcher was "rebranded" under the WebSphere umbrella and is now part of the WebSphere Edge Server. Load sharing in Network Dispatcher is supported by a user-level manager process that monitors the load on the servers and controls the connection allocation algorithm in the kernel extension. At many Internet sites, the workload required of various services has grown to the point where a single system is often unable to cope. Offering the same service under a number of different node names is a solution that has been used by a number of sites—for example, by Netscape for its file transfers. Round-robin DNS can also be used for much the same purpose and allows the servers to publish a single name by which the service is known.

Neither of these approaches spreads the load evenly as the number of servers providing the service increases. In the Netscape approach, clients manually pick from a list of names providing the service. In all probability, the names at the top of the list, or with some other feature, will be chosen. Whatever attribute is chosen, it is unlikely to be related to the current load on the servers. The list can be manipulated to put lightly loaded nodes in a place where they are most likely to be picked, but this cannot necessarily be done on a timely basis. Round-robin DNS eliminates psychology as an issue but replaces it with IP address caching. Once a particular client or proxy has received an IP address for a service, that IP address may be cached for hours, or even days. Intermediate name servers and gateways also frequently cache IP addresses and usually ignore any time-to-live limits suggested by DNS. With either method, control over the number of requests per server is limited, which causes the load to be unevenly spread.

Network Dispatcher is one solution to the problems of keeping the load evenly spread or balanced on a group of servers. Network Dispatcher is integrated with the TCP/IP stack of a single system. It acts as a dispatcher of connections from clients who know a single IP address for a service, to a set of servers that actually perform the work. Unlike other approaches, only the packets going from the clients to the server pass through Network Dispatcher; packets from the server to client may go by other routes, which need not include Network Dispatcher. This reduces the load on Network Dispatcher, allowing it to potentially stand in front of a larger number of servers.

Network Dispatcher has been used to spread the load as part of several large-scale Web server complexes—for example, the 1996 and 2000 Summer Olympic Games, Deep Blue vs. Kasperov, and the 1998 Winter Olympic Games Web sites. These systems handled millions of requests per day, going to dozens of servers at a time. Network Dispatcher has also been successfully used with many different types of Web servers.

The Weighted Round Robin (WRR) connection allocation algorithm is efficient and maintains the following conditions: All servers with the same weight will receive a new connection before any server with a lesser weight receives a new connection. Servers with higher weights get more connections than servers with lower weights, and servers with equal weights get an equal distribution of new connections. Finally, an eligible server will be returned on each invocation of the allocation function or an indication that there are no eligible servers.

The service time and resources consumed by each type of TCP connection request varies depending on several factors. These factors include request-specific parameters (type of service, content) and currently available resources (CPU, bandwidth). For example, some requests may perform computationally intensive searches, while others perform trivial computation.

A naïve distribution of TCP connections among the back-end servers can (and often does) produce a skewed allocation of the cluster resources. For example, a simple round-robin allocation may result in many requests being queued up on servers that are currently serving "heavy" requests. Obviously, such an allocation policy can cause underutilization of the cluster resources, as some servers may stay relatively idle while others are overloaded. This condition, in turn, will also produce longer observed delays for the remote clients.

Load-balancing and load-sharing are two main strategies for improving performance and resource utilization. Load-balancing strives to equalize the servers' workload, while load-sharing attempts to smooth out transient peak overload periods on some nodes. Load-balancing strategies typically consume many more resources than load-sharing, and the cost of these resources often outweigh their potential benefits. IBM found this to be the case in allocating TCP connections for HTTP requests; therefore, the first Network Dispatchers implemented a load-sharing allocation policy.

Network Dispatcher can be configured with multiple network interfaces and can be optionally co-located with a server. Packets from Network Dispatcher to the servers can follow a separate network path than the response packets from the servers to the clients. This enables better use of network bandwidth. For example, the 2000 Olympic Web site had a Network Dispatcher node on the public network, a separate token-ring network between Network Dispatcher and the servers, and the servers attached to the public network via ATM. This takes advantage of the fact that Web responses are generally much larger than the corresponding requests.

Network Dispatcher has explored two approaches for providing high availability: integrating Network Dispatcher with an existing recovery infrastructure and providing HA for smaller stand alone sites. The high-availability infrastructure consists of a set of daemons running on all nodes, exchanging heartbeat messages, and automatically initiating a recovery program via a recovery driver when a specific failure event (software or hardware) occurs. The recovery infrastructure described here is available as a product called High Availability Cluster Multi-Processing for AIX (HACMP) on the IBM SP-2, where multiple heartbeats on distinct networks are used to tolerate network adapter failures. The Web recovery programs have been prototyped and demonstrated on the SP-2 using HACMP with multiple Network Dispatcher nodes co-located with servers.

Figure 14-5 illustrates a scalable Web server with a set of Network Dispatchers routing requests from multiple clients to a set of server nodes. While a single Network Dispatcher node can handle the routing requirements of a large site, multiple Network Dispatchers may be configured for high availability and nondisruptive service. Recovery programs consist of multiple steps, and each step, which can be a single command or a complex recovery script, can execute on a different node of the system. Recovery steps can be forced to execute in sequence via `barrier` commands that are used for synchronization.

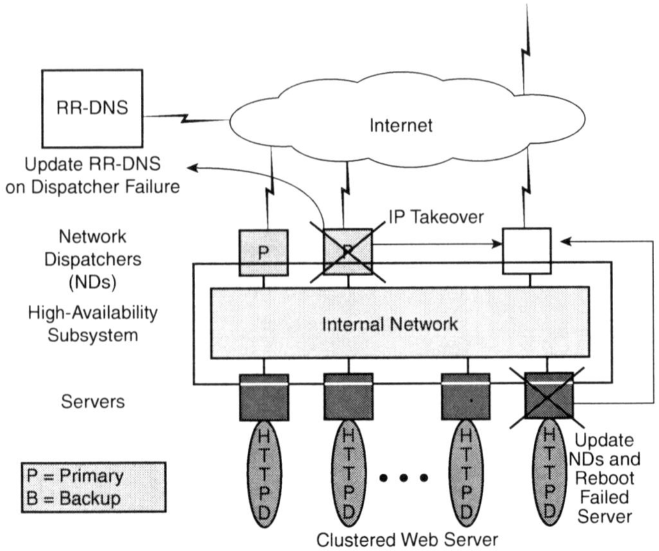

Figure 14-4 High availability in a scalable Web server using Network Dispatchers.

If there is no recovery infrastructure, then Network Dispatcher would be a single point of failure. To address this issue, Network Dispatcher has been augmented with primary backup function. Two Network Dispatchers, ND1 and ND2, with the same LAN connectivity can be configured to serve the same set of servers: One is designated as primary (ND1), and the second as backup (ND2). When the backup Network Dispatcher initializes, it requests replication of the primary Network Dispatcher's state. After the primary has replicated its state to the backup, it runs a cache consistency protocol with the backup. All changes made to the state of the primary are replicated to the backup. If the primary fails, the backup takes over the address of the primary using a gratuitous ARP.

Many Internet sites rely on DNS-based techniques to share load across several servers (e.g., NCSA). These techniques typically include modifications to the DNS BIND code. There are several drawbacks to all DNS-based solutions. All DNS variations may disclose up to 32 IP addresses for each DNS name, due to UDP packet size constraints. Knowledge of these IP addresses may create problems for clients (e.g., unnecessary reload of HTML pages in cache) and also for network gateways (e.g., for filtering rules in IP routers). Any caching of IP addresses resolved via DNS creates skews in the distribution of requests.

DNS-based solutions are very slow (or unable) to detect server failures and additions of new servers. Furthermore, while the server host may be running properly, the specific server software (e.g., httpd) may have failed. In a pathological case, a load-balancing DNS tool may see that a server is under-loaded (because the httpd daemon failed), and give it even higher priority.

Several packet-forwarding hardware products have been introduced (e.g., Cisco's LocalDirector). These devices translate (rewrite) the TCP/IP headers and recompute the checksums of all packets flowing between clients and servers in both directions. This is similar to Network Address Translation (NAT), adding the choice of the selected server. Notice that this device quickly becomes the main network bottleneck.

IBM's Experience Implementing High-Availability WebSphere Servers

Here are details on IBM's experience in implementing two HACMP WAS Cluster projects at our server farms. The first case study is for an e-Commerce Application and is based on WAS V3.5 without load balancing. The second case study is based on WAS V4 with load balancing.

Case 1—An e-Commerce Application with HA but Without Load Balancing

To satisfy the high-availability requirements associated with an Internet facing e-commerce application (which we'll refer to as "ECAPP" here), we implemented an HACMP cluster of two AIX servers in a hot-standby configuration with both servers having WAS 3.5.2 installed. The WAS setup included the configuration of a DB2 database for WAS configuration data only. The application data that resided on the cluster servers was kept in a shared logical volume group on the AIX filesystem (most of the application data was immediately pushed to internal data processing servers). The WAS configuration database was not shared because of certain difficulties, which we'll discuss.

The following diagram (Figure 14-5) shows the configuration of the HACMP cluster for the ECAPP implementation.

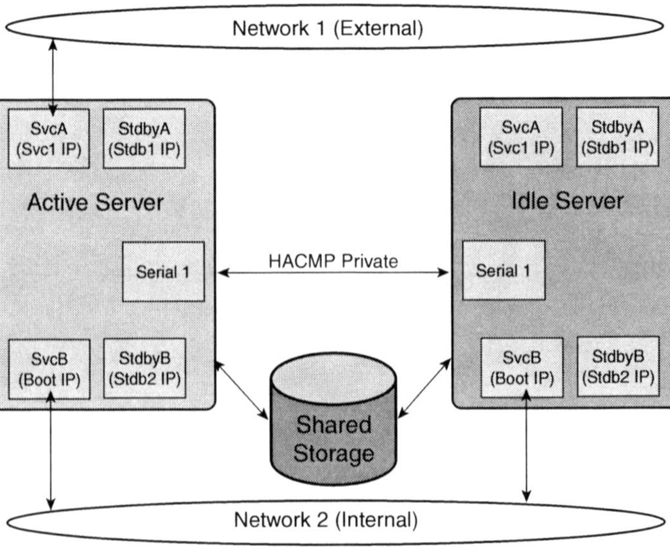

Figure 14-5 HACMP WAS cluster for an e-commerce application.

Let's try to clarify some items in the diagram. Both networks shown are "public" networks to the HACMP function (no HACMP cluster configuration data flows on either network). For each network, there is a "service" IP address defined, which provides external access to the cluster. The service IP remains the same regardless of which server is active. Each server must have a unique IP address on which the operating system can be accessed. This is referred to as the "boot" IP address. When HACMP is started on the active server, it reconfigures the service adapter for each network and replaces the boot IP with the service IP.

The hardware configuration for the ECAPP server cluster consists of the following:

ECAPP Servers Qty. 2, AIX 4.3.3, HACMP

- 7025 F50 RS/6000 Model F50
- 1×16-bit PCI SCSI SE adapter to 6 bays cable
- 2×9.1GB Ultra SCSI HDD for operating system
- 2×18.2GB Ultra SCSI HDD for logging
- 4×10/100Mbps Ethernet adapters
- 8×256MB SDRAM DIMMS (2GB total)
- 2×2 way RS64 II 332MHz processor card

- 1×SCSI Hot Swap 6 Pack
- 1×RS-232 serial cable

External Drive Array Qty. 1

- 7133-D40 Advanced SSA Subsystem
- 6×18.2GB 10K Advanced HDD
- 1×50/60Hz AC 300 VDC Power Supply
- 12×2.5m Advanced SSA Cable

Note the use of four separate Ethernet adapters (NICs) per single server. We choose to make both the Internet facing and internal network interfaces redundant on each server and to configure HACMP adapter failover for each NIC pair. The RS-232 serial cable is used for the HACMP serial network (the intra-node "heartbeat"). The external SSA subsystem provides the shared storage among the clustered servers.

The HACMP cluster is configured in "hot-standby" mode—that is, one server is always active in the cluster while the other server sits idle but automatically can be made active should the primary server fail. The automatic switch of the idle server as the active server is referred to as node "failover" or "fallover." On failover, the idle server is dynamically reconfigured to assume the identity and resources of the primary server. Specifically, the following resources are "moved":

- The service IP network address/adapter binding
- The shared resources (in our case, the SSA backed mount points)

In addition, on failover, the WAS/DB2 configuration is started on the idle (now active) server and takes over the ECAPP processing.

The fact that the TCP/IP network addresses and hostnames are changed on failover complicates the WAS configuration in our ECAPP cluster. This is because WAS ties itself to the server's hostname on a normal install; that is, WAS uses the server's hostname as its node name and writes this host/node name to its configuration database. Thus, even if WAS were installed on the primary server as a shared resource, when WAS is started on the secondary server on a failover, the host name of the secondary server would not match the node name of the WAS installation. For this reason, we chose to install WAS separately on each cluster server. Since our WAS database is only used for configuration data, we choose to create separate database instances on each server as well.

Another complication arises because of the changing hostname to IP address associations on failover (both node and adapter). For various reasons, it is beneficial to have a fixed association between a server's TCP hostname and its IP address. This is important not only for server applications such as WAS, but also for system management functions such as backup and monitoring. One example of the difficulty with HACMP dynamically changing the adapter to IP associations (and thus the hostname to IP association) is when WAS is started on the failover node, its node name (hostname) resolves to a different IP address (the common service IP) then when it was installed (the server's boot IP) and results in error messages about the intra-WAS communication protocols.

To overcome this changing adapter/IP address complication, we defined a third IP address for each network interface point. This third IP address was made distinct from the boot IP and standby IP addresses although placed within the same subnet. The third IP is established as an alias IP for whichever adapter is active for the network connection. HACMP pre- and post-event scripts were written to configure the third IP as an alias for the service IP (using the AIX `ifconfig` command with the `alias` option). Thus, whenever HACMP is started or an adapter failover is performed, the third IP address is associated with the active (service) adapter and thus is maintained as a fixed IP address for that network connection. (Note that the third IP address is not made known outside of the server itself; external access to the server is still via the service IP address.)

We chose to configure the servers in a "rotation" type of HACMP node failover. This means that when the failed server reenters the cluster, it assumes the idle server role. We chose this approach to avoid the overhead of fallback to the restored server as would occur in a "cascade" type of failover. It also allows the entire failover process to be completely automatic.

Case 2—WAS V4 Web Servers with Both HA and Load Balancing

This section discusses an example of a high-availability WAS V4 implementation with load balancing that is based on a system IBM designed and implemented at one of our server farms. This WAS system supports a large Siebel (CRM) application. The implementation uses WAS V4.0, HACMP in active/active mode, and also uses Network Dispatchers at the front end. The design concepts for this system are typical of the large scale, high-availability WAS implementations that IBM has recently been designing for customers at our server farms. Most of IBM's early WAS high-availability implementations are based on WAS V3.5x and use AIX HACMP in active/standby mode. The example in this section is based on the following requirements and WAS high-availability design concepts. These requirements are typical of our recent WAS V4.0 high-availability designs.

General WAS high-availability requirements at IBM server farms:

- High-availability—99.9 percent availability (about eight hours of unscheduled downtime per year)
- Performance—High performance, capacity on demand
- Scalability—Technology platform that scales for CPU, Memory, Disk
- Reliability—Internal redundancy, fault tolerance, and automatic failover

Technology

The servers are IBM pSeries Nodes with high-availability, scalability, reliability, and performance features. The servers typically used for WAS V4.0 are the pSeries model 660 which are Midrange Enterprise Servers (6H0, 6H1, 6M1) with four to eight processors (450MHz to 750MHz), expandable from 32GB to 64GB of Memory (RAM) with internal or external disk storage and redundant internal system components.

HACMP clustering and WAS clustering provides workload management and failover capabilities. There are the dual benefits of high-availability and workload management with WAS clustering and HACMP. HACMP provides high availability from a hardware perspective, and WAS V4 Clustering provides software failover. These technologies work in conjunction to provide fault tolerance and high availability.

The environment can start out small and grow to the full capacity of servers. It is very scalable with HACMP in an Active/Active configuration since both servers in a cluster can provide continuous processing and failover. The theory behind high availability is to reduce the single points of failure where possible and build redundancy within the infrastructure to mitigate risk. The redundancy is automatic and seamless throughout the infrastructure to reduce the impact on performance.

For the enterprise systems, HACMP (High-Available Cluster Multi-Processing) is used to help build high-available systems from the hardware level. On the software side, WebSphere Clustering is used to build a high-available WAS environment in either an active/active or active/standby mode. By coupling both HACMP and WAS Clustering, you can introduce the benefits of both hardware and software failover to build a robust and highly available node.

While HACMP or WAS Clustering by themselves provide decent HA solutions, both technologies need to be combined to mitigate the risks of a failure. Using both HACMP and WAS Clustering within a WAS Node, the two technologies work together to allow continuous processing. Shared Disk configuration provides redundant access to application data and log files to ensure that all messages are processed. In an active/active configuration, also known as mutual takeover configuration, multiple nodes can simultaneously be running workload.

Recommendations on the Best Ways to Achieve WAS High Availability

High availability including failover and recovery in WebSphere is a very broad subject, and there are a lot of options. This section provides some advice, tips, and best practices specific to using WAS (including V4 or V5) to provide high availability. Generally it is best to first consider using WebSphere's built-in failover capability. Second, consider application code best practices for failure related exception handling and recovery. Third, consider using other IBM and third-party products in conjunction with WebSphere such as HACMP, MSCS, and Network Dispatcher. Here are some details:

- WebSphere's built-in failover capability is achieved using the Work Load Management (WLM) facility at the Web Module, EJB Module, and Administrative Server levels. WAS V4 with "Server Groups" or WAS V5 with "Clusters" provides a WebSphere administrator with the ability to create any number of application server instances, or clones, of the server group or cluster. These clones can all reside on a single node or can be distributed across multiple nodes in the WebSphere Domain. Clones can be administered as a single unit by manipulating the server group or cluster object. WebSphere clones can share application workload and provide failover support. If one of the clones fails, work can continue to be handled by the other clones in the server group or cluster, if they are still available.

- Application code best practices for failover exception handling and recovery means that the application code should be able to handle JDBC exceptions and rollover to a "healthy" database server. This is achieved by ensuring that the application code is "failover-ready." It is important to remember that even in a WebSphere environment that employs high-availability mechanisms such as HACMP, MSCS, or Network Dispatcher, there is still a disruption in service while the database is switched from a failed server to an available server. Details on how best to code for the application to recognize that the database is not responding to requests and to easily reconnect to a database server once it has recovered from a failure are given in the IBM Redbook listed later in the chapter.
- In addition to WebSphere's built-in failover and load balancing capabilities and application code best practices for failover exception handling and recovery, there are a variety of IBM and third-party products that can be used in conjunction with WebSphere. These products include the OS clustering software discussed earlier in this chapter, such as AIX HACMP and Microsoft OS Clustering (MSCS). Most of the OS clustering products can be used in either an active/passive (hot standby) or active/active configurations. OS clustering (also called hardware clustering) provides a cluster manager process that periodically polls (checks the "heartbeat") the other software processes in the cluster to determine if the software and hardware it is running on is still active. If a "heartbeat" is not detected, then the cluster manager moves the software process to another server in a cluster. The movement of the process from one machine to another is not instantaneous but is usually accomplished within a minute or two (thus, the term "high availability" rather than the ideal "continuous" availability). OS clustering techniques rely on a shared disk array that is shared by all servers in the cluster—hence, OS clustering is not applicable for failing over between two sites over the wide area network.
- IP Sprayers such as Network Dispatcher (WebSphere Edge Server) can also provide high availability and load balancing—without the need of shared disk arrays. So with IP Sprayers, you could have failover across the wide area network. Also, IP Sprayers such as Network Dispatcher are needed for Web sites that require massive scalability since IP Sprayers provide for load distribution and failover for incoming HTTP requests.

These options for WebSphere high availability are shown in Figure 14-6. Notice that the diagram includes Network Dispatchers, WAS clusters (V4 server groups or V5 clusters), HACMP options, etc. A good document to reference for all of these options is the IBM Redbook, “IBM WebSphere V4.0 Advanced Edition: Scalability and Availability,” located at `http://www.redbooks.ibm.com`, document SG24-6192, May 2002.

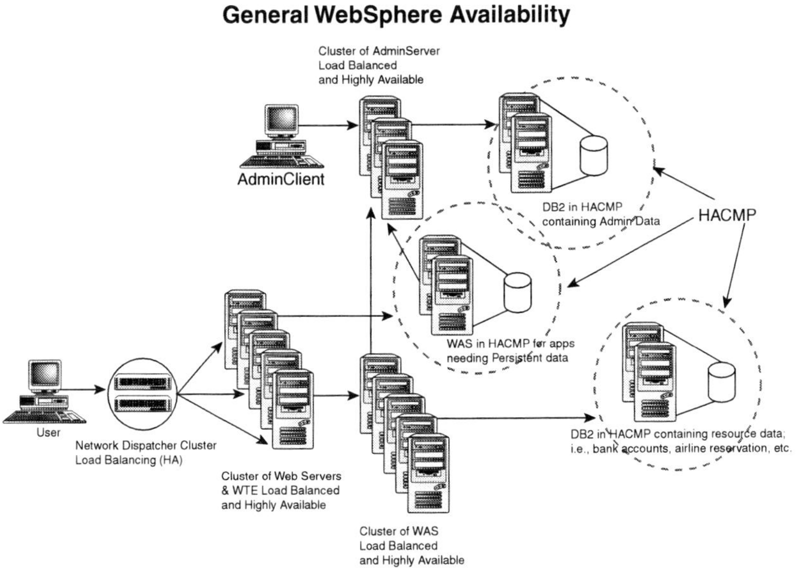

Figure 14-6 General WebSphere availability options.

The basic ideas for the different WAS load balancing and failover options are also given in the following table. The type of failover and load balancing is dependent on the high availability requirements. If you only need failover without load balancing, an HACMP implementation with an active/passive configuration is sufficient. IP Sprayers such as Network Dispatcher are required for very high scalability for your Web servers or where failover between two sites is required.

Table 14-2 WAS Load Balancing and Failover Options

High-Availability Requirement	WAS V3.5x	WAS V4	WAS V5
Failover only; Active/passive (i.e., no need for high volume).	OS Clustered Servers in Active/Passive mode using HACMP.	WAS "server groups" to allow software load balancing with HACMP for system administration server failover.	WAS "clusters" for load balancing with HACMP for system administration server failover.

continues

Table 14-2 Continued

High-Availability Requirement	WAS V3.5x	WAS V4	WAS V5
Failover & Load Balancing. Single Site.	WAS with HACMP active/active Net Dispatcher (WebSphere Edge Server).	WAS "server groups" with HACMP active/active or Net Dispatcher.	WAS "clusters" with active/active HACMP or Net Dispatcher.
Failover & Load Balancing between two or more sites.	Net Dispatcher	Net Dispatcher	Net Dispatcher

Conclusions—So What's the Best Way to Provide High Availability for Your WebSphere Servers?

The first thing to do in providing high availability for your WebSphere servers is to first leverage the built-in failover capabilities in WebSphere. Those capabilities vary with the version of WebSphere used, with V4 and V5 providing the greatest capability (e.g., leverage "server groups" in V4 and "clusters" in V5). Next, the application code should be designed for failure-related exception handling and recovery. After those two steps, then consider use of the two basic hardware approaches to high availability for WebSphere servers discussed in this chapter. The first approach is to use some sort of OS clustering (e.g., AIX HACMP or Microsoft MSCS) that provides failover with or without load balancing. The active/passive approach works well if high availability is required without high transaction rates. OS clustering systems that allow load balancing in addition to failover allow for greater transaction rates. Since OS clustering includes a disk array that is shared by Web servers, OS clustering is not an appropriate consideration for failover between different server farms.

The second hardware approach relates to the case of having a high-volume system where a very scalable system is needed (e.g., the Web server for the Olympics); then some sort of Web sprayer (e.g. Network Dispatcher) provides a very good solution. It is a very scalable solution since it's easy to add Web servers to group of servers under the control of Network Dispatcher. For failover and load balancing between two widely separated server sites, IBM has used techniques such as Network Dispatchers and Cisco Distributed Directors.

However, as discussed in this chapter, high availability can be solved with a continuum of techniques. A very basic way is to use a fault-tolerant server that has two of everything (Tandem computer or IBM zSeries). Highly fault-tolerant computers can be an expensive solution since half the components are always in standby mode. These days, almost all large Web servers do have many elements of fault-tolerant computers, such as RAID hard drives, redundant processors via Symmetric Multiprocessing (SMP), and redundant power supplies. Another basic solution would be to use round-robin DNS instead of devices such as Network Dispatcher.

Since very high availability for Web servers (especially those related to e-commerce) is becoming more and more important, we can continue to expect to see significant advances in the technology to support high availability for WebSphere servers.

Net Dispatchers and HACMP for WebServer High Availability

IP sprayers like Network Dispatcher provide a good method for both failover and load balancing for WebSphere systems that require high volume (high number of "hits" a minute) and high availability with failover. OS clustering such as AIX HACMP is a good way to have HA usually without load balancing. This section describes a high-volume, high-availability public access Web server system that IBM hosts for a customer at our Schaumburg, Illinois, server farm call an IBM Universal Server Farm (USF). IBM's USFs are very large server farms that host Web servers for many major IBM customers. The Web server system described here uses both Net Dispatchers and AIX HACMP. The diagram for this system is shown in Figure 14-7. The system consists of eight AIX servers accessed via the Internet plus numerous backend servers.

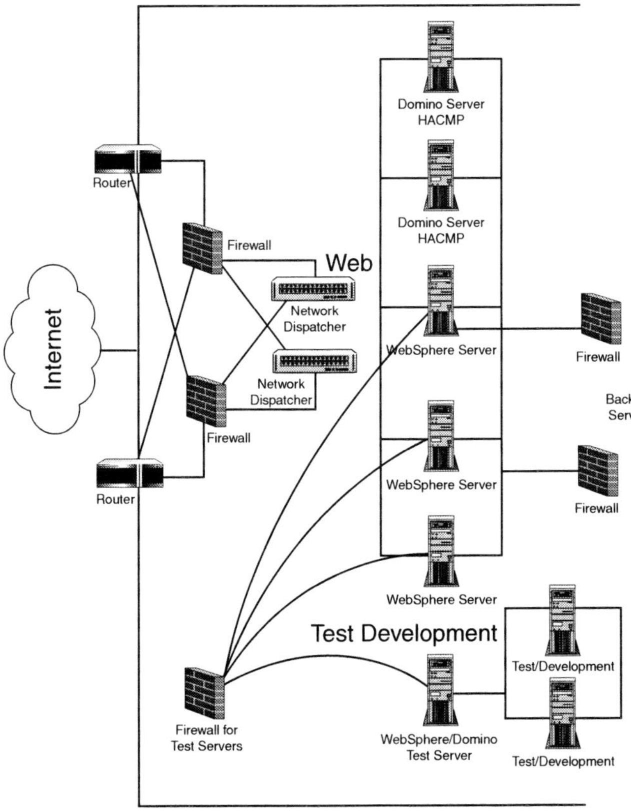

Figure 14-7 Net Dispatchers and HACMP for Web server high-availability design.

As indicated in the diagram, the eight AIX servers accessed via the Internet are made up of two Network Dispatchers for failover and load balancing: three large WebSphere servers that serve as the main public Web server system, two Domino servers that share a hard drive system via HACMP, and one large test server for both WebSphere and Domino. The configurations for the AIX servers are given here. Notice that the Net Dispatchers are relatively small servers (only 256MB RAM, 1×200MHz processor), while the WebSphere servers are quite large (e.g., 8GB RAM, 4×500MHz processors).

Net Dispatcher Qty. 2

- 7043 RS/6000 Model 43P
- 1×200MHz processor
- 256MB RAM
- 2×4.2GB HDD (OS—Mirrored)
- 3×10/100Mbps Ethernet adapters
- Software: AIX 4.3.3, WebSphere Edge Server V1.0, and eNet Dispatcher 3.6.0.8

WebSphere Servers Qty. 4

- 7026-M80 RS/6000 Model M80
- 4×500MHz processors
- 8GB RAM
- 2×9.1GB Ultra SCSI HDD (mirrored) for operating system
- 4×18.2GB Ultra SCSI HDD (mirrored) for data
- 6×10/100Mbps Ethernet adapters
- 2×Power Supplies
- Software: AIX 4.3.3, IBM HIS v1.3.12.0, and WebSphere 3.5 Advanced Edition (both with Service Pack 2)

Domino Servers Qty. 2

- 7025 RS/6000 Model F50
- 2×332MHz processors
- 256MB RAM
- 2×9.1GB HDD (OS – Mirrored giving 9.1 GB usable)
- 2×9.1GB HDD (logging – Mirrored giving 9.1 GB usable)
- 3×10/100Mbps Ethernet adapters
- 2×Advanced SerialRAID Adapter
- Software: AIX 4.3.3 , HACMP 4.3, and Lotus Domino Server R5.0.8

External Drive Array (shared by both Domino servers) Qty. 1

- 1×7133-T40 Advanced SSA Subsystem
- 4×9.1GB HDD (customer data—mirrored giving 18.2GB usable)
- 6×5 Meter Advanced SSA Cable

Users accessing this very high transaction rate system are load balanced across the three large WebSphere servers via the two Net Dispatchers. Three large WebSphere servers are required to handle the heavy load. The three WebSphere servers in turn access the Domino system to handle such things as customer registration. The Domino servers are not required to be a high-volume system but are required to have high availability. Therefore, the two Domino servers use AIX HACMP for failover but do not have load balancing. Notice that each Domino production server is a relatively small machine with 512MB of RAM, 9.1GB (mirrored) hard drive for the OS, and 9.1GB (mirrored) hard drive for logging. The shared disk drive system for the Domino Application and Data consists of 4×9.1GB drives (mirrored), for an effective size of 18.2GB. The mirrored (RAID 1) aspect of the shared drive system gives the shared drive very high availability and reliability, which is necessary since the shared drive system is effectively a single point of failure.

Because of HACMP clustering, there is only one Domino server name. Most of the time only the primary Domino server in the cluster is running. The secondary Domino server comes up only when the primary Domino server fails. Under failure conditions, the secondary server will come up, use the shared disk subsystem, restart the respective applications, and continue with the processes. It will use the same disk where the Domino data resides and the same IP address that the Domino server is using. Thus, the failover function is transparent to the WebSphere servers that access the Domino system. This Web server system using both Net Dispatchers and HACMP has been in place for over three years and has worked very well.

CHAPTER 15

Implementing High-Volume Web Sites

"The Network is the Computer."

—Sun Microsystems advertisements for the last several years

This chapter looks at the design aspects of high-volume Web sites—especially those based on WebSphere and Lotus Collaboration tools. IBM's experience designing the Web sites for the Olympic Games, which included Domino for collaboration, will serve as a way to give the history of the evolution of these Web site design concepts. Many of the design concepts for high-volume, high-availability, high-profile sites that included collaboration features went through the intense proving ground of Web sites for the Olympics and other international sporting events.

IBM and other computer services companies have long hosted Web servers on the Internet for corporations as part of "outsourcing" agreements. For example, IBM hosts the information Web site for the New York Stock Exchange at `www.nyse.com` and the transactional Web site for Victoria's Secret woman's underclothing on-line sales at `www.victoriassecret.com`. Both of these Web sites are very high-volume, high-profile sites, and the customer in each case needs to know details on any problems with the availability of the site in addition to the details on the use of the site (number of "hits," hotspots, network bandwidth required, etc.). Detailed "data mining" on the use of a Web site is often included as part of the outsourcing agreement. However, there are two new aspects of Web services that companies such as IBM are starting to offer. These are in the areas of "collaboration" services and "on demand" options. The need for collaboration services has been driven largely by the significant growth in Web commerce between companies or business-to-business (B2B) commerce. B2B commerce requires much greater need for relationships and collaborative technologies (such as Web conferencing, instant messaging, and Web teamrooms) to facilitate purchasing for large, complex transactions between teams of people. Also sites with just transactional capabilities are realizing that they could increase the amount and frequency of purchasing by assisting transactions. On demand services allow companies to pay for only the amount of computing power they use (as you would pay for electricity from the power company). On demand computing allows companies to immediately and automatically ramp up power to their Web site during required periods (e.g., the Christmas season or after advertisements at the U.S. Football Superbowl for a site such as Victoria's Secret).

This chapter will discuss the design of high-volume Web sites along with new features such as collaboration and on demand options—all of which can be implemented with WebSphere and Domino systems. Often the experience gained as a best practice for solving the problems of high-volume Internet sites carries over to Internet sites of all sizes. For example, the use of hardware SSL accelerators to offload the encryption function from your high-volume Web servers is usually a good idea even when you only have one moderate size Web server. Hardware SSL accelerators can usually handle at least 10 times the SSL sessions that can be handled by a general purpose Web server, so it's a simple way to allow for significant scalability for your Web server—even when you're just starting out. IBM's high-volume sport Web sites make use of both WebSphere and Domino, with many WebSphere and many Domino boxes. However, the IBM lessons learned using WebSphere and Domino together for these large sport Web sites carry over to Web sites of almost any size.

History of IBM's Experience with High-Volume Sport and Event Web Sites

Much of the information in this section was provided by Paul Dantzig, manager of the High-Volume Web Serving group at IBM Research in Hawthorne, New York. Professor Dantzig has been involved with high-volume Web serving design for the IBM Sport and Event Web sites since before the 1996 Summer Olympics in Atlanta, Georgia. He has also applied the "lessons learned" from these high-volume sport Web sites to commercial sites such as `www.ebay.com`, where he has served as a design consultant. Copyright 2002 ACM, Inc. is included here by permission (see the Paul Dantzig reference in Bibliography for details of the ACM paper, which is the basis for much of the material in this section).

Architecting and designing high-volume Web sites has changed immensely since 1996 when IBM hosted the Olympic Web Site for the Atlanta Olympics. These changes include the availability of inexpensive Pentium based servers, Linux, Java applications, commodity switches, connection management, caching engines, bandwidth price reductions, content distribution services, and many others. This section describes the evolution of the best practices within IBM in "architecting" sites that handle millions of page views per day. Discussed is the transition to multi-tiered architectures, the use of publish/subscribe software to reduce Web site hits, the migration from static to dynamic content, and techniques for caching of dynamic and personalized content.

Domino played a significant role in IBM's design of the Olympic Web sites and other sports and events sites. Domino's role was in the collaboration areas of Guestbooks, Feedback, Kids (mail, stories, pictures), and Digital Postcards. Domino replication was used to distribute the Domino content to each of the sports and events server complex, and then Domino replication was used again to distribute the content throughout the servers within a complex.

Evolution of Design for Very High-Volume Web Sites

The initial design of high-volume sites consisted of clustering a set of Web servers together and either using round-robin Domain Name Servers or routing at the TCP level. All the Web servers were fully replicated mirrors of each other (containing the same database, application software, Web serving software, etc.). Users of such systems often experienced inconsistencies in a site's content due to lags or failures in replication as different servers served each page. The cost of replicating databases on the serving nodes consumed cycles needed for serving, resulting in inconsistent response times.

To reduce the consistency and performance issues, sites began implementing two-tiered architectures. The first tier consisted of a set of systems containing just Web servers, and the second tier consisted of back-end servers such as databases, payment systems, and other legacy systems. This type of architecture provided significant relief from the above problems. Reduced numbers of database replications improved consistency and consumed significantly fewer resources. Response time for static requests was not impacted by the dynamic requests, which were now run in the back-end systems. Some high-volume sites grew to contain hundreds of front-end nodes. Scaling the backend usually consisted of installing bigger and bigger machines to avoid the database replication problems. It was quickly realized that the primary difficulty of running hundreds of nodes was not the hardware cost but the personnel cost to manage them.

Designing Web Sites to Support Over One Million "Hits" Per Minute

In order to help reduce the number of "front-end" Web servers for high-volume Web sites, many companies entered the Web server accelerator market (Cisco, Cacheflow, Network Appliance, IBM, and others) to market new devices to reduce the number of servers. These devices typically handle an order of magnitude more traffic than a single Web server running on a general-purpose operating system. These devices run as reverse proxy servers with large amounts of memory and disk. In addition, specialized switches and routers came on the market to distribute connections to the Web servers or the Web server accelerators. Newer devices on the market function both as proxies and connection routers (load balancers), which use heuristics such as the number of connections, server load, and content to distribute connections to multiple servers.

In addition to specialized hardware, new software techniques were developed to improve the performance of dynamic pages. One technique that was quite radical at the time was to cache dynamic pages the first time they were created. Methods for constructing pages from individual fragments allowed expensive content to be generated once and shared among many pages. However, successful deployment of these techniques requires much stricter cache consistency. In order to maintain the consistency of the cached data, the server needed to explicitly manage the cache contents. Algorithms for keeping cached dynamic data consistent were developed.

Dynamic data was realized to be nothing more than rapidly changing static data. To keep certain types of content up-to-date required reducing the expiration times to tens of seconds. Some Web sites adopted automatic refresh of pages thus creating an enormous numbers of hits. The new problem was how to reduce the number of hits without sacrificing the better response time to which users had become accustomed.

New Designs with Sports Consoles, Persistent Sessions, and Java Applets

For some types of content, in particular sports scoring results and financial market data, even better response times were considered necessary. Sport and Financial consoles have been created to invoke Java Applets, which open a long-term connection to a "publish and subscribe" system using either private protocols or published protocols, such as Java Message Service. As content changes, the publish and subscribe system pushes changes to those users interested in the change. These systems have become enormously popular with hundred of thousands of active consoles being supported.

Many high-volume Web sites require 100 percent availability. Geographically redundant sites and content distribution services (such as Akamai) place servers and proxy servers closer to the user's ISP connections. In addition to improving availability, using multiple geographic sites or content distribution services can reduce the latency for delivery of graphic and streaming content.

Role of WebSphere and Domino in IBM's High-Volume Sport Web Sites

Both WebSphere and Domino played significant roles in the design of IBM's High-Volume Sport Web sites. For the 2000 Olympics at Sydney, Australia, almost half of the Web server role was Domino. Collaborative—interactive features for sending mail, postcards, guestbook entries, and feedback based on Domino could be used both by the athletes and the "virtual spectators" by accessing the Web Site. Domino replication and the use of NotesPump provided the distribution and publishing of Domino content.

The following sections give examples of how the IBM Sport and Event Web sites were impacted by the high-volume Web site technologies discussed previously.

1996 Atlanta Summer Olympic Web Site

Like many first of a kind Web projects during the early years, the 1996 Atlanta Games Web Site was architected, designed, and programmed in a hurry. The project started in September of 1995. Figure 15-1 shows the basic architecture for the part of the Atlanta site for handling results on the Web.

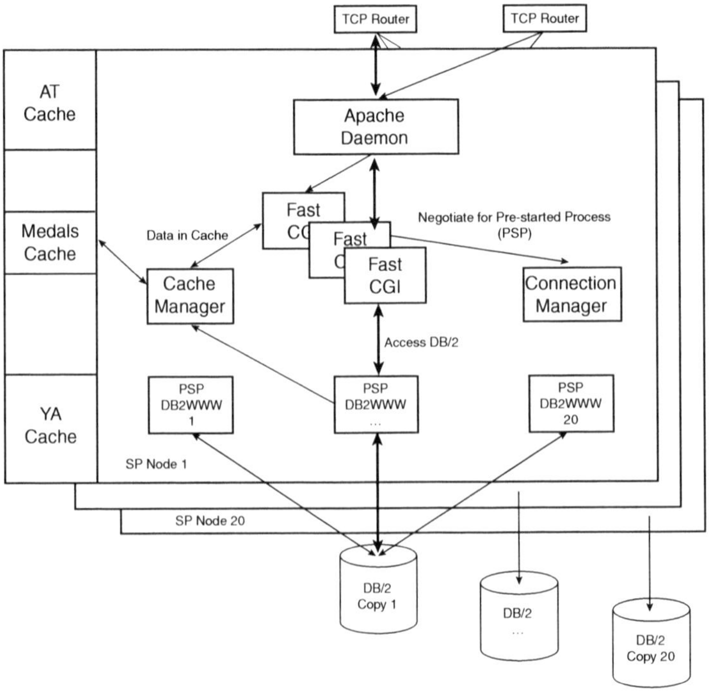

Figure 15-1 1996 Atlanta Summer Olympics Web site architecture.

The basic system setup was a cluster of 20 IBM RS6000s rack mounted in two SP2 Frames and two experimental TCP/IP router boxes for routing the incoming HTTP traffic. Each node contained identical software including the Apache HTTP server, a Fast CGI Apache extension, a connection manager daemon, a cache manager daemon, a database, and 20 prestarted processes reconnected to the database that performed SQL queries in the database if the memory cache did not contain the prebuilt page. Each sport and medals information section was allocated a separate cache memory. If a node were restarted, the caches would come up completely empty.

The first request for a page would invoke the following actions:

1. The HTTP requests were distributed to a server node based on system load from the TCP/IP Router to the Apache Process.
2. Apache called the Fast CGI Apache Extension.
3. The Fast CGI Apache Extension queried the Cache Manager with the unique cache ID (part of the URL).
4. If the page was cached in the appropriate Sport or Medal Cache, the cache manager returned the data, which was then returned to the user.
5. If the page was not cached; the connection manager was called to negotiate for a prestarted database process if available.

6. The prestarted process negotiated for and retrieved information from the database and generated an HTML result page. The Cache Manager was invoked to store information in the appropriate cache. Finally, the page was returned to the user.

In order to maintain cache consistency, whenever an update was made to one of the database tables related to a sport or medal, a database trigger invoked a process to delete all entries in the specific cache for the affected sport.

Since the system depended on keeping 20 nodes in sync (which was difficult to do), node-to-node consistency and performance problems were both quite visible to the user community. The characteristics of sporting events made the performance problem even more obvious. Users would note when an event would end and inundate the system with requests. This was precisely when the result arrived in the databases, causing the entire sport cache to be invalidated. All the requests would have to be resolved from the database until that data finally arrived in the cache. There was more than an order of magnitude difference between a cache hit and a database hit in response times.

Over the 17 days, the 1996 Atlanta site received over 190 million hits. 114 million of those hits were result requests. Peak result server hits reached nine million on August 1, 1996. Cache hit ratio averaged 80 percent, and a peak of 45 cache transactions per second per node was achieved.

1998 Nagano Winter Olympic Web Site

By 1998, IBM had signed up to sponsor several major sporting and other events. These new events included the U.S. Open Tennis Tournament, Wimbledon, the French Open Tennis Tournament, the Grammy Awards, the Tony Awards, among others. It was decided to invest in building a permanent infrastructure that could last several years to handle the Sport and Event Web sites. This new infrastructure was to be architected, designed, built, and tested before the Nagano games.

To achieve a high level of redundancy, three permanent geographic locations were chosen close to major Internet backbone sites. One east coast and two midwestern United States sites were constructed. In addition, a fourth temporary site for Asian traffic was established in Tokyo. Each site consisted of either three or four racks of 11 RS6000 nodes (one SP2 Frame). Figure 15-2 shows the configuration of one rack of nodes.

One of the criticisms of the Atlanta Web site was that finding information required too many page fetches. The Atlanta site was organized by results, news, photos, and so on—and by sports and events. Although country information and athlete biographies were provided, results corresponding to a particular country or athlete could not be collated. Consequently, the Nagano Web site organized results, news, photos, etc. by countries and athletes as well as by sports and events. The Web results displayed all of the information related to an event (the Atlanta site in many cases only had summaries). Some of the pages now were complex enough to require up to 30 seconds to produce the HTML tables (Ice Hockey results were a notable example). To reduce computational load, Country and Athlete pages were only generated every 15 minutes.

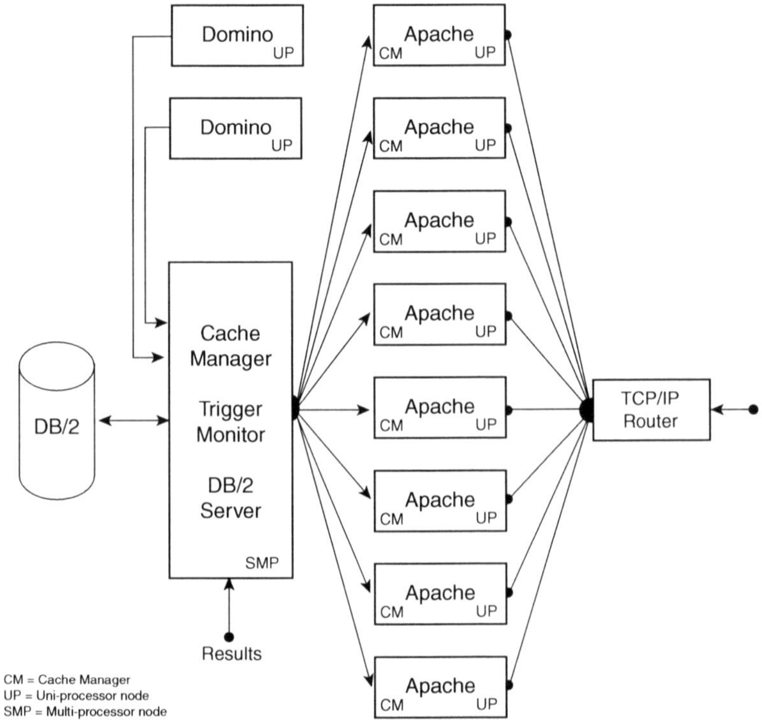

Figure 15-2 1998 Nagano Winter Olympics Web site architecture.

Shown in Figure 15-2 is the Nagano architecture. This architecture is a combination of two architectures: one to pregenerate content and perform content distribution and the other to serve Web requests. Each time a change was made to news, photos, and related content, the Domino server triggered a change. Each time a schedule, bio, or result changed, the database triggered a change. The pieces of information flowing into the system (each news item, result, etc.), when converted to a Web object (HTML, image, etc.), were now considered a fragment of information and were usable anywhere in the site. A new technology, Object Dependency Graph (ODG), was developed to keep track of the relationship between fragments and pages.

The Cache Manager on each node took on the role of a distributed file system with no persistence. In theory, a 100 percent cache hit-ratio could have been achieved with this type of system. Because it was a memory-based cache, pages infrequently requested, such as athlete biographies and country information pages, were frequently invalidated. Except for the country and athlete pages, the new model became a pro-active page replacement in which a page is replaced in cache and never invalidated. Using fragments and the ODG, only the pages rendering from page serving each server node now had consistent response time. Note: When an SMP node goes down, the cache information is lost for all the nodes; to avoid performance problems, key pages in the site were primed before the nodes were put back on-line.

Based on the lessons learned from Atlanta, database replications were minimized. Each site contained no more than four databases (one for each rack). Users were directed to one of four sites by using standard router algorithms (e.g., Japanese traffic to Tokyo or British traffic to the East Coast). If one site failed or needed to be down for maintenance, the routes automatically shifted load to the other sites. The only data inconsistencies were between sites, so users were unaware of the fact that the Asian site and East Coast site were not consistent with each other.

Nagano was an immense success. All of the goals for the site were achieved: 100 percent availability, good and consistent response times, every country and athlete treated equally, and no single points of failure. At the end of the event the publishing system contained approximately 50,000 pages in English and Japanese. The site had built 10,645 news article fragments, 14,355 photo fragments, 5,988 result fragments, and total combinations of the previous into 23,939 Sport, Country, and Athlete pages. The average number of pages rebuilt each day was 20,000, and on the peak day, 58,000 pages were rebuilt.

The Lotus Notes/Domino collaboration statistics were also impressive: 116,540 digital postcards; 12,990 kids' stories, pictures, and mail; 45,914 guestbook entries; 31,265 feedback entries; and 5,511 newsletter subscriptions.

The traffic to the site broke two Guinness records: *The Most Popular Internet Event Ever Recorded*, based on the officially audited figure of 634.7 million requests over the 16 days of the Games, and *The Most Hits On An Internet Site In One Minute*, based on the officially audited figure of 110,414 hits received in a single minute around the time of the Women's Figure Skating Free Skating (see Figure 15-3). Hits on the peak day were 55.4 million. Note, however, that Guinness records based on Web site activity are short lived. Figure 15-8, later in the chapter, shows that this 1998 Guinness record was surpassed the next year (1999) by about a four-fold increase in hits at the Wimbledon Tennis Web site, where 430 K hits per minute were recorded. Then at the 2000 Sydney Summer Olympics, an officially audited figure of 1.2 million hits a minute was recorded. This interest in growth in hits at Web sites for sports and event sites is also very evident in commercial Web sites, where companies are very interested in how many people are visiting their Web site and which areas of the site are most popular. How high-volume commercial Web sites have leaned from the earlier high-volume Olympic Web sites is discussed in the second half of this chapter.

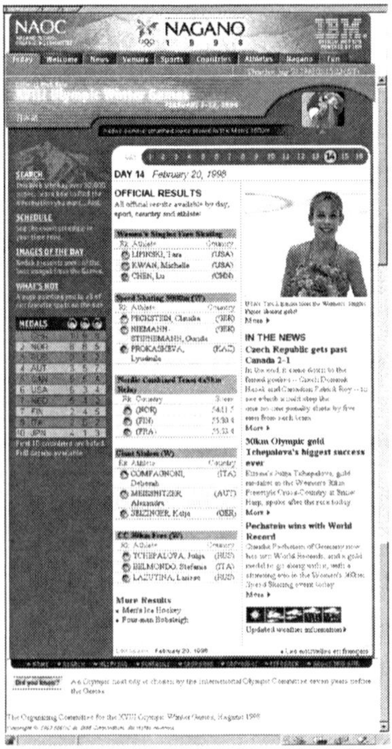

Figure 15-3 1998 Nagano Winter Olympics results page.

1999 Wimbledon Tennis Web Site

Even before Nagano, many of the large Web sites were already looking for cheaper and faster ways of serving static content. By 1997, many of them had switched from using expensive proprietary Unix systems (SUN, IBM, HP) to Intel-based systems running operating systems such as Free-BSD and Linux with Apache to reduce costs and improve performance. Also in 1997, IBM Research prototyped a combination router and proxy server cache that could support an order of magnitude more hits than a server on the Sport and Event Web sites could handle. The product was transferred to development and made available to the Sport and Event environment after Nagano was complete.

After a number of trial runs during events in 1998, four production Web server accelerators (IBM 2216s) were put into production for the 1999 Wimbledon event. Figure 15-4 shows the Web site architecture for that event.

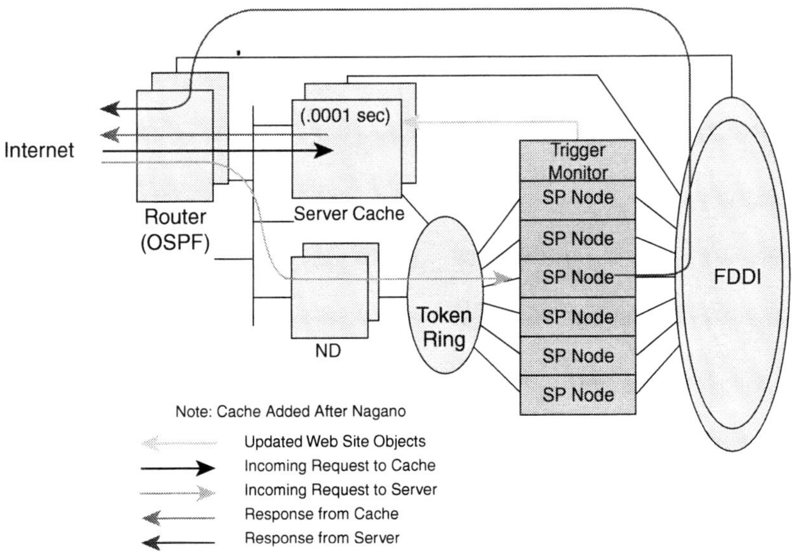

Figure 15-4 1999 Wimbledon Tennis Web site.

To get high-cache hit ratios (93 percent), result-scoring content was pushed into the Web server accelerators through a proprietary interface whenever content changed using the Trigger Monitor daemon. The proxy server built-in to each Web Server Accelerator requested other site content. Each Web server accelerator contained 500MB of memory, and the working set for Wimbledon was less than 200MB of memory.

The 1999 Wimbledon event ended with the officially audited figure of 942 million requests over the 14-day event. The peak hit rate of 430K hits per minute was received on June 30, 1999. Also received were 71.2 million page views and 8.7 million visits to the site. This event proved that inexpensive Web Server Accelerator hardware could reduce the size and cost of the infrastructure even with the enormous rate of growth of these events.

Many Web Server Accelerators (including the IBM 2216) produce no Web server logs, either because the logging function does not exist, or because logging impacts the performance of the accelerator. However, most do support SNMP interfaces, which can be used to record hits, misses, memory usage, and so on. This was the process IBM used to obtain information on the performance of the accelerators.

2000 Sydney Summer Olympic Web Site

Based on what we had learned from the previous events, the key to the success of the Sydney Olympic Web site would be making Web Server Accelerator cache misses low-cost. Two critical pieces were already prototyped. Learned from Nagano was how to make a system for prebuilding pages (Trigger Monitor) and distributing pages. Learned from Wimbledon was how to make a really good edge-serving infrastructure using Web server accelerators. Both of these systems needed critical improvements to handle the bigger loads anticipated in Sydney.

Installed after the 1999 Wimbledon event were 65 Web server accelerators. A fourth temporary server site in Sydney was planed to augment the three permanent server sites, which now had one East Coast, one Midwest, and one West Coast United States locations. An agreement was made with a telecom to collocate a large number of the Web server accelerators at five major Internet hubs in the USA. The installed capacity of the accelerators was between three and four million hits per minute. Figure 15-5 shows the server and Web server accelerator locations. The East Coast would handle the European traffic, and the West Coast the Asian traffic. The origin servers were located at IBM Global Services Hosting locations nearby. Experiments had been run during other events, which demonstrated that if *all* of the content (HTML, GIFs, JPEGs, etc.) were given *correct expirations*, content would not have to be pushed to the caches; the accelerators would simply proxy the content on a miss. The cache-hit ratio typically dropped to about 75 percent, which was within the hit capacity of the server sites.

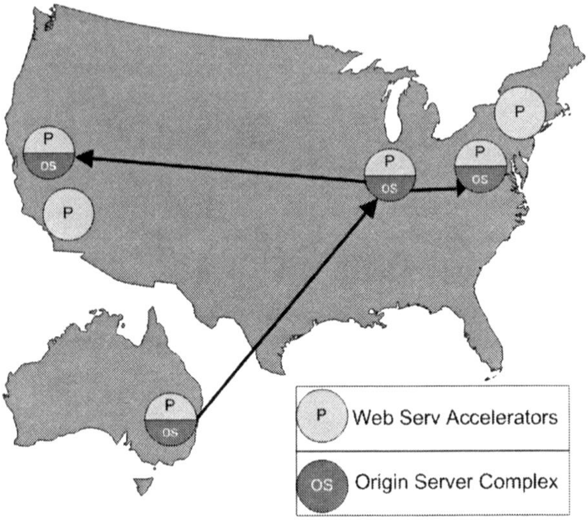

Figure 15-5 2000 Sydney Olympics origin server and Web server accelerator locations.

Because of the experimental nature of the dynamic page generation system, both the Atlanta and Nagano sites were designed and implemented as two separate sites—one of which served relatively static content and the highly dynamic results delivery system. Shortly before the games started, an ad-hoc effort was made to merge the sites and present the illusion of a single, unified site. It became quite clear during the merge phase of the Nagano site that the Sydney site would need to be constructed from the beginning as a single, unified site.

The combination of building a complete, unified site based on fragment technology, as well as improvements in the technology itself, produced an unforeseen problem: how to manage a very large number of fragments. Early estimates for the size of Sydney content were based on simply taking the number of sports, events, counties, athletes, etc. and multiplying Nagano times three (75K pages and 150K rebuilt pages a day). No more than four fragments per page and no imbedded fragments were supported in the Nagano publishing system.

There was a large shift between Nagano and Sydney from many expensive replicated engines at every site to a single, large, centralized multiprocessor engine for content management and many inexpensive Web server accelerators to serve the site. Also, in Sydney, there was only a single instance of any database, finally eliminating all replications.

Figure 15-6 shows a Sydney 2000 Web site page. It illustrates the depth of the information available to the Web site users. All of this information typically was available in less than one minute after the result system triggered a new result.

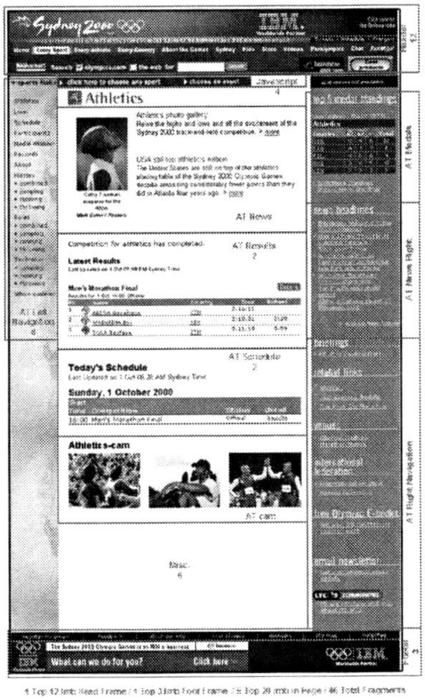

Figure 15-6 2000 Sydney Web page.

More than seven terabytes of data was distributed by the content distribution system to the 57 nodes. Scoreboard updates were delivered from Sydney to the East Coast, serving nodes in less than six seconds. Given automatic delays built into most television broadcasts of 30 seconds, users discovered that the site had scores on the Web before the broadcasters communicated them. Even the complex result updates with hundreds of pages changed were delivered on average in less than 40 seconds to the serving nodes.

It should be no surprise that every event record was broken again. The traffic to the site broke the most hits on a sport event site in a single minute based on an officially audited figure of 1.2M hits. Total traffic for the event over the 16 days was 11.3 billion hits. The Web server accelerators handled less than half (5.5 billion) of the hits; the ISP proxy caches and the user browsers handled the rest (due to the expiration policy). The Web server accelerators had an 80-90 percent cache-hit ratio during the entire event. Thirty percent of the log entries on the serving nodes were UC.GIF records. Other significant statistics include the peak traffic day of 875M hits, 2.8M visits, and 18M page views and a total for the event of 35M official visits and 222M page views. Sydney was completely successful and fulfilled the promises of being bigger than any of the previous events.

2001 Wimbledon Tennis Web Site

Sports fans are a difficult audience to satisfy. For Atlanta, almost all of the scores were delivered after the event was over. For Nagano, many events provided periodic updates. For Sydney, all the major sports had sport consoles with 30-second updates. With Wimbledon 2001, technology finally caught up with the fans desire for real-time scoring consoles.

With the new content distribution system, Web objects could be reliably distributed to all the geographic server sites in seconds. This enabled very short expirations of 30 seconds to be set on data for the scoreboard applets. With scoring data delivered to the most distant serving node in about six seconds and the scoring data's 30-second expiration, the scoring console on the user's workstation was nearly always up-to-date. New technologies based on "publish and subscribe" were added to the even infrastructure to support a real-time scoreboard. Figure 15-7 shows a sample tennis console from Wimbledon 2001. The console is a Java Applet and uses Java Message Service (JMS) as the protocol to subscribe with the message broker. Each user acquires a permanent TCP/IP socket to a message broker. The trade-off of the cost of hits to a Web server accelerator or a permanent connection to a message broker is based on update rate.

Using current technology update rates of less than 30 seconds, messaging tends to be cheaper (although research into the exact "break-even" point is needed). For the 2001 Wimbledon finals between Rafter and Ivanisevic, over 230K consoles were active. Two other sporting event records were broken during this event with peak hits per minute at 1.4M and peak page views in a day at 24M.

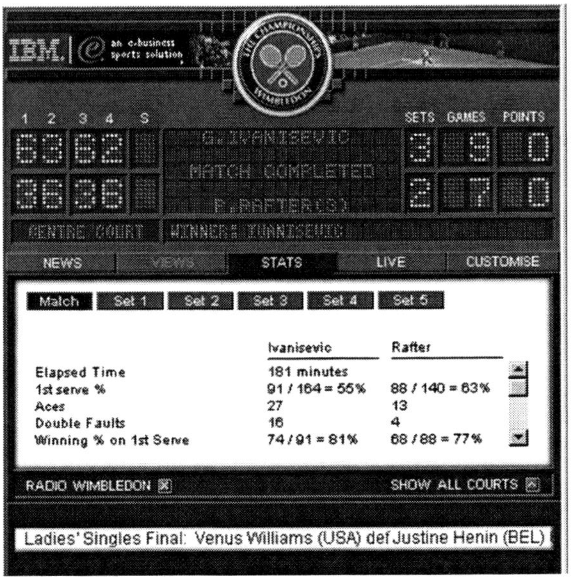

Figure 15-7 Wimbledon 2001 tennis console "real-time scoreboard."

Current Design for IBM Sport and Events Web Sites

Based on "lessons learned," all of the hardware and telecommunication equipment for the IBM Sport and Events system has been replaced in the last few years. The servers are now IBM Pentium-based processors running Linux. The Web server accelerators were replaced with commercially available equipment. Multiple vendors provide telecommunication bandwidth directly (OC-12s) into each of three geographic sites. Figure 15-8 shows the huge growth the sites have received. As the Web audience has increased worldwide, the sites have continued to receive more hits, page views, visits, etc., year after year. Notice that the 1998 Nagano Guinness record of 110,414 hits in a single minute was quickly eclipsed. The 2000 Sydney Summer Olympics at 1.2 million hits per minute was more than 10 times that Nagano Guinness book of records mark.

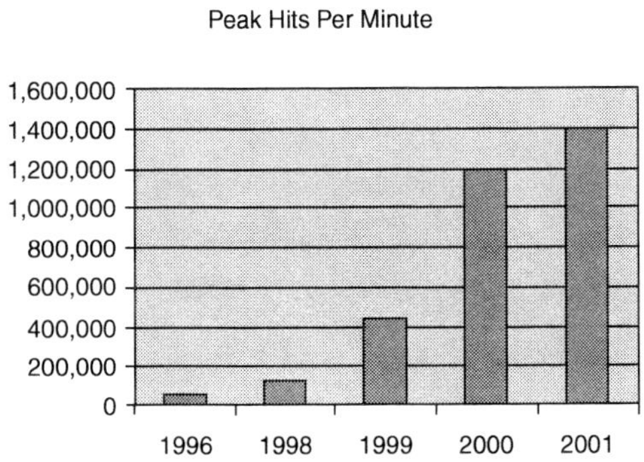

Figure 15-8 Growth of sport and events peak hits per minute.

Going from High-Volume Sport Web Sites to Commercial Web Sites

How has IBM gone from the valuable experiences of hosting high-volume sport Web sites to incorporating the best practices "lessons learned" into the design of large commercial Web sites? Some examples of high-volume commercial Web sites that routinely use the best practices learned from the high-volume sport Web sites are as follows:

- New York Stock Exchange Public Site, `www.nyse.com`
- eBay auction site, `www.ebay.com`
- Victoria's Secret woman's clothing, `www.victoriassecret.com`
- News/weather sites

First, let's look more in-depth at one of the most important aspects discussed for the Olympic Web sites in the previous sections: the use of Web server accelerators for the Olympics. These Web server accelerators fall into the general description of Web hardware appliance and are based on caching technology.

Web Caching Alternatives

IBM's experience with the Olympic Web sites and other high-volume sporting event Web sites helped show the significant benefits of using caching technology to increase the performance and scalability of Web sites. This section will discuss the hardware appliance, dynamic caching, and content delivery network (CDN) solutions available, explaining their differences, the types of content each serves best, pricing structures, and how they can act as complementary technologies as opposed to competitive ones.

Caching Theory

Caching is a fundamental computer science technique that involves the storing of frequently accessed content. It is used to address the Web site performance and scalability issues brought forth by ever-growing Internet traffic. Businesses tend to seek caching technology for three primary reasons:

1. Provide quality of service to end users
2. Protect from Web server stresses that cause site outages
3. Reduce performance-bogging site traffic

Caching theory suggests that when a particular resource is accessed, it is likely the same resource will be accessed again and again. As with the World Wide Web, one page request indicates that an identical one will be made shortly thereafter. Caching allows otherwise resource-intensive pages to be served without any burden to the Web server.

Hardware appliances, dynamic caching and CDN solutions, use caching theory as a basis to improve Web site performance, scalability, and uptime at a layer between the user's browser and the Web server. However, they all do it differently.

The Different Approaches

Hardware and CDN solutions increase Web site performance by handling page requests before they reach the server. Dynamic caching solutions improve Web site performance by reducing the overall number of resources necessary to compile and execute database-dependent pages and eliminate the need for page re-executions.

One way to think about it is that hardware and CDN solutions take pages that are easy to serve, like graphics and streaming media, and place them on multiple servers closer to end users. Dynamic caching solutions make hard-to-serve pages such as Active Server Pages, ColdFusion, and Java Script Pages easier for Web servers to handle.

Because hardware, dynamic caching, and CDN solutions specialize in different file types, they can be used together to produce better performance for the entire array of Internet file types. Contrary to what many believe, these technologies are complementary, not competitive.

Hardware Caching Solutions

Hardware solutions are caching servers deployed locally, either in the rack with the Web servers or at co-location centers with the Web servers (see Figure 15-9). The actual caching devices are basically computers with preinstalled software. The difference between this type of hardware and a reverse proxy server is minimal, although hardware usually offers alternative means of purging the cache.

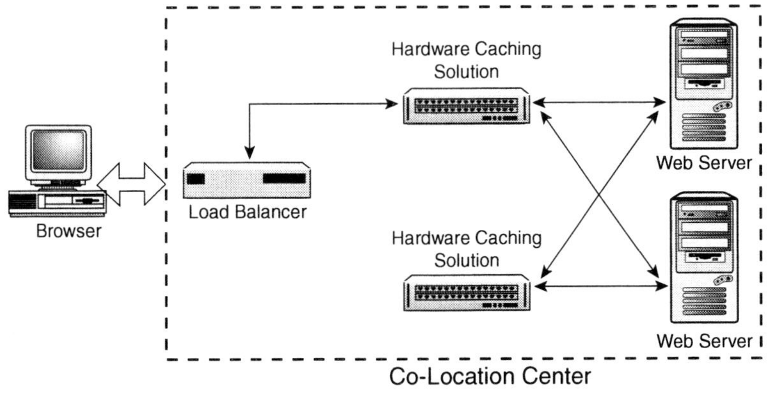

Figure 15-9 Hardware caching solution.

Caching hardware handles page requests before they get to the Web server, so if a requested page is found in cache, the hardware fulfills the request without communication with the server. This caching technique removes server load, freeing it to perform other tasks. To implement a hardware solution, each load balancer must point to the IP address of the device as opposed to the actual Web site. More than one caching device is typically needed for redundancy—you wouldn't want your Web site going down should you have to reboot the hardware cache.

Pages stored in a hardware cache are purged when they reach their expiration time, which is based on headers returned from the Web server. Alternatively, pages can be purged based on a maximum duration configured by the administrator of the hardware cache. Administrators also can decide which files should be cached and which should not. Hardware caching solutions are best suited for graphics and streaming media. Dynamic pages are not typically cached on hardware because there is no way for the Web server to update them on content or database changes. CacheFlow, Volera, and Application Server are some of the hardware vendors for this type of solution.

Dynamic Caching Solutions

Dynamic caching solutions are designed to cache database-dependent pages. Typically these are software products that are installed directly on the Web server. Unlike hardware and CDN solutions, dynamic caching solutions reduce the overall resource load on the Web server.

Two things distinguish dynamic caching solutions from hardware and CDN solutions:

1. When changes are made to the database, the same changes reflect in the cache.
2. The ability to perform partial page caching.

Partial page caching (see Figure 15-10, based on use of XCache Technologies as the dynamic caching provider) enables administrators to cache parts of pages while leaving selected elements such as stock quotes, personalized content, and rotating banners dynamic. This greatly reduces the number of server roundtrips and system resources required to fulfill requests. To execute uncached page pieces, most dynamic caching solutions must reside on the Web server, which is the primary reason hardware and CDN solutions do not perform partial page caching.

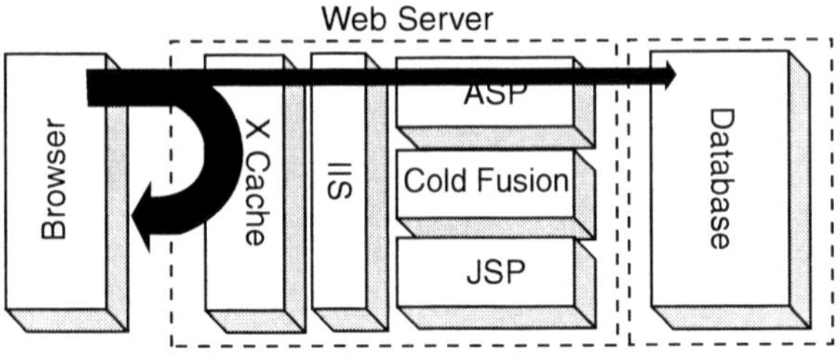

Figure 15-10 Partial page caching.

Administrators can also decide which files should be cached and which should not. Hardware caching solutions are best suited for graphics and streaming media. Dynamic pages are not typically cached on hardware because there is no way for the Web server to update them when content or database changes are made. XCache Technologies, Persistence Software, and Chutney Technologies are some of the dynamic caching solution providers.

Content Delivery Networks

Content Delivery Networks (CDNs) are caching servers deployed around the world to form a network (see Figure 15-11). They usually reside in co-location centers and have access to large amounts of bandwidth. CDNs provide a way to offload Web traffic and are primarily used for graphics and streaming media. Unique to CDNs is the ability to deliver content from servers that reside closer to end users. This is to say that when a graphic is requested from the network, the server sitting closest to the particular request serves the graphic.

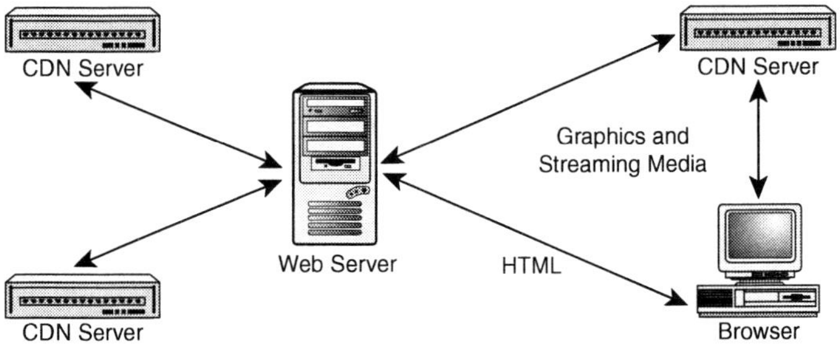

Figure 15-11 Content Delivery Network.

CDN servers often do not have the ability to "talk" to one another. This means if a user in Singapore requests a graphic, the CDN server sitting closest to the Singapore user will cache and serve the graphic, but a New York server belonging to the same network will not know to behave in the same manner. In other words, performance improvements are not realized until a second user in Singapore requests the same graphic, and a New York user will not realize any performance improvements at all. Implication: The more CDN servers there are, the more traffic a Web site has to have before performance improvements are achieved. This scenario assumes traffic is evenly spread across the CDN network.

CDNs know how to retrieve, cache, and serve graphics. How this works is the browser requests graphics from the CDN who then retrieves them from the Web server. Once the graphics are cached on the CDN, they are served directly to end users without any further server communication. An important concept to remember is that the HTML returned from the Web server must contain links to the graphics cached on the CDN for all this to happen, which typically requires code revisions. There are products available that can help change the HTML in dynamic Web pages so that they reference graphics located on CDNs.

Removing images and streaming media traffic from the origin Web server can greatly reduce load, a CDN specialty. On the other hand, dynamic pages that must be generated on the Web server must be served from the Web server—once they are placed on a CDN, they become static. CDNs do not have the ability to know when Web server data changes and therefore do not have the capability to update page content. CDNs typically charge monthly service fees for their service and look for customers who have sites attracting a million plus page views per month. CDN solution providers include Akamai, Mirror Image, and SolidSpeed Networks.

Choosing the right caching solution depends on a Web site's design and the areas needing improvement. If there is a need to cache secure information, hardware caching solutions can keep all caching in house. If offloading traffic from a facility is the goal, CDN solutions can cache graphics and streaming media on the edge. If database-dependent pages are slow to deliver, dynamic caching will dramatically improve site performance.

Design of a Public Information Site for an Exchange

This section gives information on the design of an Exchange public site, which is hosted by IBM at an IBM server farm. This Web site brings in many of the design concepts for large Web sites discussed in the previous sections. This site receives over a million hits per hour during business hours (as opposed to a million hits per minute for the Olympic Web sites). Figure 15-12 shows the design schematic for this Web site.

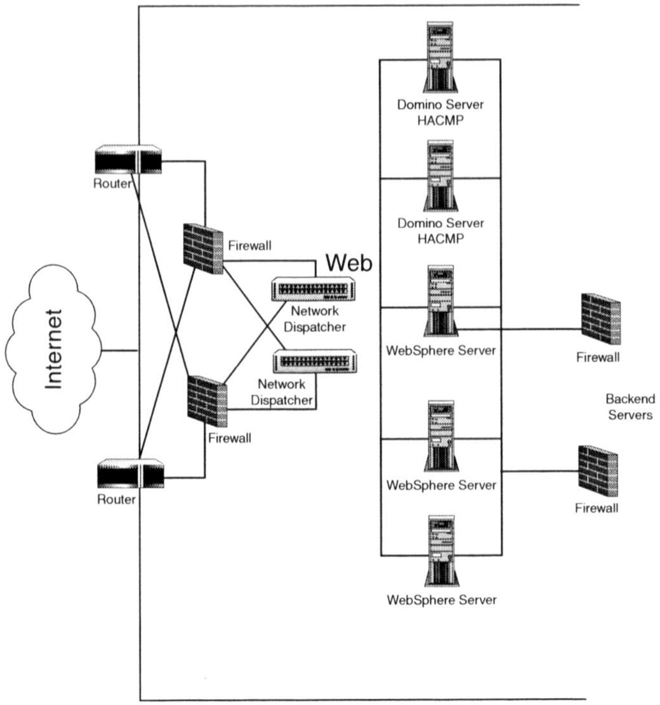

Figure 15-12 Design of a public information Web site.

This design for a high-volume Web site uses IP Sprayers for failover and load balancing provided by IBM Network Dispatchers. WebSphere Application Server is used for the three large Web servers. Domino servers with HACMP Active/Passive failover are used to register users and keep track of their portfolios. The Domino servers are accessed by the WebSphere servers, not by end users.

New Design for the Exchange Public Site

In order to provide for very even higher availability and disaster recovery, IBM proposed to the customer the use of multiple sites—in case of a disaster at one site, the second site would immediately be available to take all of the load. Instead of multiple sites, the customer chose to use a Content Delivery Network (CDN) as discussed earlier in this chapter. Akamai Content Distribution (`www.akamai.com`) was chosen. Akamai has hundreds of server sites worldwide, and users are now sent first to the closest Akamai site, which improves end-user response time and helps off-load the large Web servers. For some content the end users must still be sent on to the large Web servers. The Akamai servers constantly receive new information from the large Exchange Web servers, which is cached by the Akamai servers. The Akamai servers have features that can help prevent Denial of Service (DoS) attacks—an added benefit to use of a Content Delivery Network.

A Global Web Conferencing Offering on the Internet

This IBM Web conferencing offering gives another good example of leveraging "lessons learned" from the design of IBM's high-volume Web sites for the Olympics to commercial projects. IBM's On Demand Workplace (ODW) Web Conferencing is a hosted offering designed to allow customers to securely share Web conferencing resources. ODW Web Conferencing was also discussed at the end of Chapter 13 since it involves the use of WAS clusters.

This offering also has redundancy for all components. The system is built around the Lotus Web Conferencing 3.1 product (a.k.a. Sametime 3.1) including the Enterprise Meeting Server (EMS) option. Sametime 3.1 runs on a Domino 6 platform, so this is a Web application that is primarily about Lotus Collaboration and the role of WebSphere as that of a support role (as opposed to most Web server sites where the collaboration role is secondary).

The core software implementing the solution functionality is based on a set of IBM software products. In particular, the solution will leverage the following IBM products:

- Tivoli Monitoring
- Tivoli Storage Management (TSM)
- IBM Directory Server (used as LDAP directory)
- WebSphere Application Server (WAS)
- IBM/Lotus Web Conferencing 3.1 with the Enterprise Meeting Server (EMS) option
- WebSphere Everyplace Subscription Manager (WESM)
- DB/2
- Domino 6

When this book was written, this IBM offering was being upgraded to Lotus Web Conferencing version 6.5.1 and Domino was being upgraded to Domino 6.5.1. However, the overall design of the Web conferencing offering at an IBM server farm as given in Figure 15-13 is the same, although several software levels have been upgraded and will continue to be upgraded as new versions of software appear.

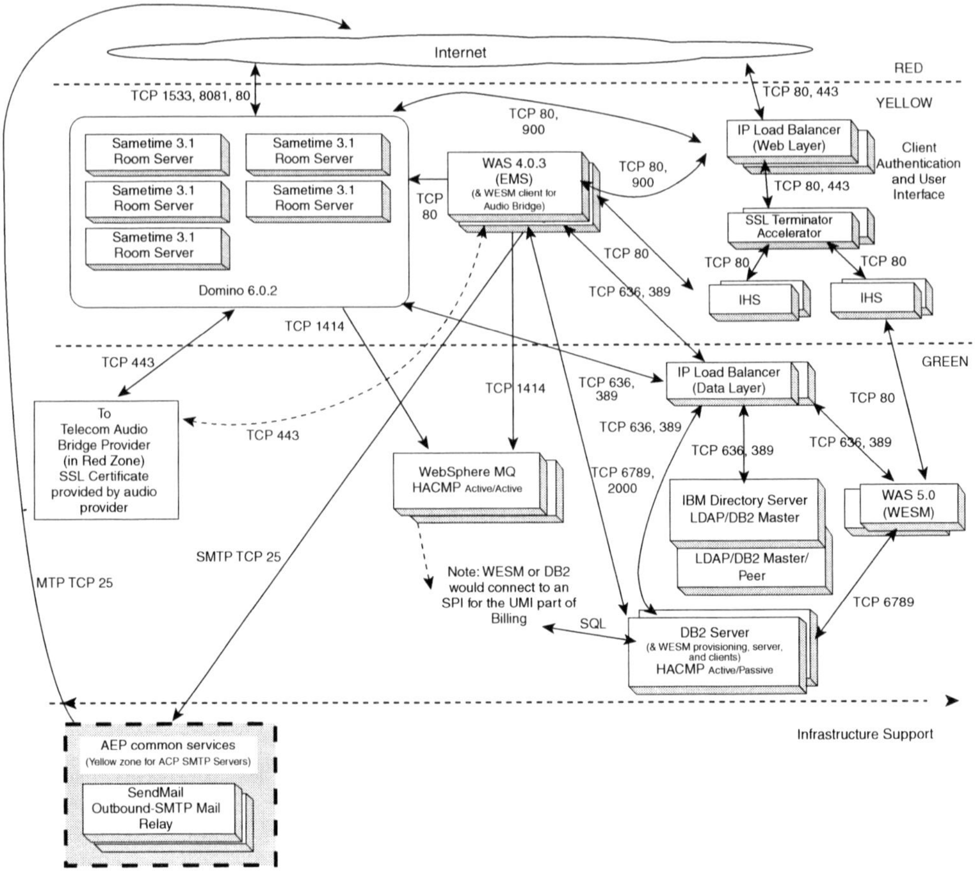

Figure 15-13 Web Conferencing Logical server architecture.

The use of SSL Accelerators for the IBM Web Conferencing offering comes from IBM's experience supporting high-volume Internet sites.

SSL Accelerator/Terminator

SSL accelerators are network appliances that off-load the CPU (Central Processing Unit)-intensive encryption and decryption associated with SSL from the servers. When a server is running SSL by itself, a majority of the available CPUs are consumed with just the cryptographic functions, leaving few resources for actual server functions. With an SSL accelerator, the encryption functionality is off-loaded, and the server is free to do its other requirements.

Although SSL is encrypted, every SSL connection has a session ID that is not encrypted, which allows both sides of the connection to know which encrypted transaction is in session. However, some versions of Internet Explorer (5.0 through 5.5) renegotiate the SSL session ID every two minutes, thus making the SSL session ID ineffective as a way to differentiate between users. SSL accelerators can solve this problem. Because they decrypt traffic before it hits the Web servers, a load balancer between the SSL accelerator and the Web server facilitates cookie-based persistence.

SSL cards were once fairly popular items, but the advantages of SSL accelerators have started to win out. SSL cards address the need to off-load the work of encryption, but, because the demarcation point for the SSL traffic is the server itself, one cannot use cookie-based persistence from a load balancer. The only option for persistence is source-IP address. For all of these reasons the use of a hardware SSL accelerator for ODW Web Conferencing is a good idea and something that has been recently accepted as a "best practices" design at the IBM commercial server farms.

IBM's Internal Use of Web Conferencing and IM Collaboration Tools

This section gives examples of how IBM has used the Domino- and WebSphere-based collaboration tools of Web conferencing and Instant Messaging (IM) internally to significantly increase productivity. In a sense, this section discusses how IBM "eats its own cooking." Of course, large companies like IBM have always used extensive collaboration between employees, suppliers, and customers as part of the business process. However, the need for collaboration tools in large companies has never been greater. Only a few years ago, the employees in most IBM departments were all located in the same building and collaboration occurred when those employees frequently met in conference rooms, in individual offices, or even around the water cooler. Now, at IBM and other large companies, the employees in a department and certainly the employees working on the same project are usually scattered around the country and frequently around the world. An "ad hoc" meeting in an office or around the water cooler is no longer a collaboration option. Virtual meetings using Web conferencing and TeamRooms or virtual water cooler collaboration via instant messaging is now the requirement. Without the new electronic collaboration tools available to IBM employees internally, the business process would be far less productive.

IBM Web Conferencing Internal Design and Utilization

IBM's internal design and architecture for Web conferencing is very similar to the global Web conferencing offering described in the previous section. IBM has had an internal Web conferencing system (to which customers and vendors could be invited) since January 2000. The system was always based on the Lotus Sametime web conferencing product, but the code to manage and allow meetings to be spread over many room servers was developed internally. IBM's earliest web conferencing systems allowed customers and vendors to join IBM web meetings from an Internet Sametime server that was "invited" to join the meeting on a server on IBM's internal network. However, toward the end of February 2004, IBM migrated to a system based on Lotus Sametime 3.1 and the Lotus Enterprise Meeting Server (EMS) used to manage meeting distribution across nine Room Servers.

The usage statistics for the first three months of operation of this new system are shown in Table 15-1. The usage information shows that on average, there are only about around 4 to 5 users per meeting. Thus, for IBM internally, the meetings tend to be small but long (three-hour meetings). Meetings on IBM's commercial web conferencing also tend to be small, but are considerably shorter in duration. One difference, of course, between IBM's commercial offering and IBM's internal system is cost. With the internal IBM system, the moderator's department does get charged, but the charge is modest so the meetings might tend to be longer. IBM's internal system is worldwide, and many meetings include customers and vendors (about 10% of attendees are not IBM employees). All Web conferencing servers are at one IBM server farm in Southbury, CT. There are nine Sametime 3.1 room servers, two EMS servers running on a clustered WebSphere Application Server base, two MQ servers, two LDAP servers, and two DB2 servers. Statistics for Table 15-1 were taken from the DB2 server. All users (IBM and customer/vendor) must be authenticated. The IBM Unified LDAP Directory with over 2 million entries (including 340,000 IBMers) is used for authenticating Web conferencing users for the internal system.

Table 15-1 IBM Internal Web Conferencing: Usage for the First Three Months on New System

Month 2004	Total Users	Average # Users per Meeting	Largest Meeting Size (# of Users)	Number of Meetings	Total Connect Hours	Average Hours per Meeting
March	74,223	4.03	281	18,424	53,829	2.92
April	78,515	3.93	297	19,960	65,798	3.30
May	80,556	4.67	170	17,263	68,516	3.97

The EMS server does a good job load balancing meetings over the nine Sametime 3.1 Room Servers. Table 15-2 shows the distribution of meetings for the first three months of operation. The use of this Lotus EMS web conferencing system has served as a "proving ground" for IBM's commercial offering. When this section was written, it was the largest deployment of Lotus EMS web conferencing to date. As with IBM's commercial Web conferencing offering, the internal deployment will be upgrading to Sametime 6.5.1 and Domino 6.5.1.

Table 15-2 IBM Internal Web Conferencing: Usage by Room Server for the First Three Months on New System

Lotus Room Server	Total Users	Average # Users per Meeting	Largest Meeting Size (# of Users)	Number of Meetings	Total Connect Hours	Average Hours per Meeting
D02XDB054	25,029	4.11	214	6094	20,693	3.40
D02XDB055	26,749	4.27	125	6265	21,696	3.46
D02XDB056	27,016	4.05	85	6676	22,255	3.33
D02XDB057	24,351	4.08	113	5973	20,328	3.40
D02XDB058	24,555	4.21	98	5835	20,213	3.46
D02XDB059	26,389	4.41	170	5980	20,070	3.36
D02XDB060	26,351	4.01	145	6576	21,637	3.29
D02XDB061	26,487	4.23	297	6262	20,988	3.35
D02XDB062	26,367	4.40	281	5986	20,264	3.39

IBM Instant Messaging Internal Design and Utilization

IBM internally uses instant messaging based on the Lotus Sametime 3.1 product. All Sametime 3.1 servers for IBM worldwide are based at IBM's server farm in Boulder, Colorado. There are 285 thousand registered users with 150 thousand concurrent users. The typical volume is five million chat messages a day. Instant Messaging (IM) is ubiquitous within IBM. The current IM deployment consists of four Sametime 3.1 servers that are supported by 10 MUX servers. IBM plans to implement a second "hot swappable" site at Southbury, CT, for disaster recovery. Also, Sametime 3.1 will be updated to IBM Lotus Instant Messaging 6.5.1. Most IBMers would be lost without the IM capability to anyone in IBM worldwide. One area that is lacking in the current implementation is easy IM access to customers, vendors, and others outside IBM. Currently, IBM's Sametime link to AOL Instant Messaging is the method often used to communicate with people outside IBM. However, use of the SIP Gateway and the use of SIP and SIMPLE standards by all vendors will soon dramatically change IBM's instant messaging capabilities.

SIP and SIMPLE

The key to allowing everyone to communicate easily with Instant Messaging is the use of standard protocols and standard encryption methods. *Session Initiation Protocol (SIP)* is a standard protocol defined by the Internet Engineering Task Force (IETF). SIP is an application-layer signaling protocol with rules that govern interactive, multimedia sessions, including presence and instant messaging. SIP is used by the Lotus Instant Messaging product (Sametime) and uses existing transport protocols, like TCP, to initiate a session. In addition, SIP can modify or terminate a session. SIP is a control protocol that doesn't care about content. With SIP, users are identified by a Uniform Reference Identifier (URI). When you connect with a SIP server, you send a registration to the Lotus Sametime server that includes your URI. A SIP Registrar stores your registration information on the SIP server. If another user wants to create a session with you, the Sametime server locates you using your URI. The same thing happens when another user subscribes to your presence or sends you an instant message—the server uses your URI to locate you. The SIP URI uses the format sip:username@domainname, which is the same format as an Internet email address. Typically, SIP uses a user's email address as a Uniform Reference Identifier. Within IBM and for most companies and communities a user's email address is also used for instant messaging.

SIP for Instant Messaging and Presence Leveraging Extensions (SIMPLE) is an emerging standard based on the Session Initiation Protocol. SIMPLE is an extension of SIP that enables awareness and instant messaging. The SIP Registrar uses your registration information for call routing. SIMPLE extends call routing to online status to ask for presence and to provide instant messaging. The Sametime SIP Gateway supports SIMPLE, and as other instant messaging vendors, such as AOL and Microsoft, support SIMPLE, you will be able to connect your Sametime community with third-party SIP-enabled communities.

To connect your Lotus Web Conferencing (Sametime) server to a separate Sametime community, you need the Sametime SIP Gateway and the SIP Connector. The other community that you are connecting with needs a SIP Gateway and a SIP Connector as well. Together the SIP Gateway and SIP Connector act as a SIP proxy server. The SIP Connector creates the connections to other SIP-enabled communities. It is responsible for both networking and the preliminary parsing of instant messages. The SIP Gateway also processes messages, but the primary responsibility of the gateway is to provide translation between the SIP network and your Sametime community.

Security

Once you start connecting your SIP-enabled community to other SIP-enabled communities, security becomes a big concern. You can determine which communities you connect with when you configure the SIP Gateway. In addition, the Sametime SIP Gateway addresses server-to-server authentication through IP addresses, which is how Sametime currently authenticates servers. When a SIP Connector connects with another server, it first checks whether or not the external community is enabled. Whether or not a community is enabled is determined when you configure the documents in the Sametime Configuration database. When you enable a connection to another SIP-enabled community, you list the DNS domains of that external community. If the community is enabled but the SIP Connector can't find a proxy server for the external community, the SIP Connector uses the DNS to create a connection. When an external server tries to connect to your Sametime community, your SIP Connector attempts to find the name of the proxy server using DNS. If the IP address of the proxy server matches the IP address that the message came from, then the Connector accepts the connection. If the two IP addresses don't match, the Connector rejects the connection.

Encryption

The latest Web browsers support the Transport Layer Security (TLS) protocol and so does the SIP Gateway. TLS is an Internet protocol that prevents third parties from listening in or tampering with a message communicated between a server and client. Because the SIP Gateway supports connections to other communities, Lotus Sametime 3.x uses the TLS protocol to encrypt messages sent between servers, between servers and SIP Gateways, and between SIP Gateways.

The Ultimate Solution

When everyone is using SIP and SIMPLE and the SIP Gateway, you'll be able to see when any contact you have in the world is online and communicate with that person instantaneously. This could be construed as good or bad—but it will certainly aid in collaboration. Users with such ubiquitous instant messaging capabilities have to be willing to be quite "interrupt driven."

Providing Information on Performance and Use of Web Sites

This section shows how companies such as IBM provide customers with information on performance and use of the Web sites hosted by IBM. This is related to the "lessons learned" from hosting the Olympic and other sporting and event Web sites. Customers are very interested in how their Web sites are performing and who is accessing what parts of the sites.

Details on Server Utilization, Network Utilization, and Problems with Availability

One area that must be addressed first is how to provide customers with the information they need on server utilization, network utilization, and problems with system availability. The following graphs (Figures 15-14 and 15-15) show information for the Exchange public site described earlier. This server measurement data at IBM server farms is measured with standard tools including IBM's Server Resource Management (SRM) and BMC Patrol. These tools are used to report historical trends of key server resources, such as CPU, memory, and disk usage.

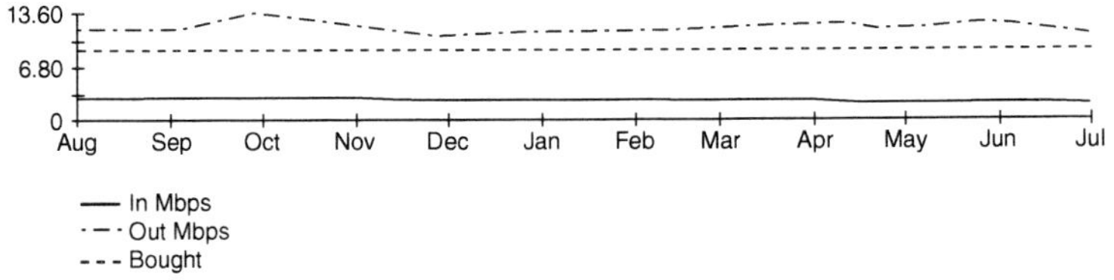

Figure 15-14 Providing front-end (Internet) bandwidth utilization information for companies.

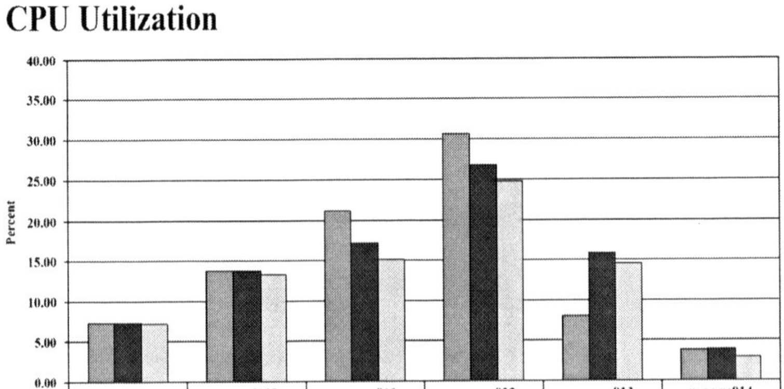

	nysend001	nysend002	nysews010	nysews012	nysews013	nysews014
May	7.32	13.75	21.12	30.76	8.04	3.73
Jun	7.29	13.76	17.11	26.88	15.83	3.89
Jul	7.21	13.29	15.08	24.75	14.58	2.88

Figure 15-15 Providing server utilization information for companies.

How Web Sites Are Being Used

Customers also want to know how their Web sites are being used (e.g., number of hits, hotspots, number of new visitors, which pages are most popular, etc.). The following graphs (Figures 15-16 and 15-17) show information for the Exchange public site discussed earlier. These graphs were produced by "Surfaid" a service IBM provides to their Web customers.

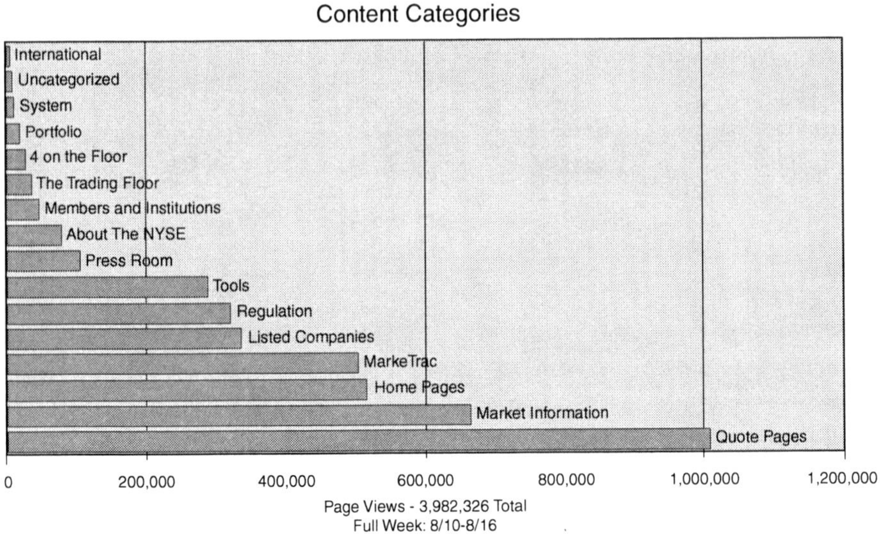

Figure 15-16 Providing information on which pages are most popular.

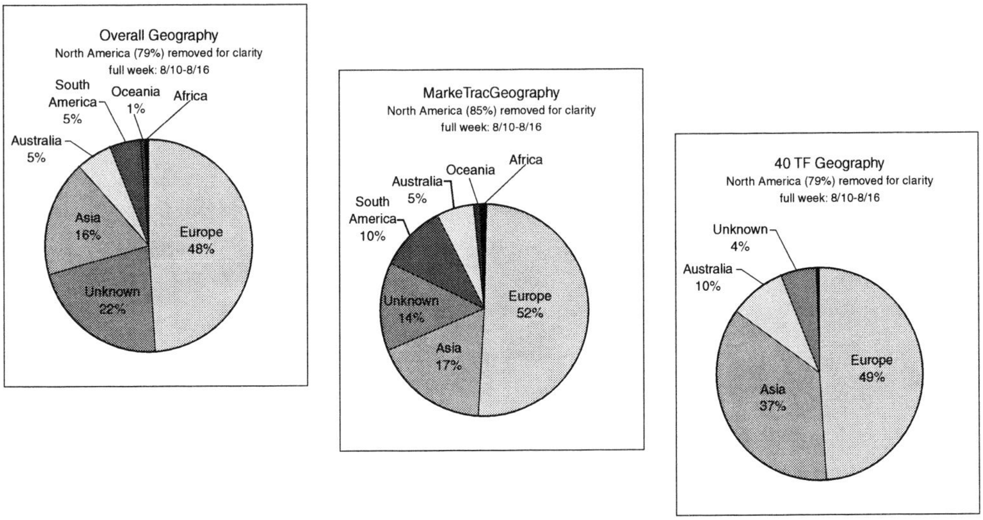

Figure 15-17 Geography of users (outside NA).

New Options for Business-to-Business (B2B) Web Sites

Companies are now using Web sites to conduct business with their suppliers (business-to-business). Collaboration techniques (Web conferencing, instant messaging, teamrooms, and so on) are becoming options. Figures 15-18 and 15-19 show the type of Web conferencing and instant messaging options that can be added using Lotus collaboration tools. The screen shots are from the demo server for a Web conferencing service IBM offers out of a San Jose, California, farm based on Lotus Web Conferencing and Lotus Instant Messaging tools.

Collaboration can significantly help facilitate purchasing for large, complex transactions between teams of people.

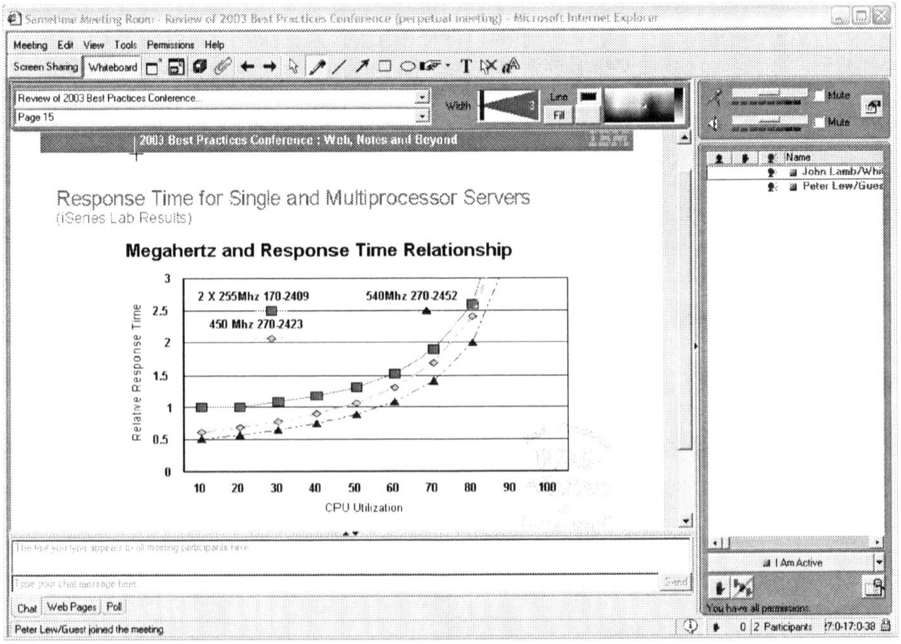

Figure 15-18 Collaboration with Web conferencing.

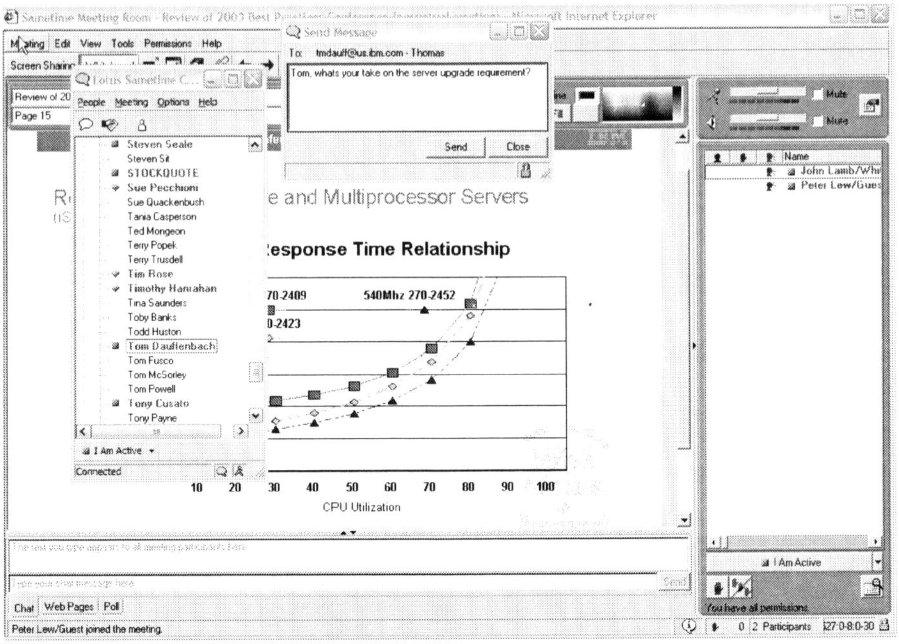

Figure 15-19 Collaboration with instant messaging and Web conferencing.

New On Demand Concepts for Web Sites

The idea of on demand computing is to allow companies to pay for only the amount of computing power they use, as they would pay for electricity from the power company. IBM's strategy for "on demand" offerings is to create a computing infrastructure using servers, storage, middleware, and distributed networks that is seen by users as a flexible, integrated common resource that can span traditional enterprise boundaries. In an on demand environment, IT resources are gathered from multiple physical locations to create an integrated "virtual" computing environment. Traditional IT environments typically dedicate IT resources to applications. Substantial economic and competitive efficiencies are gained using this strategy, since virtualized resources can be shared between applications to achieve much higher utilizations. On demand deployments have already been initiated during 2003 and 2004 in large enterprises, including the IBM environment to save billions of dollars—e.g., supply chain management applications. An on demand infrastructure is the basis for business transformations, including being a services delivery platform for e-hosting, web portals, and mission-critical applications. Grid computing is one of the first logical steps to creating an on demand environment.

Virtualizing Resources: VMware and IBM Virtualization Engine

VMware was founded in 1999 to offer server management software that allows mainframe-class virtual machines to be created using industry-standard computers (see `http://www.vmware.com`). VMware is one of the proprietary and evolutionary means to manage or virtualize Windows and Linux servers for "on demand" offerings to gain efficiencies with clusters of servers. An example is that IBM can often consolidate eight to 10 customer Domino application servers onto a larger one using VMware. A simple "rule of thumb" for migrating to VMware is to add up the processing power of the source servers and compare that to the processing power of the target server. If the target xSeries 445 servers have four 2.4Ghz processors, the total Ghz of the target server is 9.6Ghz. Based on that simple guideline, we could consolidate about 10 single processor NT boxes with average processor power of 1Ghz onto one IBM xSeries 445 using VMware. Announced in April 2004, IBM's Virtualization Engine (VE) software suite is a major industry advancement in managing servers, storage, and network resources across multiple locations and heterogeneous platforms (see `www.ibm.com/servers/e-server/about/virtualization`). At one end of the suite, the software allows "micro-partitioning" of a single CPU into ten pieces, thus allowing a 4-way system to have 40 CPUs, each an independent fault-tolerant partition to maximize system utilization. IBM VE suite includes TotalServer and TotalStorage technology to enable enterprise optimization of heterogeneous pools of servers, storage file systems, and disk volumes across networks in a secure and efficient environment, something not easily done before. To the user who runs applications in the IBM VE environment, the resources are "virtualized" as a common resource, and this offers powerful new control over a heterogeneous infrastructure.

Grid Computing

Grid computing is a major evolutionary step that virtualizes an IT infrastructure. It's defined by the Global Grid Forum (www.gridforum.org; see also www-fp.mcs.anl.gov/~foster/Articles/WhatIsTheGrid.pdf) as distributed computing over a network of heterogeneous resources across domain boundaries and enabled by open standards. While the industry has used server cluster technology and distributed computing over networks for nearly two decades, these technologies cannot in themselves be grid computing. What makes grid computing different is the use of open source middleware to virtualize resources across domains. Middleware makes it possible to apply policy-based control to the management of those resources. Key factors that differentiate grid computing from previous distributed computing technologies are as follows:

1. The computers and resources involved can be heterogeneous. They can run different operating systems and use different server and storage platforms that may even be owned by different individuals, companies, labs, or universities. Grids may include clusters, individual servers, or entire data centers as part of the resources that are virtualized for sharing with other members on the grid.
2. Grids are based on open standards such as Globus and Global Grid Forum based on protocols like SOAP and WSDL (Web Services Description Language) to define all resources as services, for discovery, management, security, etc., in a layered OGSA industry-approved Open Grid Services Architecture (OGSA).
3. Grids are capable of massive scaling to share and manage applications, data, and resources between local, campus, regional, or international locations.
4. Grids are based on policies created to define job scheduling, security, resource availability, and workload balancing across multiple applications and user domains.

Grid computing can be thought of as a services delivery platform. Multiple users can launch jobs into the grid environment running simultaneous applications with access to multiple databases over networks. Grid computing offers individuals, national labs, universities, and commercial enterprises a previously unavailable flexibility to perform internal IT computing tasks faster and at much lower costs, to manage critical capital expenditures, to leverage open standards, and to access new markets. Grids enable new or improved services that were simply not possible in the past. For example, small bio-genetics companies can now have easy access to massive grid computing services (IBM Deep Computing) or a large international bank or telecommunications company can grid-enable their customer web portals or large call centers to optimize costs and to win in a competitive environment.

Around 2001, it became apparent that the mix of grid software, high-performance computing applications, proprietary solutions from vendors, and increasing need for individuals to collaborate across a grid had become a real challenge for interoperability and infrastructure management. It was also apparent that while many grid efforts were remarkably successful, a new stage of common development and open standards would be required to make grid computing infrastructures a commercial success. In the commercial world, applications have been typically "pinned" to dedicated resources within administrative domains under layers of security. Expensive IT resources are typically underused in anticipation of peak capacity loads and the launch of new commercial services is gated by the availability of capital funding to buy new platforms. A powerful yet flexible framework would be needed—but one based on open standards.

IBM is an acknowledged industry leader in grid computing and a key collaborator in defining open standards. To enable grid computing and on demand, open standards have been and are being adopted to create flexible, reusable components that can be inserted into a layered "services-oriented architecture" (SOA). An example of a services-oriented architecture is the well-known Open Services Grid Architecture (OGSA) now adopted by the industry through the Global Grid Forum. IBM's own on demand architecture is itself a SOA that will embrace the OGSA. The key concepts are that open standards allow applications to ride on a layer of middleware (OGSA) which, because of flexibility and reliability needs, calls for resources in the physical infrastructure below. Web Services hold the key to grid computing for commercial applications and business applications that will take advantage of the grid through XML, UDDI, SOAP, and WSDL. A major advancement in early 2004 occurred with the public announcements by GGF to "refactor" the OGSA grid standards to merge with Web Services standards to create a WS-RF Web Services Reference Framework and corresponding family of Web Services standards through year 2004. Adoption and development of standards are underway by the international grid community comprised of more than 100 companies in the Global Grid Forum, researchers, and commercial software providers to extend existing grid standards.

An important observation is that most commercial applications—e.g., in banking, life sciences, manufacturing, and telecommunications—have not been written specifically to run on a grid. Yet since 2003, IBM has successfully launched nearly 200 grid engagements with 35+ public grid references and has found that many stable UNIX-based applications are capable of running in an OGSA grid environment without major software porting. Specific knowledge of the application, workflows, architecture, and platform performance is required to extract the most business value from specific grid implementations.

Grid Computing—The Reason

The adoption of grid computing by commercial enterprises for mission-critical operations is rising quickly in 2004 and is seen by business and national leaders as a way to effectively transform their enterprises and organizations. Grid computing is a practical first step to creating a scalable, responsive "on demand" infrastructure. There are powerful business reasons to build or join a grid services infrastructure:

1. Dramatically improve % utilization of dedicated IT resources that are now typically "pinned" to an application or platforms, either within or between enterprises.

2. Enable new services to customers at lower cost and faster time to revenues.
3. Create a highly resilient infrastructure (an inherent feature of grids) for disaster recovery or mission-critical "business continuity" services.

An example of the first business reason to deploy grid computing is emphasized by the following table (see Figure 15-20) derived from market studies from 35 large global companies (see IBM Scorpion Study 2001) that shows a low average % utilization for IT resources. The results have remained fairly consistent for different enterprises and reflects some of the realities in which capacity is overprovisioned and sometimes guarded with anticipation of peak demands. In addition, other studies have shown that people and administration, maintenance, and service comprise some 60% of the overall cost of typical data center operations, while the other 40% of costs is from all hardware/software platforms. One can easily observe that low average % utilization of hardware/software platforms has a corresponding cost in effective use of skilled people. Grid computing offers a powerful solution to breaking down the barriers created by IT resources that are held in silos and have low utilization. Sharing the resources with grids has shown dramatic (2 to 200 times) improvements in process cycle times with huge reductions in costs for those steps. Once higher value business applications are put onto a grid, they typically expose the next weakest link in business application processing. A grid computing infrastructure creates a services delivery platform that will lower costs, help fund the next round of improvements, enable flexible launch of new services, and create a "virtuous circle" of valued improvements. See Figure 15-21.

Asset Utilization

	Peak-hour Utilization	Prime-shift Utilization	24-hour Period Utilization
Mainframes	85-100%	70%	60%
UNIX	50-70%	10-15%	<10%
Intel-based	30%	5-10%	2-5%
Storage	N/A	N/A	52%

Figure 15-20 Grid computing—the reason.

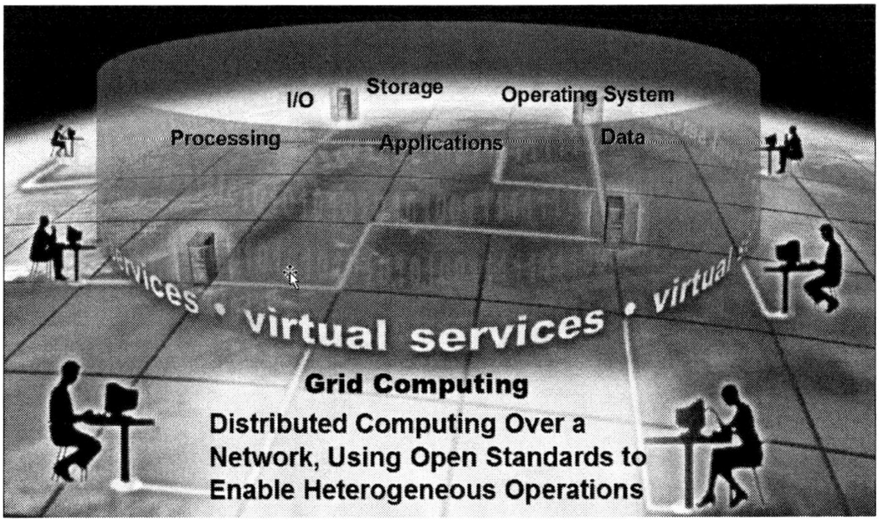

Figure 15-21 Grid computing—the concept.

Grid Computing Is Becoming Reality

Commercial adoption of grid computing during 2003 and 2004 has been rapidly lead by six active industry global segments: R&D, life sciences and pharmaceuticals, government services, finance services, industrial and manufacturing, and most recently, telecommunications. Leaders in grid computing like IBM have focused on building grids for mission-critical or highly-valued categories of applications. These are:

- "Business analytics acceleration" of financial derivatives portfolio, insurance actuarial calculations, seismic and petrochemical calculations
- Engineering and design for CAD/CAM in aerospace, automotive, and electronics
- R&D in life sciences for modeling bimolecular components, agricultural chemicals
- Government and organizational collaboration by diverse users across grids
- Enterprise optimization of business processes in petrochemical, finance, and telecommunications

Grid computing is in an early commercialization stage but is already providing impressive business value. OGSA and Web Services have addressed lower layers of the architecture to drive interoperability and these standards have been launched into the final approval process during the first quarter of 2004. Other goals of the OGSA standards development are given in great detail on the Global Grid Forum site (`www.gridforum.org`), in which the family of Web Services standards will be merged. Web Services enables better interoperability and improved distributed system integration. Finally, grid computing needs to develop a base of deeper commercial experience to address future opportunities in which grid services providers can offer services to unknown third parties, allow secure access for their services, and accurately bill for services rendered, all in an automatic manner. The technology underpinnings for this are several years away.

The History of Grid Computing

Grid computing for commercial applications has been successfully adopted in 2004 by executive leaders in companies in Europe, Asia, and the Americas. Now it has been extended well beyond the mix of distributed computing technologies originating from various universities, supercomputing centers like San Diego, national labs like Argonne, and international scientific consortia like CERN during the late 1980s through the 1990s. One of the origins of grid computing can be traced to cluster computing. Clustering was made possible by the advent of powerful desktop workstations in the 1980s. These machines, which could deliver nearly one million operations per second, were easily affordable and connected by high-speed networks. It wasn't long before researchers began to make these workstations work collaboratively.

To gain access to new computing power that would be created by sharing resources, various researchers wrote their own software and collaborated to support the many specific implementations of early grid computing. A critical enabler was the rise of new Internet standards and protocols (see the seminal reference on grids, "The Grid: Blueprint for a New Computing Infrastructure" by Foster & Kesselman, 1998). This activity resulted in the development of the earliest-known grids of computers and enabled large-scale, resource-hungry problems to be solved cost-effectively. Several early grid initiatives have attempted to provide a more formal model for sharing of resources in wide area cluster environments.

The Business Benefits of Grid Computing

Grid computing is a first logical step that can be taken by individuals or an organization to building a more powerful, lower total cost infrastructure. After business process re-engineering, workload consolidation, and simplification of the IT environment, executives should logically consider grid computing as part of their strategy to transform their businesses. Grid computing environments have proved valuable in a variety of application domains. Industry leaders have chosen grid computing to win commercial benefits in applications for business analytics acceleration, as well as enterprise optimization (see the previous section). Although grid computing infrastructures conform to the OGSA and emerging standards, they can be implemented and optimized for their intended applications. One can think of a "style" of grid computing implementation represented by business analytics acceleration that is more computationally intensive, while government organizations require more collaborative use of common applications and synchronized databases to create a "collaborative grid." Some applications require a hybrid grid environment combining computationally intensive and collaborative styles to address "enterprise optimization." An example of the latter might be a grid to support massive billing systems used by telecommunications companies.

Examples of grid application domains and their applicability to business environments include the following: High-performance distributed computing applications leverage large numbers of network-connected computing resources so users can collaborate to solve complex problems. By distributing the work, you can reduce the amount of computing time needed to complete an application. Grid computing dramatically increases the number of computing resources available to a distributed computing application. In a business environment, this capability enables an enterprise to improve the quality of service of its distributed applications, even when unplanned events that affect the enterprise occur.

Collaborative support has been traditionally used for activities such as enabling geographically dispersed teams to simultaneously analyze information contained in scientific datasets and displaying this information graphically for ease of analysis. In a business environment, companies can leverage the mechanisms grid computing provides for collaborative support across multiple departments and across multiple companies while protecting intellectual property and confidential information. Furthermore, these collaborative capabilities support virtual collaborative organizations (i.e., multiple businesses pooling their resources to solve a mutual problem), in which the servers and storage devices contributed by each participant work together seamlessly and optimally.

Grid computing also helps to enable the development of on demand utilities. Through the use of advanced resource sharing capabilities provided by the grid, an on demand utility can offer computing power and storage resources to companies that subscribe to these services, and these companies use only what they need. This approach can save companies substantial IT investments while giving them access to supplemental computing power and storage resources during peak usage periods.

Remote access to resources is an inherent feature of grid computing environments in which the scientific community already has demonstrated access and use of remotely located telescopes and data mining of medical databases over the Internet. In a business environment, the same mechanisms that make these operations possible also can be used to securely access and exchange distributed sources of business information in real-time and at high-speeds.

A Real-World Business Grid Application

To better illustrate the applicability of grid computing to business applications, we'll examine a real-world private exchange application that benefits from the grid's capabilities. Let's assume that our private exchange application enables product suppliers to make available information about their products (e.g., product description, availability, and prices) to buyers who browse catalogs to find products of interest. Also, the exchange is considered private, as it's accessible only to registered buyers and suppliers. Let's also assume that we have thousands of registered buyers and suppliers and a product catalog that contains millions of products.

In this environment, each supplier has its own enterprise information system where it stores its product information. However, registered suppliers must also store all their product information in a large global exchange repository. Furthermore, because each supplier stores its product information in a proprietary format, each supplier is responsible for converting its format to the format supported by the exchange repository. Additionally, the supplier is responsible for publishing the product information in the product catalogs provided by the exchange; this requires yet another transformation of the information in the supplier's information systems. Suppliers also typically need to synchronize their product information with the global exchange repository to update pricing and description information. Registered buyers can then browse the product catalogs created by the suppliers to identify the items they wish to purchase.

Because the registered buyers and suppliers need a mechanism for implementing the private exchange, they set up a grid computing environment where they can contribute servers and associated storage. To manage the computing and storage resources, these buyers and suppliers use grid service middleware so that they can share resources in a more seamless and transparent fashion.

Following are the benefits the grid service-based computing environment provides for our private exchange application:

Sharing of storage resources allows the cost of storing the global exchange and the product catalogs to be distributed among the buyers and suppliers that have contributed storage to the grid computing environment. How the buyers and suppliers choose to distribute the cost of storage requires some negotiation between the parties involved. The ability to share the cost of storage is significant, as both of these repositories have massive storage requirements. Moreover, even though the storage has been distributed throughout the grid, the grid service infrastructure virtualizes this storage support such that it appears available as a monolithic resource from a centralized hosting site in the environment.

Sharing of processing resources enables several different servers connected to the grid to share and perform the work of transforming product information into the formats supported by the exchange repository and the product catalogs. The grid service infrastructure handles the processing at each server, and as a result, it appears to each supplier as if it has a much larger cluster of servers available for performing this computing-intensive task.

Improved business resiliency, reliability, and quality of service for accessing product catalogs results from the availability of a larger pool of servers. This greater number of servers allows the hosting of product catalogs at a variety of sites in the grid, improving reliability should an unplanned outage occur at some of the locations hosting portions of the grid. In addition, during peak product catalog usage periods, numerous resources can be dedicated to the hosting of product catalogs on servers, thus improving the quality of service for those accessing and searching the catalogs.

OGSA and WebSphere

Open Grid Services Architecture (OGSA) is a services-based architecture that builds on Web services technologies. IBM is working with Web services standards organizations to ensure there's a single set of standards that meets the needs of both conventional applications and applications that are optimized for a grid environment. The WebSphere software platform will provide tools and runtime support that leverage these standards. IBM will also use this platform to provide a high-quality implementation of OGSA that will be available on IBM and other vendors' server hardware.

WebSphere will likely become the runtime environment of choice not only for the grid infrastructure but also for applications as they evolve to leverage the benefits of a grid environment. Because OGSI and OGSA are open standards, application developers can use any vendor's implementation of the hardware and software infrastructure to build and deploy their applications.

IBM's Grid Strategy

IBM is addressing Grid Computing across four dimensions—storage virtualization, global file systems, database management, and system management. The company's strategy concentrates on five focus areas—research and development, engineering and design, business analytics, enterprise optimization, and government development—for customers in the aerospace, automotive, financial, government, and life sciences industries. IBM is strongly committed to the Open Grid Services Architecture (OGSA—a set of specifications and standards that form a common framework for building commercial grid solutions. The company also actively participates with the Global Grid Forum in the development of the standards. IBM's grid offerings, as well as their server, storage, and middleware products, will support OGSA.

Over 100 Grid Customers Up and Running

Although grid technology is still very much under development for corporations, IBM has more than 100 grid customers up and running worldwide. The latest grid offerings are for the financial services industry. IBM and software provider, SAS, have teamed to provide a new Grid offering in the banking industry. Using the offering and its own network and IT resources, a bank can use its data to better understand its customers and their banking needs, reducing its cycle time for executing statistical models and providing more sophisticated and accurate customer analysis. This should help banks better identify, acquire, and retain customers.

The other grid offering, developed in conjunction with DataSynapse, is designed to help risk managers implement a grid infrastructure to support real-time credit limit monitoring. The offering can also form the foundation for a credit risk application infrastructure that complies with banking organization and Federal government reporting requirements. Examples include the Patriot Act, which provides U.S. law enforcement officers with information and tools to investigate and prevent acts of terror, and recently toughened New York Stock Exchange reporting requirements.

Real-World Examples of IBM's On Demand Implementations

Here are some examples of IBM's On Demand implementation for customers.

On Demand at the U.S. Open Tennis Site

Figure 15-22 shows the home page of the U.S. Open Tennis Site. The U.S. Open infrastructure, which is run by IBM, when not being used for U.S. Open Tennis, is used to explore protein folding—as a preview to the Blue Gene Project.

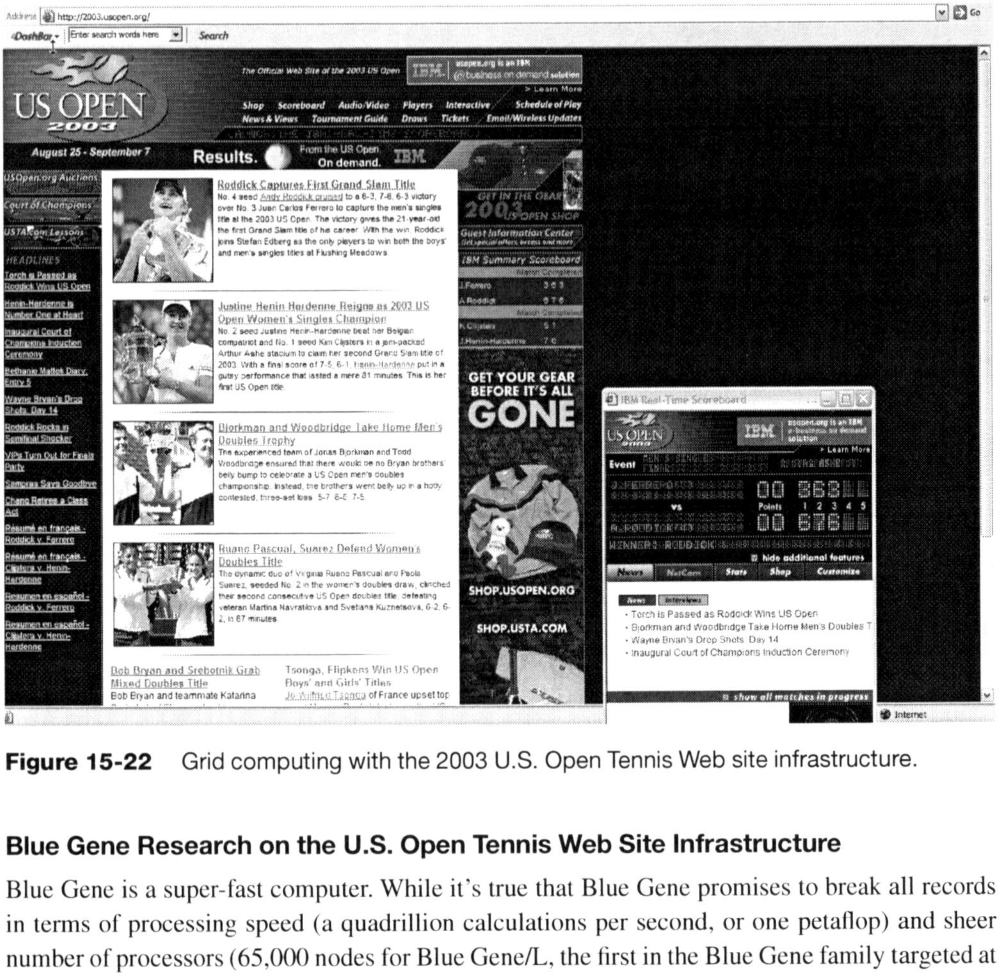

Figure 15-22 Grid computing with the 2003 U.S. Open Tennis Web site infrastructure.

Blue Gene Research on the U.S. Open Tennis Web Site Infrastructure

Blue Gene is a super-fast computer. While it's true that Blue Gene promises to break all records in terms of processing speed (a quadrillion calculations per second, or one petaflop) and sheer number of processors (65,000 nodes for Blue Gene/L, the first in the Blue Gene family targeted at 360 teraflops), an interesting aspect of the story involves Blue Gene software. The U.S. Open infrastructure has been used as a grid to help run protein folding simulation experiment. The Computational Biology Center at IBM Research is running Blue Gene science applications today using the U.S. Tennis Open infrastructure for a grid before the actual research version of the Blue Gene supercomputer is available.

IBM scientists in labs around the world have been developing applications that will allow life sciences researchers, for the first time, to explore the intricacies of protein folding. The Blue

Gene science program is already in full swing, well in advance of the hardware. When the technologies are not needed for tennis, this same infrastructure is being used for other workloads—in this case, running applications for protein folding. IBM researchers are using capacity at the U.S. Open for three major projects, as discussed next.

Understanding Protein Aggregation

Blue Gene software will help researchers look not just at how proteins fold up to their correct biological structure, but also what happens when folding goes awry. Misfolded proteins can clump together, or aggregate. Different proteins have different propensities to aggregate, and these aggregates can be completely disordered (like pieces of spaghetti tangled together) or show some degree of order (like Lego bricks stuck together). Amyloid fibrils, a class of protein aggregates, are involved in a number of major diseases, including Alzheimer's, Creutzfeld-Jakob Disease (prion), and transthyretin amylodosis. A research group at UCLA has recently reported the discovery of a tiny protein fragment that can form amyloid-like aggregates. IBM researchers are using the U.S. Open infrastructure to carry out simulations on concentrated solutions of these four amino acid peptides to help understand the process of protein aggregation and to give a physical picture of the interactions that stabilize the aggregated state.

Folding Simulations of the Villin Headpiece

The villin headpiece is a 36-amino acid alpha-helical protein and is one of the fastest folding proteins known (it folds in less than 10 microseconds, which is ten-millionths of a second). As part of the Blue Gene project, IBM is trying to understand why this protein folds so quickly. Since protein folding is a statistical process, the researchers need to simulate numerous folding events that may occur by a number of different pathways in order to determine the rate and the overall process of protein folding. That type of simulation takes enormous computing power.

Simulations of Protein-Protein Interactions

IBM researchers are also doing simulations of a protein-protein interaction module composed of 35-40 amino acids. This module is a subject of general interest because several signaling complexes that the domain mediates have been implicated in human diseases including Muscular Dystrophy, Alzheimer's Disease, Huntington Disease, Liddle Syndrome of hypertension, and, recently, in cancer.

Grid Experiment Turns into Real-World Healthcare Business

Another example of a real-world grid and on demand system is a digital medical archive—funded by IBM and the National Library of Medicine. The archive of digitized mammograms, which draws on the shared computing resources of hospitals in Chicago, North Carolina, and Toronto, is expected to create a comprehensive records system that will facilitate near-instantaneous access to breast cancer screening data and transform the way hospitals deliver healthcare information to doctors and patients.

The National Digital Mammography Archive (NDMA) is another example of on demand technology. Factors such as shared computing resources, open standards, and "autonomic" self-managing servers are making it possible for institutions across North America to distribute critical information quickly, cut waste out of supply chains, and reduce administrative costs associated with phone or paper-based processes. In the case of NDMA, server clusters, or grid technology, are putting high-resolution medical images, records, and clinical history in the hands of radiologists at three hospitals, letting them compare an individual's mammography records to make fast, reliable diagnoses.

On Demand Is Still Evolving

As some 200 medical centers in the U.S. and Canada log onto the scalable, secure, and economically feasible grid, breast cancer data will reach more radiologists as soon as they need it. Patients will be the ultimate beneficiaries. So although grid and on demand technology are still evolving, there are many real-world examples in the works. Grid and on demand technology both provide an excellent base for collaboration tools for the enterprise.

Lessons Learned from Implementing High-Volume Sport Web Sites

This section summarizes the "lessons learned" by IBM implementing high-volume sport Web sites. As discussed throughout this chapter, these lessons have been a great help in designing high-volume commercial Web sites—including those Web sites that include collaboration techniques.

The lessons learned from high-volume sport Web sites are

- Many techniques have been incorporated into commercial products.
- Less cost handle more hits!
- Web site requirements are not static.
- Web architecture is not static!

Commercial groups want to know how their Web site is being used (like the sports Web sites):

- Who are their users, and where are they from?
- How many hits a day are occurring, and is volume growing?

Companies (and Web site hosting providers) are looking at new ways to improve Web site function and reduce costs, including the following:

- The use of Linux and Intel based servers
- The use of collaboration tools for B2B transactions
- Pay only for the computing power you use
- On demand like you pay for electric power
- Exploit server virtualization and grid computing techniques

CHAPTER 16

The Future for WebSphere and Lotus Collaboration

"What's past is prologue."

—William Shakespeare, *The Tempest*

"IBM's holistic approach is refreshing. It gives customers a number of options in terms of where to start in creating what it calls an 'on demand' environment. While it does not dictate specific technology, it focuses on positioning technology as the tool and business process as the enabler."

—IDC, November, 2003

Where IBM and Lotus Are Heading

The future direction for IBM and Lotus will be to address the changes in the needs of many corporations, from basic Web hosting to new services that would include collaboration and on demand capacity. The continuing enhancements to WebSphere Portal are particularly interesting in this context. Over the past couple of years, the Lotus range has expanded to embrace technologies such as instant messaging and chat; in effect, IBM has recognized that these new technologies of presence offer the possibility of incremental collaboration. The use of WebSphere and Domino together and use of the improving WebSphere Portal is the payoff. The WebSphere Portal incorporates the Lotus-derived Collaboration Center, which offers "portlets" integrated with Lotus technology for messaging and team workplaces. Lotus technology has long carried the stigma of its products being for smaller companies (e.g., the 1,000-seat-at-a-time per server imposed solution). Including Lotus Collaboration under the WebSphere cover could be an excellent way to shed this image and introduce the WebSphere community to the joys of collaboration.

With the Lotus group embracing DB2 and J2EE, most of the recent traffic between the Lotus and WebSphere communities has been one-way. Now the WebSphere technologists might have something to learn from the fuzzy logic of the collaborative world.

On Demand Business

One major buzz phrase of 2003 and 2004 was "On Demand Business." The range that this phrase covers is gradually becoming clear. The on demand era isn't about trading goods and services over the Web. On Demand Business is a set of Web-based strategies for increasing efficiency and driving down costs internally, as well as in an organization's dealings with its partners.

On Demand Business is an ambitious program, embracing everything from hardware to business consultancy. For a services company, On Demand Business is about rapid and, as far as possible, automated application development. IBM wants to sell this approach to the world at-large, including Windows users. For a software company, On Demand Business means component-based systems, easy to assemble and reconfigure, and portal-based hybrids—in practical terms, with the horsepower of WebSphere and the flexibility of Domino. For a hardware company, On Demand Business means transparent end-to-end integration of the type promised by autonomic computing. IBM's CEO, Sam Palmisano, defined an on demand company as "an enterprise whose business processes—integrated end-to-end across the company and with key suppliers and customers—can respond with speed to any customer demand, market opportunity, or external threat." He also said that "it is not about gizmos and gadgets, but business processes and people...It's about eliminating the gaps and seams and spaces." So On Demand is a strategy for all departments of IBM—and all parts of IBM's user base. Any business can implement at least part of the On Demand Business vision and take steps toward implementing it more fully. WebSphere together with Domino are key platforms for On Demand Business.

The role of the portal server is continuing to take on more and more importance. Enterprise portals that present Web-enabled applications alongside collaborative workspaces, search facilities, e-mail, and news are a familiar concept. International Data Corporation (IDC) had divided portals into four groups: the basic enterprise information portal (EIP), the groupware-enabled enterprise collaborative portal (ECP), the personalized enterprise expertise portal (EEP), and the enterprise knowledge portal (EKP), which incorporates knowledge management technology. Portals have been around for a long time but have never quite lived up to the early hype. Now, however, with IBM and BEA both firmly behind the technology, there seems little doubt that we'll all soon be speaking the language of portlets and e-workplaces. After several years, portals are finally going to make it big.

But why now? One answer is that it's taken until now for people to realize what a great idea they are. Portals were discredited by the early implementations, because they gobbled up screen area for little benefit (how often do you need a weather forecast?). Later, less flashy implementations delivered more tangible benefits: always-on access to key applications, e-mail, chat, and a browser. People have caught on, and now portals are going to be big. However, there's more to the story.

A more complete explanation is that the portal is the logical next step for IBM and other companies providing solutions for Web sites. These companies have already provided a J2EE application server, in-depth support for XML, and application integration tools. The offerings themselves sharpen the need for portal services. Server-side Java means assembling applications from reusable interoperable objects, XML and Web services making it possible to combine remote and local components, and legacy integration promising to open a whole new range of data sources. Applications built under these conditions will be significantly more open to extension and customization than the ones most of us are used to. Portals mirror this flexibility on the client side, offering users a personalized combination of application components and data sources.

The similarity isn't coincidental. Portals are built on the same open standards used by application servers. Using WebSphere Portal Server, you can either plug in an application with a Web service interface as a remote portlet or bind it to an existing portlet, for example, to return information from an Internet search to a local application.

Benefits of the Portal Model

IBM's overall software strategy favors portal development. The company has grouped its software under four key middleware brands: WebSphere, DB2, Tivoli, and Lotus Domino. IBM will continue to integrate the four more closely. The portal model demonstrates the benefits of this approach, yoking the WebSphere platform to Tivoli system management tools and a range of Lotus collaboration and knowledge management technologies, including QuickPlace, Sametime, and Lotus Extended Search.

WebSphere covers a lot of ground. Across the four eServer platforms, there are at least 10 current versions of the WebSphere Application Server, including Linux versions for xSeries and zSeries. As for the broader WebSphere product range, IBM's WebSphere Developer Domain Products Page (`http://www7b.boulder.ibm.com/wsdd/products`) introduces you to WebSphere Portal Server, WebSphere Voice Toolkit, WebSphere Host Integration, WebSphere Host On Demand—and more.

IBM explains this profusion of WebSphere products in this way: "The WebSphere software platform for e-business consists of a Foundation, Foundation Extensions, and Application Accelerators." The core of the foundation layer is the WebSphere Application Server (in its two main editions, Advanced and Enterprise). Also in the foundation layer is WebSphere MQ. This isn't the MQSeries messaging middleware but a whole family of messaging products, including products such as the WebSphere MQ Integrator (an "information broker" offering intelligent message routing and information flow modeling).

The foundation extensions are a range of software packages for building WebSphere applications (development), making them look good (presentation), and keeping them running (deployment). Development extensions include WebSphere Studio and WebSphere Business Components (of which there are many). Presentation extensions include WebSphere Voice Server and WebSphere Everyplace Suite. Then there are the deployment extensions; these include WebSphere Site Analyzer and WebSphere Host Integration.

At the top level are the application accelerators: tools and templates for application development. This group includes products such as WebSphere Commerce Suite (in four editions: Start, Pro, MarketPlace, and Service Provider) and WebSphere MQ Workflow, which offers message-based automation of business processes. The Crossworlds Enterprise Application Integration (EAI) toolset, acquired by IBM in 2002, was transformed into a WebSphere application accelerator under the WebSphere umbrella of "Business Integration Accelerators."

The WebSphere Approach

There are overlaps and some gray areas in the WebSphere architecture, and it's not always clear how it maps to actual commercial products. What's more significant is the overall approach that it represents. First, IBM is determined to cover all the bases. Consequently, the WebSphere range is not so much a suite but more the raw materials for a huge range of potential, user-designed suites.

Second, IBM clearly wants to ensure that the product line stays current. With mobile e-commerce in its infancy and B2B e-business struggling, the company is unlikely to make it big in the short term out of the WebSphere Business Integrator ("provides a complete and consistent model for both EAI and B2B integration") or the WebSphere Transcoding Publisher ("adapts, reformats, and filters content...giving companies better access to customers, business partners, and mobile employees"). Nevertheless, if and when e-commerce and the B2B model start to deliver on their early promise, IBM and the WebSphere brand will be there.

The third and perhaps the most significant point follows from the first two. While we might tend to concentrate on the WebSphere application server, WebSphere is also a brand—indeed, it's now regarded as one of IBM Software's four main brands, alongside DB2 and the two great acquisitions, Lotus and Tivoli. At the core of WebSphere is a strong product, but the strength of the product does not necessarily transfer to the brand. If anything, the uniformity of the WebSphere branding draws attention to the diversity of the products underneath. However, when WebSphere is combined with the collaboration strengths of Lotus and the convenience of portals, most companies will have the platforms they need for all their Web server needs.

Appendix A

lication

on IBM WebSphere Application Server
and enhanced capabilities abound. You'll
t the new browser-based administration
Management eXtensions (JMX)-based
about the performance of your applica-
oving applications from earlier versions
ng new application development on V5,
erformance enhancements you can't live

—*e-Pro Magazine*, May, 2003

ation Server (WAS) 5.0. Also see Appendix D
phere Application Server, WAS 6.0. At the time
ed but features were available for evaluation.

new release of the WAS product. V5 includes
different approach to product packaging and
installation of single and multiserver configurations; an all new server administration architecture (including an XML-based administrative repository, a Web-based Administrative Console, and a completely new scripting facility); substantial security, availability, and performance enhancements; and many new programming extensions.

Feature Details

WAS V5 brought major changes to IBM's strategic application server. This section gives details on WAS 5.0 features, along with some advice on using those features.

WAS Product Packages

- **WAS Express** is a lightweight server that supports static content, servlets, and JSPs—but not EJBs. Its features simplify installation and administration.
- **WAS is the core application server.** It provides full J2EE 1.3 support and many additional features. WAS is licensed for a single system and doesn't support workload management. This V5 product is the successor to WAS Advanced Single Server Edition (WAS AEs) in V4.
- **WAS Network Deployment (WAS ND)** is the core WAS product plus the Network Deployment add-on module, which enables multiserver configurations and workload management. This product is the successor to WAS Advanced Edition (WAS AE) V4.
- **WAS Extended Deployment (WAS XD)** is WAS ND plus the Extended Deployment module, which adds more configuration, runtime, and administration features for large-scale, multiserver environments.
- **WAS Enterprise** is WAS XD plus the Enterprise Extensions module, which includes a set of additional programming features. This product is the successor to WAS Enterprise Edition V4.
- **WAS for z/OS** is a platform-specific implementation of WAS Enterprise (except for those features that don't make sense on z/OS).

WAS Enterprise, WAS XD, and WAS ND are delivered as a set of modules that you install on top of the base WAS product. Unlike in V4, you don't have to uninstall one edition of WAS to upgrade to a higher-level edition. For example, you can initially install WAS V5 on a machine and later install the Network Deployment module to upgrade to WAS ND. This packaging should simplify upgrades and provide more consistent behavior among different editions.

New Administration Architecture

WAS V5 no longer requires a relational database management system (RDBMS) as the administrative repository for multiserver configurations. All editions now use XML files to store configuration information. IBM says this change is a response to market resistance to the requirement of an RDBMS just for the repository. The Java application used for the WAS AE Administrative Console in V4 has been replaced by a Web application, which all editions use. The V5 Web-based console is much expanded and improved over the Web-based console used for WAS AE in V4.

The WebSphere Control Program (WSCP), available in WAS AE V4, was replaced by a more flexible and coherent facility (known as wsadmin) that's available in all V5 editions. You can use wsadmin in interactive or batch modes and perform any operation that you can do with the Administrative Console. The wsadmin utility is based on IBM's Bean Scripting Framework

(BSF) and supports a variety of scripting languages, including JavaScript, JPython, and Jacl (a Tcl derivative). The new V5 administrative architecture uses 30 or so Java Management Extensions (JMX) managed beans (MBeans) to control WAS servers, and these MBeans are accessible in wsadmin scripts. The V5 support for JMX is important not just for wsadmin; JMX also lets independent developers use a standards-based interface to develop WAS administrative tools.

Workload Management

Multiserver configurations and workload management (WLM) have also been redesigned in V5. In WAS ND and higher, distributed configurations are organized as cells (comparable to a V4 administrative domain). A cell is a set of nodes, and a node is a set of managed servers. A node cannot span multiple machines, but a machine can have multiple nodes. In WAS terminology, a managed server means an instance of WAS—that is, a single JVM running in its own process.

A cell has one master administrative repository file, and each node and server also has its own local administrative repository file. Changes to a node or server repository file are temporary and while in effect override specifications in higher-level repository files. Changes to the cell repository file are permanent and are propagated to the node and server files when the next synchronization action occurs (e.g., when a server starts).

A cell manager process runs on one of the systems encompassed by the cell and communicates with one or more node managers, each of which runs in its own lightweight (i.e., not a full J2EE environment) JVM process on the same machine as the associated node. Having independent cell manager and node agent processes increases the independence of node and server operations, while maintaining coordination among them. For example, you can start a node's servers and run applications on them whether or not the cell manager is running. Any necessary synchronization occurs among servers and nodes whenever a server, a node agent, or the cell manager is started (or possibly at other defined synchronization points). You also can attach an administrative client (such as the Administrative Console) at any of the three levels and manage the server(s) encompassed by that level. This lets you administer parts of a cell even when the cell manager is down.

Within a cell, a cluster is a workload-managed set of servers. (Clusters replace the V4 server group feature.) A cell can have multiple clusters, and servers in a cluster can all be on the same node (vertical cluster) or in multiple nodes (horizontal cluster). New in V5 is the ability to define a backup cluster, which mirrors and provides failover for an entire (primary) cluster. V5 WLM also supports a new server-weighted, round-robin routing mechanism that adds better control over load distribution. Application deployment is greatly simplified in a V5 cluster because you need only install the application on a single node, and WAS will replicate the application to other nodes as needed.

The WAS ND module also bundles the DMZ CD, which includes WebSphere Edge Server components, such as the IBM HTTP Server, the Network Dispatcher router, and a caching proxy. WAS XD adds features that allow dynamic and custom tuning of WLM, asynchronous application distribution, Simple Network Management Protocol (SNMP) support, and proximity-based (i.e., tiered) name lookups.

Security

V5 supports the full Java 2 and J2EE 1.3 security architecture, including the Java Authentication and Authorization Service (JAAS). JAAS provides for plug-in authentication modules, and the base WAS product supports Simple WebSphere Access Mechanism (SWAM), which is equivalent to V4 LocalOS, as well as Lightweight Directory Access Protocol (LDAP) and custom user registries. When you add the ND module, you also get Lightweight Third-Party Authentication (LTPA) support. The WAS XD module will add Kerberos support in a point release to follow V5. Version 5 also supports the OMG Common Secure Interoperability (CSI version 2) standard for communication within a distributed environment.

WAS administration now supports roles that have different sets of capabilities. The base WAS edition provides Administrator, Operator, Configurator, and Monitor roles, and WAS XD adds other, finer-grained roles. You authorize users to appropriate roles so they can perform the administrative operations related to their job responsibilities. In a point release to follow V5.0, WAS XD will bundle the Tivoli Access Manager (formerly called Policy Director), which provides an integrated user administration facility and allows access policies to be changed dynamically. For Web services, WAS V5 supports the Web Service Security (WS-Security) standard developed by IBM, Microsoft, and Verisign. WS-Security supports digital signatures and identity propagation.

Performance

In addition to server code optimization, several other enhancements have been made to improve performance under WAS V5. There's a new memory-to-memory option for sharing session data among multiple WAS instances, which lets you avoid the cost of using RDBMS persistence for this type of sharing. The existing Dynacache facility has been expanded to support edge-side includes using the Akamai ESI product. Also available are new caching policies and cache replication across nodes.

The V4 Performance Analyzer tool was enhanced and renamed as Tivoli Performance Viewer. A "technology preview" for Performance Viewer will provide tuning advice, a first step in what IBM plans as automated WAS performance tuning. The Performance Monitoring Interface (PMI) adds new instrumentation for session times, Dynacache, WLM, JDBC, and other WAS components. The new information available from PMI can be exploited by third-party and locally written tools, as well as by IBM products.

Problem Diagnosis

WAS V5 has several enhancements that should aid in problem diagnosis. Diagnostic messages have been improved to include help on probable causes and suggestions for corrective actions. All internal exception handling (i.e., catch blocks) has been equipped with a system known as First Failure Data Capture (FFDC). FFDC makes it easier to diagnose the root cause of a problem, and WAS provides an analysis engine that matches FFDC data against a knowledge base of common errors and fixes to provide help in correcting a problem. V5 also includes a new end-to-end transaction trace facility to examine application execution flow.

J2EE 1.3 Support

Perhaps the most important aspect of V5 is the support for J2EE 1.3, especially EJB 2.0. The EJB persistence manager has been completely redesigned. Not only does the new facility support EJB 2.0 features, but much of the implementation support has been moved out of the code-generation tools that are in VisualAge for Java and WebSphere Studio Application Developer and into WAS runtime components. This change should substantially reduce the need to regenerate code as the persistence manager evolves.

Another notable change brought about by J2EE 1.3 is that access to data in an RDBMS is through new relational resource adapters (RRAs) that are part of the Java Connector Architecture (JCA). (Older-style, direct JDBC access is still supported, but it's not the recommended approach.) This change will impact application coding and administration tasks. The base WAS product and WAS ND both contain a limited version of WebSphere MQ as the Java Messaging Service (JMS) provider. This limited version doesn't support non-WAS applications or many of the management functions of the full-blown WebSphere MQ product, which is bundled with WAS XD and WAS Enterprise.

Web Services

WAS V5 expands on the previous version's support of Web Services. Version 5 will use the new AXIS engine (aka SOAP 3.0), which should improve performance. WAS ND includes a private UDDI registry that companies can use to store service definitions. WAS ND also provides a Web Services Gateway that acts as an intermediary between services available on an enterprise's intranet and those on the Internet. This feature lets an organization control the services it exposes and how to make public services available to local applications. The gateway also provides for uniform deployment of Web services delivered by WAS and by other means regardless of which protocols are used (e.g., SOAP/HTTP, SOAP/JMS, or Java over JMS). Support for the new JAX-RPC and JSR 109 (enterprise Web services) standards will appear in additional "technology previews."

Programming Extensions

WAS Enterprise includes a raft of interesting programming extensions, including the following enhancements to EJBs:

- SQLJ support for CMP beans (this feature is also in WAS XD)
- Specifying EJB access intent (concurrency strategy, type of access, read-ahead, and pre-fetch)
- Defining CMP beans over procedural code rather than a DBMS
- Defining EJBs by composition of other EJBs
- Runtime creation of EJB (QL) queries

- Ability to incorporate nontransactional, or one-phase commit, resources into a transaction
- Container-managed messaging

WAS Enterprise V5 also supports asynchronous execution of EJBs—essentially enabling a "batch" subsystem for executing code. Building on this feature, WAS Enterprise also includes a scheduler for running work within WAS, along with support for callback notification of execution events or status. These features take WAS further along the road to providing more and more of the capabilities traditionally provided by enterprise operating systems.

A significant feature in WAS Enterprise is the WebSphere Workflow facility. This feature lets you use a graphical workflow editor to assemble a variety of services, rules, and lower-level flows into a large-scale, transactional, potentially long-running business process. One of the key underlying concepts is that individual processes within a workflow define a "compensation" process that the workflow engine can invoke to undo work if a downstream process fails. It will be interesting to watch how the WebSphere workflow capability impacts the Lotus Domino product that has traditionally been IBM's strategic product in this area.

JDK 1.4 Performance Improvements

WebSphere Application Server V5 supports Java Developer Kit (JDK) 1.4. One of the hottest features of JDK 1.4 is its new server optimization feature. If you use the `-server` option, the JDK will scan loaded classes as they are being run, eliminate methods by inlining them, turn methods into their native code equivalents, remove content evaluations from loops, and perform a number of other code optimizations. Those optimizations can potentially yield a ten-fold performance improvement in CPU-intensive sections of code.

`wsadmin`, WAS 5.0's Replacement for `WSCP` and `XMLConfig`

WAS V5 replaces both the XMLConfig and WSCP administration utilities with `wsadmin`. So, with WAS V5, there is one utility to become familiar with rather than two. In addition, `wsadmin` is based on Java Management Extensions (JMX). JMX requires the use of MBeans, which are Management JavaBeans. JMX and MBeans follow a standard API, allowing tool vendors such as IBM and BEA to develop sophisticated interfaces to their Web application servers. Furthermore, developers also can easily create custom utilities for managing their application servers. `wsadmin` uses a JACL interface to WebSphere's JMX MBeans. JACL is basically a Java/TCL language. You can directly access WebSphere's MBeans from within `wsadmin`'s shell or with JACL scripts. The main MBeans available for WAS 5.0 are `AdminControl`, `AdminConfig`, and `AdminApp`. Those three MBeans each have a set of methods that you can access from `wsadmin` and JACL commands. All this JMX, MBean, JACL, and TCL material can be overwhelming, so let's experiment a bit. For example, to add a WAR file to WAS, you could use the following commands from `wsadmin`'s shell.

First, start `wsadmin` with wsadmin.bat or sh wsadmin.sh; then enter the following from the `wsadmin` prompt key:

```
wsadmin> $AdminApp installInteractive C:\dir\to\some\test.war {-
contextroot testuri}
```

Note that the dollar ($) symbol is not a Unix symbol but a JACL symbol. `$AdminApp` refers to the MBean, `installInteractive` is a method name, and the `test.war` and curly braced `contextroot` setting are parameters to the `installInteractive` method. The tool will then prompt you for several items:

```
Pre-compile JSP:  [No]: No
Directory to Install Application:  []:
Distribute Application:  [Yes]:
Use Binary Configuration:  [No]:
Deploy EJBs:  [No]:
Application Name:  [test.war]:
Create MBeans for Resources:  [Yes]:
Enable class reloading:  [No]:
Reload Interval:  []:
```

After the installInteractive tool finishes, tell `wsadmin` to save the changes and exit by entering the following:

```
wsadmin> $AdminConfig save
wsadmin> exit
```

Some of the `wsadmin` command sets get somewhat verbose, but you can put them in a file and reference the file with the hyphen "`f`" option:

```
c:\wasroot\bin\wsadmin.bat -f importwar.jacl
```

If you added that test WAR file to your WebSphere server, here's how to remove it:

```
wsadmin> $AdminApp unin`stall test.war
wsadmin> $AdminConfig save
```

To show you a bit of the strengths of `wsadmin` and JACL, here's a JACL file that lists all your servers and the applications hosted by those servers:

```
<!--text or script-->

# Begin listapps.jacl
set cells [$AdminConfig list Cell]
foreach cell $cells {
  set nodes [$AdminConfig list Node $cell]
  foreach node $nodes {
     set cname [$AdminConfig showAttribute $cell name]
```

```
        set nname [$AdminConfig showAttribute $node name]
        set servs [$AdminControl queryNames
        ➥type=Server,cell=$cname,node=$nname,*]
        puts "node $nname server count: [llength $servs]"
        foreach server $servs {
            set sname [$AdminControl getAttribute $server name]
            set ptype [$AdminControl getAttribute $server
            ➥processType]
            set pid   [$AdminControl getAttribute $server pid]
            set state [$AdminControl getAttribute $server state]
            set jvm [$AdminControl queryNames
             ➥type=JVM,cell=$cname,node=$nname,process=$sname,*]
            set osname [$AdminControl invoke $jvm getProperty
            ➥os.name]
            puts "  $sname process Id: $pid State: $state O/S:
            ➥$osname"
            set apps [$AdminControl queryNames
            ➥type=Application,cell=$cname,node=$nname,process=$sname,*]
            puts "  $sname application count: [llength $apps]"
            foreach app $apps {
               set aname [$AdminControl getAttribute $app name]
               puts "    $aname"
            }
        }
    }
}
# End listapps.jacl

<!--text or script-->
```

The following is what will appear on your test server when you run the `listapps.jacl` script:

```
c:\wasroot\bin\wsadmin.bat -f listapps.jacl
Number of running servers on node DefaultNode: 1
  server1 (UnManagedProcess) has pid 3789; state: STARTED; on
Linux
  Number of applications running on server1: 2
    IBMUTC
    insuraceEAR
    test.war
```

Exploiting Dynamic Caching in WAS 5.0

At the heart of all caching features in WAS is the WebSphere Dynamic Cache. This caching feature provides a declarative-based caching service for J2EE applications. Dynamic Cache works within a WAS instance, intercepting requests to cacheable objects. If the object resides in the cache, it's served immediately; otherwise, the request is processed as normal, and the output is used to populate a cache entry. Cached content is generally served in a fraction of the time it normally takes to complete a noncached request.

Web Page/Fragment Caching

WAS Dynamic Cache provides caching for static and dynamic content of servlet and JavaServer Pages (JSP) pages and fragments making up Web pages. You cache servlets and JSPs in WAS declaratively and configure the caching by using a cache policy defined as an XML deployment descriptor included in your Web application. The cache policy file—named *Cachespec*.XML—is located in your Web module's WEB-INF directory along with the standard Web.XML file. When configuring the cache policy for Dynamic Cache, you must consider which servlets or JSPs to cache, what makes an invocation of that servlet or JSP unique, and when to remove content from the cache.

Identifying Servlets or JSPs to Cache

WAS Dynamic Cache parses the Cachespec.XML deployment descriptor on application startup and extracts from each `<cache-entry>…</cache-entry>` element a set of configuration parameters. Then, every time a new servlet or JSP is initialized (e.g., when the servlet is first accessed), the cache attempts to match that servlet/JSP to each cache-entry element to find the configuration information for that servlet.

You can indicate which servlets or JSPs to cache by either specifying the class name of the servlet or using its relative URI. For example, if you want to cache the servlet class `com.ibm.samples.MyServlet.class`, use the `<name>` and `<class>` tags, as the following code shows:

```
<cache-entry>
     <name>com.ibm.samples.MyServlet.class </name>
     <class>servlet </class>
</cache-entry>
```

Whenever a servlet of the specified class is initialized, WAS Dynamic Cache will match that servlet with the configuration for this element. The `<class>` tag identifies the type of object being specified in the `<name>` tag. For servlets and JSPs, the `<class>` tag always has a value of servlet. Other types of cacheable entities (e.g., commands) use different values for the `<class>` tag.

You can also identify a servlet or JSP using its Web URL path. For example, the following specification defines a cache policy for a servlet with a mapping in Web.XML of /action/view:

```
<cache-entry>
     <name>/action/view</name>
     <class>servlet</class>
</cache-entry>
```

The specification of the Web path is relative to the Web application's context root; therefore, the context root isn't included in the servlet's name.

Identifying What Makes a Cache Entry Unique

Most servlets and JSPs make use of a variety of inputs to generate unique dynamic content (otherwise, static HTML is used). This variable input usually comes in the form of request parameters, browser cookies, request headers, session attributes, path information, and request attributes from parent or peer servlets. To *correctly* cache a servlet or JSP, the cache policy author must identify the minimum set of input variables that makes a servlet's or JSP's output unique. Dynamic Cache supports declaring these inputs through the use of `<cache-id>` and `<component>` tags.

For example, the following code shows that the `viewQuote` servlet's output is dependent on a request parameter named "`ticker`" and on a style attribute that was previously stored in the user's HTTP session.

```
<cache-entry>
      <name>/viewQuote </name>
      <class>servlet </class>
      <cache-id>
             <component id="ticker" type="parameter" />
             <component id="style" type="session" />
      </cache-id>
 </cache-entry
```

When a request is made to the URL `http:/myhost/quoteapp/viewQuote?ticker=IBM` by a user with an HTTP session attribute of `style=frames`, the following cache ID is generated:

```
/quoteapp/viewQuote:ticker=IBM:style=frames
```

Each request to the `viewQuote` servlet with unique ticker and style attribute values will produce a unique cache entry instance. WAS supports many different input variable types that you can use to uniquely identify requests. Table A-1 displays a list of valid types supported for use in generating cache IDs.

Table A-1 List of Valid Types Supported for Generating Cache IDs

Input Variable Type	Valid Classes	Description
`Method`	Command	Calls the indicated method on the command or object.
`Field`	Command	Retrieves the named field in the command or object.
`Parameter`	Servlet	Retrieves the named parameter from the request object.
`Session`	Servlet	Retrieves the named attribute from the HTTP Session.
`Cookie`	Servlet	Retrieves the named cookie value.
`Attribute`	Servlet	Retrieves the named request attribute.
`Header`	Servlet and Webservice	Retrieves the named request header.
`Pathinfo`	Servlet	Retrieves the path information from the request.
`Servletpath`	Servlet	Retrieves the servlet path.
`Locale`	Servlet	Retrieves the request locale.
`SOAPEnvelope`	Webservice	Retrieves the SOAP Envelope from a Web services request; an ID attribute of Hash uses a hash of the SOAP Envelope, and Literal uses the SOAP Envelope as received.
`SOAPAction`	Webservice	Retrieves the SOAP Action header, if available, for a Web services request.
`ServiceOperation`	Webservice	Retrieves the service operation for a Web services request.
`ServiceOperation Parameter`	Webservice	Retrieves the specified parameter.

Another important element in Cachespec.XML is the dependency ID—a secondary key that describes when to invalidate a cache entry. Dependency IDs are composed of an optional static string and/or one or more input variables as used in cache ID generation. The static string and component values are appended together to form the dependency ID. One example of using a dependency ID is to invalidate a cache entry of a stock ticker when the stock price is updated, as we'll see in a moment.

Removing Entries from the Cache

After populating the cache with servlet and JSP content, the next important consideration is how to remove the content. Dynamic Cache removes content based on an explicit `timeout`, an explicit invalidation based on cache ID or dependency ID, and replacement due to being selected for eviction by the cache's replacement algorithm. Dynamic Cache manages the replacement algorithm, which uses a cache entry's priority and frequency of access to determine which entry to evict when the cache has exceeded its capacity. The cache policy in the code below sets each cache entry to have a `timeout` value of five minutes (300 seconds).

```
<cache-entry>
      <name>/viewQuote</name>
      <class>servlet</class>
      <cache-id>
       <component id="ticker" type="parameter" />
       <component id="style" type="session" />
       <timeout>300</timeout>
      </cache-id>
</cache-entry>
```

There are three ways to explicitly remove entries from the cache: programmatically by cache ID, programmatically by dependency ID, and declaratively by dependency ID.

Programmatic Invalidation

An application can programmatically invalidate entries in the cache by passing a cache entry's ID as an argument into the cache's `invalidateById` API, as indicated in the following code:

```
import com.ibm.websphere.cache.*;

public class Test {
public void invalidate(String id) {
            Cache cache = DynamicCacheAccessor.getCache();
            cache.invalidateById(id);
      }
}
```

An easier (and preferred) way to invalidate cache entries is to use dependency IDs. Each cache entry can have one or more dependency ID(s) for use as invalidation keys. As described earlier, you can create dependency IDs declaratively inside the Cachespec.XML deployment descriptor, as indicated in the following code:

```
<cache-entry>
       <name>/viewQuote </name>
       <class>servlet </class>
       <cache-id>
```

```
        <component id="ticker" type="parameter" />
        <component id="style" type="session" />
        <timeout>300 </timeout>
       </cache-id>
       <dependency-id>
       stockTicker
        <component id="ticker" type="parameter" />
       </dependency-id>
   </cache-entry>
```

When a request is made to the URL `http:/myhost/quoteapp/viewQuote?ticker=IBM`, the following dependency ID is generated:

```
stockTicker:IBM
```

To remove this entry from the cache, execute the following code snippet:

```
import com.ibm.websphere.cache.*;

public class Test {
     public void invalidate(String id) {
          Cache cache = DynamicCacheAccessor.getCache();
          cache.invalidateById("stockTicker:IBM");
     }
}
```

All cache entries that shared the `stockTicker:IBM` dependency ID are invalidated at the same time.

Declarative Invalidation

You can declare invalidation rules in the Cachespec.XML file to invalidate content based on dependency IDs. If the example in the code above had contained a servlet `/updateQuote` that updates the stock price, we could have defined the cache policy in the code below to declaratively invalidate the cached `/viewQuote` output for the updated stock ticker:

```
<cache-entry>
     <name>/updateQuote </name>
     <class>servlet </class>
     <invalidation>
     stockTicker
      <component id="ticker" type="parameter" />
     </invalidation>
</cache-entry>
```

Because we defined no cache ID rules for the `/updateQuote` servlet, its cache policy will invalidate only other cached content; its contents won't be cached.

A Declarative Caching Sample: Rich Site Summaries

Rich Site Summaries (RSS) is an XML document format that's widely used on the Internet to provide syndicated information summaries. Various Web sites make RSS files available for other sites to use to embed up-to-date information in their own Web pages. Many RSS-enabled Web sites provide current news headlines, stock quotes, and other dynamic content. The simple servlet in the following code retrieves an RSS file from a Web site, applies an eXtensible Stylesheet Language (XSL) transform to the document, and sends the result back to the client browser.

```
package com.ibm.websphere.samples.rdf;

import java.io.IOException;
import java.io.PrintWriter;
import java.net.URL;

import javax.servlet.ServletException;
import javax.servlet.http.HttpServlet;
import javax.servlet.http.HttpServletRequest;
import javax.servlet.http.HttpServletResponse;
import javax.xml.transform.Source;
import javax.xml.transform.Transformer;
import javax.xml.transform.TransformerFactory;
import javax.xml.transform.stream.StreamSource;

public class RDFServlet extends HttpServlet {

   public void doGet(HttpServletRequest request,
HttpServletResponse response) throws ServletException, IOException
{
      // Set content type for HTML.
      response.setContentType("text/html; charset=UTF-8");
      // Output goes to the response PrintWriter.
      PrintWriter out = response.getWriter();
      try {
         TransformerFactory tFactory =
         ➥TransformerFactory.newInstance();
         // Get the XML input document and the stylesheet, both in
         // the web application document directory.
         Source xmlSource = new StreamSource(new
         ➥URL(request.getParameter("url")).openStream());
         Source xslSource = new StreamSource(RDFServlet.class.
         ➥getResourceAsStream("/rss.xsl"));
         // Generate the transformer.
```

```
            Transformer transformer =
            ➥tFactory.newTransformer(xslSource);
            // Perform the transformation, sending the output to the
            ➥response.
            transformer.transform(xmlSource, new
            ➥javax.xml.transform.stream.StreamResult(out));
        } catch (Exception e) {
            // If an Exception occurs, return the error to the
            ➥client.
            out.write(e.getMessage());
            e.printStackTrace(out);
        }
    }

}
```

The simple design of this servlet won't perform well because it's retrieving the RSS file from a remote Web site on each request and transforming the XML document to HTML on every page request. We can employ servlet caching to make the servlet perform better without any additional coding on our part. We simply need to create a cache policy that specifies the *Servlet to cache*, the parameters that make the *servlet invocation unique*, and the time we want the cached response to live in the cache. Here's a sample cache policy for the servlet:

```
<cache-entry>
     <name>/RDFServlet</name>
     <class>servlet</class>
     <cache-id>
      <component id="url" type="parameter" />
      <timeout>600</timeout>
     </cache-id>
</cache-entry>
```

The cache policy indicates that the `/RDFServlet` should be cached for 10 minutes and that the request parameter URL should be used in the generated cache ID. The following code shows a snippet of a sample RDF XML document providing current news on IBM:

```
<?xml version="1.0"?>
 <rdf:RDF
  xmlns:rdf="http://www.w3.org/1999/02/22-rdf-syntax-ns#"
  xmlns="http://purl.org/rss/1.0/">


   <item>
     <title>IBM extends leadership in identity and security …
</title>
```

```
    <link>http://www.ibm.com/news/us/2003/01/301.html </link>
    <description>IBM has announced that dozens of new customers …
    </description>
  </item>
  <item>
...
```

Performance Comparison Using RSS Servlet

Employing the RSS sample servlet, we conducted a set of performance tests to compare the throughput and response time of content delivered using WAS Dynamic Cache to noncached content. For this test, we set up all the components, including the servlet and RDF XML file, on the same local server.

Caching the `RDFServlet` provides significant performance and scalability improvements. The cached servlet provides 10 times the throughput of the noncached version at the peak load of 1,000 clients. Also, the response time for the cached version is a fraction of the noncached one. Although this benchmark is fairly simple and may not reflect the results of a more complex application mix, these tests nevertheless indicate the potential for performance and scalability improvements you can achieve using WAS Dynamic Cache.

Command Caching

Command caching is a new feature in WAS that provides the capability to cache Java object instances. For example, we can cache a Java class that accesses database data or Enterprise Java Beans (EJBs) just as we cached the Web pages previously described. This provides a much finer-grain approach to caching data. Because you can combine servlet/JSP caching with Command caching, a flexible caching solution exists for almost any e-business application. Consider Command caching when the same application data is reused on multiple servlets or JSP pages. By caching the data before it's transformed into each page view, the application can avoid redundant re-retrievals or recalculations.

Cacheable commands are stored in the cache for reuse with a similar mechanism for servlets and JSP files. However, in this case, the unique cache IDs are generated based on methods and fields present in the command as input parameters. For example, a `GetStockQuote` command can have a stock ticker as its input parameter; a unique cache ID is generated from the name of the command and the value of the stock ticker.

Cacheable commands in WAS use the Command pattern programming model. Cacheable command implementations extend the `com.ibm.websphere.command. CacheableCommandImpl` base class. A command is invoked by first setting object attributes required for executing the command. This is followed by calling the `execute()` method on the Command bean. If there is a cache hit, the results of the `execute()` method invocation are served from the cache; otherwise, code within the `execute()` method is run, and the command's output properties are cached appropriately. After the `execute()` method has been invoked, results are obtained by invoking the getter methods on the Command bean. A complete example of Command caching is provided in the WAS Info Center at `http://publib7b.boulder.ibm.com/wasinfo1/en/info/aes/ae/rprf_commandxmp.html`.

Flexible, Portable Caching

WAS Dynamic Cache provides a declarative method for caching Web page content and Java object data. The key benefit of declarative-based caching is that it requires no modifications to the application source code. This advantage, combined with Dynamic Cache's page, fragment, and object-based caching, results in a flexible yet portable means to enable caching for virtually any J2EE application. Dynamic Cache works within the WAS runtime, intercepting and processing requests for cacheable objects and content. Leveraging caching features in WAS can provide tremendous performance benefits for most e-business applications.

WAS 5.1 Migration Tools

There are several migration tools that will help you get to the latest version of WAS 5. Getting from V3.5.x to Version V5 was discussed in Chapter 6, "Building a Combined WebSphere and Domino System." Here are some brief explanations on migrating from V4.x and V5.0.x to V5.1.

New Migration Tools

The WASPreUpgrade and the WASPostUpgrade migration tools in Version 5.1 are updated to work with V5.1. To migrate the configuration from another version of WebSphere Application Server, use the WASPreUpgrade tool from the migration/bin directory on the product CD-ROM. The tool from the previous release is not sufficient for migrating to V5.1. There are several scripts that the installer program copies to the V5.0.x install_root/bin directory when you use the Installation Wizard to migrate a V5.0.x release to V5.1. The pre_uninst50ws and the post_uninst50ws tools prevent removing the configuration of a migrated node from the deployment manager when you uninstall. The pre_uninst502mq and the post_uninst502mq tools prevent deleting a messaging queue manager when you uninstall after migrating a configuration of V5.0.2 to V5.1.

Version 5.0.x Migration to 5.1

Migration from Version 5.0.x to Version 5.1 involves minimal change because both releases implement the Java 2 Platform, Enterprise Edition (J2EE) 1.3 specification. Version 5.0.x supports the Java 2 SDK Version 1.3.1. Version 5.1 supports the Java 2 SDK Version 1.4.1. V5.0.x applications can run on either release without change. V5.1 applications can run on either release without change if you compile applications on a V5.1 node with the compatibility options. There are several considerations. There are new migration tools for V5.1 to help migrate federated nodes and nodes with the embedded messaging feature installed. Use the new tools when uninstalling a node you migrate or when uninstalling the V5.1 node, if you decide to keep using the previous node. One set of tools prevents the deployment manager from removing the node as a member of the cell. The other new set of tools prevents uninstalling the messaging queue manager and deleting the code for embedded messaging feature.

Version 4.x Migration

Migration from Version 4.x to Version 5.1 involves minimal change because both releases implement the Java 2 Platform, Enterprise Edition (J2EE) specification. (V4 implements the J2EE 1.2 specification. V5.1 implements the J2EE 1.3 specification.) Most V4.x applications can run without change. Despite this fact, knowledge of how the migration tools migrate Version 4.x applications is important. For example, extended messaging support service is now a component of IBM WebSphere Application Server, Version 5.1. Applications that use Version 4.x extended messaging support services require migration to use in an IBM WebSphere Application Server, Version 5.1 system.

If you are considering a migration to V5.1 from previous versions of WebSphere Application Server, Table A-2 summarizes potential migration concerns and answers.

Table A-2 Summary of Potential Migration Areas

Functional Area	Support in V3.5.x or V4.x?	Must Migrate from V3.5.x?	Must Migrate` from V4.x?	Must Migrate from V5.0.x?	Details
Enterprise Beans	EJB 1.0	Yes	Not applicable	No	Many EJB 1.0 applications can run unchanged in Version 5.1, although some changes might be required or recommended. See Migrating enterprise bean code to the supported specification.
	EJB 1.1	Not applicable	No	No	Full support for EJB 1.1 is provided. See migrating enterprise bean code to the supported specification.
Java 2 Connectors	Java 2 Connectors	Not applicable	Yes	No	The preliminary Java 2 Connector support in V4.x is completed in V5.1. Some changes might be necessary to take full advantage of this support. See J2EE Connector Architecture migration tips.

Functional Area	Support in V3.5.x or V4.x?	Must Migrate from V3.5.x?	Must Migrate` from V4.x?	Must Migrate from V5.0.x?	Details
JDBC API	JDBC API	Yes	Not applicable	No	Many applications can run unchanged in V5.1, although some changes might be required or recommended. See Migrating a Version 4.0 data access application to Version 5.1.
JavaServer Pages (JSP) files	JSP 1.0 Specification	No	No	No	JSP 1.0 ASPs are a pure subset of JSP 1.2. See Developing JavaServer Pages files with WebSphere extensions.
	JSP 1.1 Specification	Not applicable	No	No	JSP 1.1 ASPs are a pure subset of JSP 1.2. See Developing JavaServer Pages files with WebSphere extensions.
Security	IBM Security	Yes	Yes	No	Changes might be required due to J2EE security. See Migrating security configurations from previous releases.
Servlets	Servlet 2.1 Specification and IBM extensions	Yes	Yes	No	Many Servlet 2.1 applications can run unchanged in V5.1, although changes might be required or recommended. See Developing servlets with WebSphere Application Server extensions.
Servlets	Servlet 2.2 Specification	No	No	No	Servlet 2.2 APIs are a pure subset of Servlet 2.3. See Developing servlets with WebSphere Application Server extensions.

continues

Table A-2 Continued

Functional Area	Support in V3.5.x or V4.x?	Must Migrate from V3.5.x?	Must Migrate` from V4.x?	Must Migrate from V5.0.x?	Details
Sessions	IBM sessions	Yes	Yes	No	Many applications can run unchanged in Version 5.1, although changes might be required or recommended. See Developing session management in servlets and Migrating HTTP sessions.
Transactions	IBM transactions	Yes	No	No	A change in the import statement. Also, one datasource connection cannot be used across multiple user transactions. See Using the transaction service.
Web services	Apache (SOAP) 2.2	Not applicable	Yes	No	Many applications can run unchanged, although changes to use new support are recommended.
XML parser	XML 2.0.x supported	Yes	Not applicable	No	Changes to move to the supported API XML4J V4.0.6 level are required. See Configuring WebSphere Application Server after migration.
XML parser	XML4J V3.1	Not applicable	Yes	Not applicable	Recompilation is required to convert to XML4J V4.0.6.
XML parser	XML4J V3.1	Not applicable	Yes	Yes	Recompilation is required to convert to XML4J 4.2.2. See Configuring WebSphere Application Server after migration.

Functional Area	Support in V3.5.x or V4.x?	Must Migrate from V3.5.x?	Must Migrate` from V4.x?	Must Migrate from V5.0.x?	Details
XML configuration tool	XMLConfig	Yes	Yes	No	Use the JMX support provided by wsadmin. See Migrating from wscp V4.0 to wsadmin V5.0.
WebSphere Control Program	WSCP	Yes	Yes	No	Use the JMX support provided by wsadmin. See Migrating from wscp V4.0 to wsadmin V5.0.

APPENDIX B

Domino 6 Details

> **"IBM Lotus Notes/Domino 6.5 awarded 'Best Enterprise Collaboration Platform' based on 'Greatly improved interface for both Web and stand-alone clients makes Notes-Domino a win-win.'"**
>
> **—*InfoWorld Magazine*, December, 2003**

This appendix gives details on the many new features in Domino 6 and compares that release to Domino 5 where appropriate.

Overview

Notes/Domino 6 and the "point" releases of Domino 6 (Domino 6.5.1 was the latest point release when this Appendix was written) have many new features. For example, Domino 6 has many built-in client management features that allow Domino administrators to efficiently manage Notes client desktops across the enterprise. Policies are a new feature of Notes/Domino 6 that are configured using policy documents on the Domino server. Policies allow administrators to define and control Notes 6 client workstation configuration. Policies can be used to help make administering desktops easier and more efficient.

Domino 6 Feature Details

Here are details on the new features included with Domino 6.

Roaming

The roaming user feature allows users to have their own up-to-date information available on any PC. The information is stored on the server and replicated to each workstation accessed by the user. For example, users may have a system at work and one at home or may access PCs in multiple work locations. By using the roaming feature, users can have their information readily available wherever they are. This feature would be implemented by the Domino administrator. The Roaming User features work in the following manner: When registering or designating a roaming user, the user's personal information is created and stored in a separate personal directory on the user's mail server. When the user connects from a workstation, the following information is replicated to the current workstation and replicated back to the server when work is complete:

- Personal Address Book
- User ID
- User preferences from NOTES.INI
- Bookmarks
- Personal Journal

Replication and Client Performance Enhancements in Notes 6

The Notes 6 replication architecture is much more efficient than in previous versions of Notes. The following enhancements to the replication process result in greater user efficiency:

- Multitasking and Memory Usage
- Streaming Replication
- Incremental view building
- Documents replicated in ascending size
- User can specify limits for document and attachment sizes
- Client/Server Interactions
- Network compression

1. **Multitasking and memory usage.** Replication is one of many features that are "multitasked," which means users can perform other tasks while the Notes client is busy replicating in the background. Activities such as polling for new mail, detaching files, replicating, printing, monitoring alarms, subscription processing, running background agents, replacing designs, and copying databases take advantage of the new Notes 6 multitasking abilities. Another key improvement in the client is better overall use of memory and less use of graphical objects in the client itself.

2. **Streaming replication.** Streaming replication involves a single server request, which then pulls in all the data (Notes documents and their attachments) into the database. This feature dramatically reduces replication time and works in all Domino 6 client/server scenarios. Streaming replication also means that when using the Notes client you no longer have to wait until the replication is over before seeing replicated documents in folders. They appear individually, as soon as they are pulled into the system, and you can begin to work on them before the database has finished replicating. In addition, documents are replicated in ascending size order.
3. **Dynamic build of folders and views.** Streaming replication also means that when using the Notes client you no longer have to wait until the replication is over before seeing replicated documents in folders. They appear individually, as soon as they are placed in the local replica, and you can begin to work on them before the database has finished replicating. This is also referred to as incremental view building.
4. **Documents replicated by size.** In Notes 6, documents are replicated in ascending size order. This means a user can abort a long replication and still get smaller documents replicated.
5. **Document and attachment limits.** While previous versions of Notes allowed the user the option of receiving truncated documents (summary information and 40kb of rich text), Notes 6 allows users to replicate partial documents and define the size limit for documents and attachments. This is defined on the "Basics" tab of the Replication Settings dialog.
6. **Client/server interactions are also more efficient in Notes/Domino 6.** For example, an advantage of the new streaming feature is that because attachments are also streamed, Open and Save operations are more efficient. Client/server interactions are also dramatically improved by less frequent unread table exchanges, as there is a significant reduction in the number of bytes exchanged between client and server. Another performance enhancement is the use of incremental view reading, used to update design information and, in selected cases, user views. For example, when you move a document from one folder to another, the server is requested to provide just the incremental change it will take to fill the screen with the new view.
7. **Network compression.** Notes/Domino 6 network compression reduces the number of bytes sent during transactions by up to *50 percent*. Connections across heavily loaded links such as WANs and XPCs will see the most benefit. To use network compression, it must be enabled on both the client and the server. Additionally, Notes/Domino 6 incorporation of "network" streaming also reduces the overhead of large transfers.

Advantages of Using Domino 6 for Cluster Replication

This section discusses the advantages of using Domino 6 for cluster replication, and these advantages were a major factor used by IBM to encourage the customer to use Domino 6 from the beginning as they were starting the migration process from distributed NT servers to centralized Domino servers on the iSeries platform with cluster replication across the wide area network for 15,000 mail users. The customer did need some encouragement to start their project on the iSeries with the Domino 6.0 version (following the usual reluctance to use the X.0 version of any product).

Replication Notes/Domino 6 Utilizes Streaming Replication

Streaming replication involves a single server request, which then pulls in all the data (Notes documents and their attachments) into the database. This feature dramatically reduces replication time and works in all Domino 6 client/server scenarios.

To use streaming replication, both parties involved in the replication must be running Notes/Domino 6. Streaming replication is PULL only. This feature uses a single streamed RPC to read a number of documents and attachments. This is an improvement over the nonstreaming method of requesting and acknowledging one database note at a time.

Streaming replication saves time and consumes less CPU and network activity to replicate the same number of documents. Network latency is reduced, especially on slower links. Documents are also replicated in ascending size order (smallest first). This allows the client to abort a long replication but still receive some documents (for example, mail). Partial replication is also supported, allowing an aborted replication to be reinitiated and to continue where it left off. Views in folders are also updated incrementally while a replication is in progress (as opposed to the end of replication). This allows users to begin responding to initial documents (for example, mail messages) while larger documents continue downloading.

End-user benefits of streaming replication include shorter replication times, ability to start viewing and working on documents before replication is completed, and replication of smallest documents first. If replication is cancelled, the target system still has documents (especially smaller ones), which were replicated before the replication session ended. Incrementally updated folders means you no longer have to wait until the replication is over before seeing replicated documents in folders.

Advantages for IT departments (e.g., server to server replication) include faster replication, partial replication, potentially less network traffic due to single streamed RPC, and reduction of ACK (TCP/IP) responses. Lotus found that databases with 10,000 and 25,000 documents replicated 15 percent faster with streaming replication on. A database with 50,000 documents replicated 60 percent faster with streaming replication on.

Network Compression

Notes/Domino 6 network compression reduces the number of bytes sent during transactions by up to 50 percent. Network compression is an important performance feature offered in Notes/Domino 6. When you enable network compression, data is automatically compressed before it is sent over the network. This improves network performance, especially over slower line speeds. With network compression enabled on both the sending and the receiving computers, the Notes client and/or Domino server attempts to compress any data, (for example mail or replicated documents) before transmitting it over the network. Network compression works with mail routing, replication, or any other data sent via NRPC. This includes design elements in new or updated replica databases.

Notes/Domino 6 network compression offers a number of immediate benefits. For example, by reducing the amount of data being transmitted, you can improve the performance of your routing and replicating hub servers, especially if they're currently laboring under heavy workloads. Also you can enable it by default, so all your users with Notes 6 clients can take advantage of this functionality without having to select it themselves. Because network compression is a standard, out-of-the-box feature, it doesn't require any additional code, which helps simplify administration. CPU and memory resources used on the system increases by a small amount. But in many environments where it is very expensive or even not an option to upgrade the existing network infrastructure, this is a welcome trade-off.

Lotus also enhanced attachment compression in Domino 6. Attachment compression can now take advantage of the LZ1 algorithm (the same algorithm used in network compression). In R5, attachment compression used the Huffman algorithm. To use LZ1 compression in Domino 6, you must enable it in the Database Properties box. (By default, databases still use the Huffman algorithm to maintain backward compatibility.) The LZ1 attachment compression typically reduces attachment file size anywhere from 5 to 40 percent over Huffman attachment compression. The percent reduction in size depends on the type and complexity of the file attachment.

Domino 6 with Quota Mail and Cluster Replication

Here is information on how Domino 6 handles quota mail and cluster replication. The basic response is that Domino 6 is much more flexible in this area.

1. Domino R5 would not replicate databases over quota (true for both scheduled replication and cluster replication). On the other hand, the R5 router would essentially ignore the quota and route mail even if the mail database were over quota.
2. The Domino 6 default is to replicate databases over quota (i.e., ignore quotas for both scheduled replication and cluster replication). The default for the router remains the same (ignore quota). However, Domino 6 allows an INI setting to override the default and essentially go back to the R5 procedure of honoring quotas for both scheduled replication and cluster replication.

Domino 6 Policy-Based System Administration

In previous Domino/Notes releases, creating and managing a standard environment for all your users usually involved maintaining a set of written corporate policies. (An example of a policy might be a specific security configuration you want every Notes user to follow, covering things such as minimum password length and certificate expiration dates.) You would then use these policies to guide you as you entered settings for users throughout your site. If you were anything less than 100 percent familiar with every single setting, sooner or later you'd probably end up making some small, but crucial data entry error, one requiring hours or possibly days to correct. Perhaps you might identify with the plight of the administrator registering 1,000 users with the wrong Internet address format, resulting in hours spent editing each Person document containing the offending information. Those of you who know your way around LotusScript (or are on friendly terms with someone who does) might be able to create an agent to distribute at least some of this policy information corporate-wide. But wouldn't it be great if right out of the box, Domino gave you an automated way of enforcing such standards accurately and easily?

If you think so, you're in luck. Domino 6 helps you solve the problem of applying and managing standard corporate policies by providing a new technology called policy-based system administration. This feature gives you a way to centrally define, organize, and manage user settings throughout your organization. All information required for policy-based system administration is stored in your Domino Directory, giving you a single place to control administration activity, from user setup to mail archiving.

This article introduces policy-based system administration. We explain all its major features and components, how they work, and what you can do with them. And we'll provide examples of how policy-based system administration can make your job as a Domino administrator easier. For more information on policy-based system administration and all other Domino 6 administration features, see the Documentation Library.

What Is Policy-Based System Administration?

In broad functional terms, policy-based system administration is exactly what its name implies—a way for you as a Domino administrator to manage and apply corporate policies for your employees. These corporate policies define the way Domino enforces user settings, which ensures that your established corporate standards and practices control how your technology works (and not the other way around).

In Domino 6, you enforce corporate policies through individual Policy documents and Settings documents in the Domino Directory. To create a policy, you use a Policy document to specify which Settings documents to include.

A Policy document (see Figure B-1 as an example) defines a set of corporate information you want to apply to your users. It does this by specifying which Settings documents to include in the policy. Each Settings document contains detailed information applicable to a specific area of Domino administration, for example archiving or security. This information defines user settings for that area.

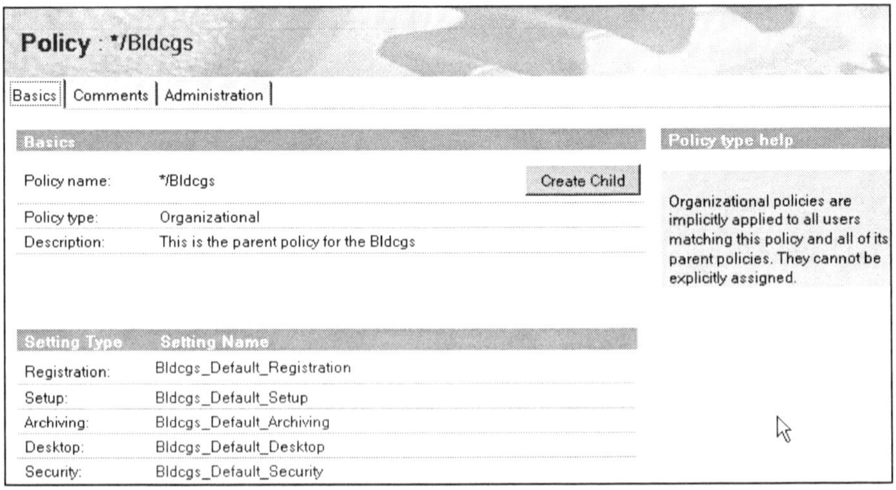

Figure B-1 Example of a policy document.

You can consider policy-based system administration as a way to create and apply rules within your user community. The Settings documents contain the rules; the Policy document organizes these rules. So when you create a Policy document, you have a single place to list these rules. You can then apply the rules to establish and enforce administrative standards by distributing them throughout your organization. And to change an existing policy, all you need to do is edit the Policy document and/or one or more Settings documents. No more running around to each user's computer, making the same change over and over—you can now do this all centrally.

You can use policy-based system administration to manage five major areas of Domino administration—archiving, desktop, setup, registration, and security—as indicated in Table B-1.

Table B-1 Five Major Areas of Domino Administration

Policy Area	Major Features	Description
Archiving	Server-to-server archiving Server-to-local archiving Folder-based archiving	Archive settings control mail file archiving. These settings determine whether or not to allow archiving, and if you do allow archiving, whether to allow Notes users to set their own private archiving criteria.
Desktop	Welcome page Deployment Bookmarks management Client upgrades	Desktop settings control the user's desktop environment. They are applied to a user's client configuration during authentication whenever a change to the policy occurs.

continues

Table B-1 Five Major Areas of Domino Administration

Policy Area	Major Features	Description
Setup	Browser Proxy Applet security Preferences	Setup settings help configure a new Notes client. They are used only once during the initial Notes client setup to populate the user's Location document and bookmarks.
Registration	Mail template Password length/quality Internet address format Certificate expiration	Registration settings predefine the User Registration options, if your policies are in place before you register users.
Security	Password expiration ECL management Password length	Security settings establish the administration ECLs and define password management options, including synchronization of Internet and Notes passwords.

Creating a Policy Document

To create a Policy document, you must have at least Author access to the Domino Directory, and you must be assigned to the PolicyCreator role. Then, from the Domino Administrator

1. Click the **People & Groups** tab and open the Policies view.
2. Click **Add Policy**.
3. In the Basics section, enter a name for the policy. If you are creating an explicit policy, enter a unique name. If you are creating an organizational policy, enter a name in one of the following formats:
   - For Organizations: */<organization>
   - For Organizational Units: */<organizational unit>/<organization>
   - For hosted organizations: */<hosted organization>
   - For hosted organizations to indicate all hosted organizations in the Domino Directory: *
4. Choose Explicit or Organizational as the policy type. Choose Explicit to create a policy that you assign to specific users and groups. Choose Organizational to create a policy to assign to all users in an Organization or Organizational Unit (OU). (See the Organizational and Explicit Policies section for more information.)
5. Optionally, enter a short description of the policy.

6. To create a child Policy document at this time, click Create Child. This creates a new Policy document that includes the name of the parent policy. You can save this new child Policy document and return to it at a later time. When you close this document you return to the parent Policy document.
7. In the Setting Type section, select the Settings documents (registration, setup, archiving, desktop, and security) you want to apply to this policy. These can be existing Settings documents you created earlier, or you can create new Settings documents on-the-fly by clicking New and completing the Settings document form.
8. Save and close the new Policy document.

Creating a Settings Document

To create Settings documents, you must have at least Editor access to the Domino Directory, and you must be assigned to the PolicyCreator role. You can create Settings documents when you create a Policy document, or you can create Settings documents by opening the Domino Administrator and following these steps:

1. Select the **People & Groups** tab and open the Settings view.
2. Click **Add Settings**.
3. Select the type of Settings document you want to create (registration, setup, archiving, desktop, or security).
4. Complete the fields appropriate to the settings type and then save and close the document.

Organizational and Explicit Policies

You can assign policies to individuals within your company two different ways, organizationally and explicitly.

Organizational policies apply settings to users or groups based on your organizational structure. An organizational policy affects all users in a naming hierarchy (*/Bldcgs, */Sales/Bldcgs, and so on) and can apply to either an Organization or Organizational Unit (OU). Organizational policies distribute settings to the broadest group of users, when the settings follow organizational lines. For example, to apply settings to all users in Sales/Bldcgs, create an organizational policy named */Sales/Bldcgs. Any user registered using the Sales/Bldcgs certifier automatically receives the settings in this organizational policy. Organizational policies can be considered wildcards and are usually the most effective method to distribute and maintain settings that apply to all users in a specific Organization or OU.

Explicit policies assign settings to people and groups across different organizations. An explicit policy is assigned in the Person document and applies to a group of users when an Organization or OU does not exist to define the group. Explicit policies are most appropriate for environments without an organizational hierarchy or when you need to assign common settings to a group that spans multiple organizations. For example, suppose your company includes contractors as temporary employees working in many different OUs. You want your contractors to inherit some settings from the applicable organizational policy such as */Sales/Bldcgs, but you also want certain settings to apply only to contractors. For instance, you can set their certificates to expire after six months. To do this, you can create an explicit policy called /Contractors that applies only to them. (You can assign explicit policies with the Assign Policy tool.)

When deciding which type of policies to use, consider the following suggestions:

- Use organizational policies to apply settings to groups in accordance with your existing naming hierarchy.
- Use explicit policies when groups to whom you want to apply settings don't match your Organization or OU structure.
- Above all, feel free to use both types of policies within your company—indeed, we expect this will be common practice.

NOTE

Both explicit and organizational policies can be configured as exception policies. An exception policy explicitly exempts a specific individual or group from one or more of your standard policy settings. For example, your executives may need to bypass certain settings. Exception policies are powerful because they override all other settings (including enforced settings) that may apply to the exempted persons. But in practice, you should be careful not to overuse exception policies. Too many exemptions might decentralize the administration of these settings and therefore could defeat the entire philosophy of policy-based system administration.

Once created, organizational policies are automatically assigned to users—based on the certifier. Explicit policies, on the other hand, must be assigned manually. You can assign explicit policies during user registration, in the Person document, or by using the Assign Policy tool. It's best to create organizational policies prior to user registration; the best time to assign explicit policies is during user registration. If you don't create your policies before you register users, you can't apply registration policy settings. Applying policies at registration time also ensures that setup policy settings are available when the registered users perform Notes client setup.

Because organizational and explicit policies are hierarchical, inheritance and enforcement automatically fit into a parent-child relationship. This model gives you the ability to define corporate-wide standards inherited by Organizational Units while allowing for the occasional exception. Just imagine, you can have an archiving policy in which your end users have no control, some control, or complete control of where and how their mail files are archived.

What Makes Up a Policy?

If you are familiar with other aspects of Notes and Domino administration, administering policies should be relatively painless. All policy information is stored on the server in the public Domino Directory. That last sentence may spark a few questions, such as, "If policies reside in the public Domino Directory, how are they enforced when my users aren't connected?" Good question—with an interesting answer. When a user authenticates with the home mail server, a process called dynamic configuration runs locally and caches all policy information for that user. This ensures that while the user is disconnected, all policies are still enforced. This also means users who remain disconnected for an extended period will not receive any changes to policies until they reconnect and authenticate with their home servers.

As stated previously, a policy consists of a collection of Settings documents, organized within one high-level Policy document containing a single set of fields filled with the appropriate settings. Let's look at this through a simple diagram as shown in Figure B-2.

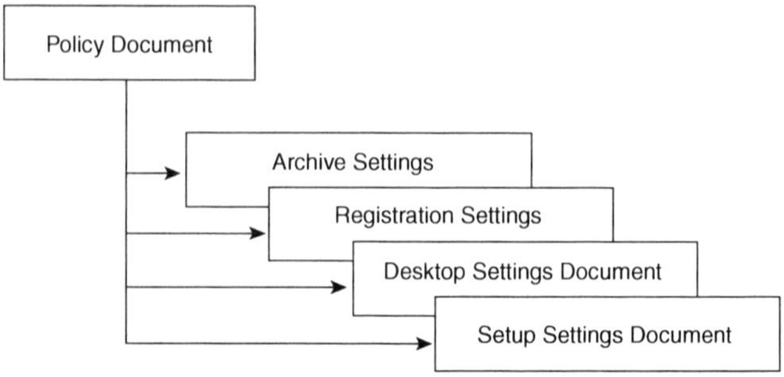

Figure B-2 Policy document.

In the illustration given in Figure B-2, the top-level Policy document represents the policy you want to enforce in your organization. The subordinate Settings documents (for archive, registration, desktop, and setup) contain the settings information that apply to this policy. You can think of these Setting documents as buckets of information containing settings that in previous releases had to be entered through many different places within the Domino and Notes user interface. The Policy document dips into these buckets to derive the settings needed to apply to the appropriate users and groups.

Knowing the relationship between Policy documents and Settings documents is crucial to understanding two important features of policy-based administration: policy hierarchies and effective policies.

Policy Hierarchies

Let's forget about Domino 6 and policy-based system administration for a moment and consider a typical large corporation. This company has rules that apply to all employees and rules that apply to some departments and not others. Some workgroups within those departments have their own rules, with some members exempt. And there are rules for particular types of employees, for example, contractors and managers. This may sound complex, but policy-based system administration can make sense of it all through policy hierarchies.

Parent and Child Policies

You construct a policy hierarchy by establishing parent-child relationships between policies. When you create a Policy document, you also can create one or more child documents for it. The original policy is then considered the child policy's parent. Through the parent-child relationship, you create a hierarchy of policies to apply across your entire corporation. In such a hierarchy, Policy documents build the relationship, and Settings documents determine the value of the fields based on their position in the hierarchy. Using either inheritance or enforcement at the field level, you can refine settings for the requirements of individual Organizational Units while maintaining control of certain standards that must be enforced across the company.

For example, if you want all users to use the same Internet mail name format, set that value in the Registration Settings document for the top-level parent policy. After you do this, you don't have to change or reenter it in subsequent child policies. You instead just instruct the child policy to inherit this value from its parent. However, if your organization includes a group of international users for whom this setting is a problem, you can create an explicit policy that applies only to this group. The combination of explicit and organizational policies together provide the control and the flexibility you need.

Inherited and Enforced Policy Settings

A major feature of a hierarchical policy structure is the ability to inherit and enforce settings. Inherited settings are derived from another Setting document. Enforced settings are set in a parent Setting document and are automatically inherited by all child documents. This setting cannot be overridden in any child document. This lets you enter a setting in a single place and then populate it throughout your corporate structure—while still allowing other settings to be set within each Organizational Unit. Note, however, that exception policies overrride enforced settings. This is another reason you should use exception policies sparingly.

Policies inherit and enforce all settings at the field level. Two checkboxes are provided for each field that can be inherited or enforced. The following screen (see Figure B-3) shows how each field in the Archive Settings document can either inherit its value from its parent or enforce a setting to all of its children.

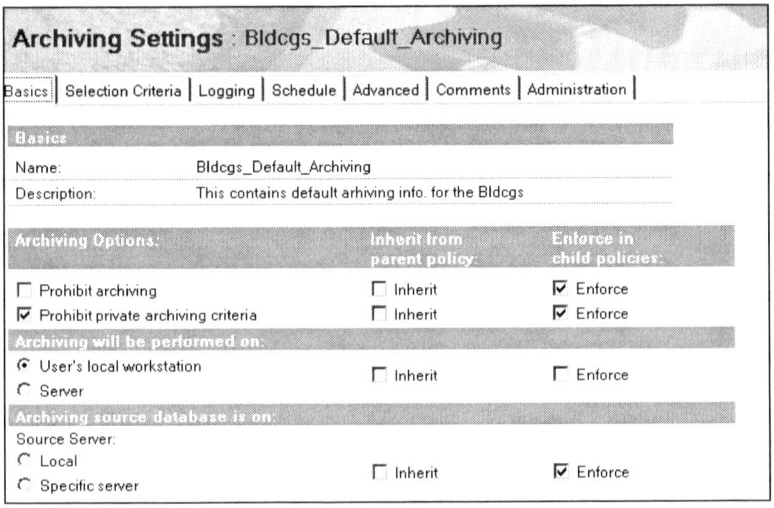

Figure B-3 Fields in the Archive Settings document.

Effective Policies

In a hierarchical policy structure, settings that apply to a particular group or user may not be from a single Policy document. Instead, policy information may be derived from multiple documents, based on inherited and enforced settings (or even exception policies if they're used). This derived set comprises the user's effective policy. The effective policy consists of policy settings dynamically calculated at the time of execution. The field values in an effective policy may originate from many different Settings documents. Each hierarchical level can have an associated policy, so users may have a combination of policy settings that include the values set at their OU level and those inherited from a parent policy. The resolution of these settings, stepping up through the organizational hierarchy, creates the effective policy for each user.

In the end, it's the user's effective policy that determines which settings apply. So even though the effective policy doesn't actually exist in the same sense a Policy document does, it's nevertheless among the most crucial concepts you need to fully understand how policy-based administration works.

Because the effective policy settings are derived at execution time, it may not always be obvious what effect changing a value of a policy setting will produce. The Policy Synopsis report shows the policy from which each of the effective settings is derived. This helps you better understand the settings of individual policies, Setting documents, and effective policies—and how they relate and affect each other.

Tools for Policy-Based System Administration

With every new technology, however beneficial, there's usually at least some additional level of complexity. While extremely flexible by nature, policies can also be somewhat unfamiliar territory to the uninitiated. And let's face it, with the way many large, global companies are structured today (national, international, local, remote, IT, HR, and so on) a well-defined policy structure can be more than a little complex to create and maintain. To help make sense of all this, Domino 6 provides three useful tools for effective policy-based system administration: Assign Policy, View Policy, and Policy Synopsis.

Assign Policy

The Assign Policy tool gives you the ability to assign explicit policies. With this tool, you can assign an explicit policy to a user or group, or you can change the explicit policy assignment. The Assign Policy tool lets you make changes to multiple users or groups, by associating an explicit policy with a person or group. When you change the explicit policy for a user or group, Assign Policy gives you the option of viewing how the change impacts the effective policy for that user or group.

For example, let's suppose your company has a group in the Domino Directory called Executives. Imagine the group contains high-level managers and VPs from different organizations and OUs within the company. Your corporate policy states that all executives must follow a strict mail archiving policy. Here's an example of a few of the people in the Executives group:

Sandy Beech/Sales/Bldcgs

Lee Smith/Bldcgs

Fran Jones/Development/Bldcgs

In this example, each person has a different organizational policy for archiving their mail, as shown in Table B-2.

Table B-2 Organizational Mail Policy Example

User	Organizational Mail Policy
Sandy Beech	* */Acme */Sales/Bldcgs
Lee Smith	* */Acme
Fran Jones	* */Acme */Development/Bldcgs

So you create a new explicit policy called /Executive and make it an exception policy. You then use the Assign Policy tool to assign the policy to each member in the Executive group:

1. From the Domino Administrator, click the **People & Groups** tab.
2. Open the People view and select the users to whom you want to assign policies or open the Groups view and select the groups to which you want to assign policies.
3. From the Tools pane, click **People** or **Groups** (depending on your selection in the previous step), and then select Assign Policy.
4. Complete the Assign Policy Options dialog box appropriately, as indicated in Figure B-4.

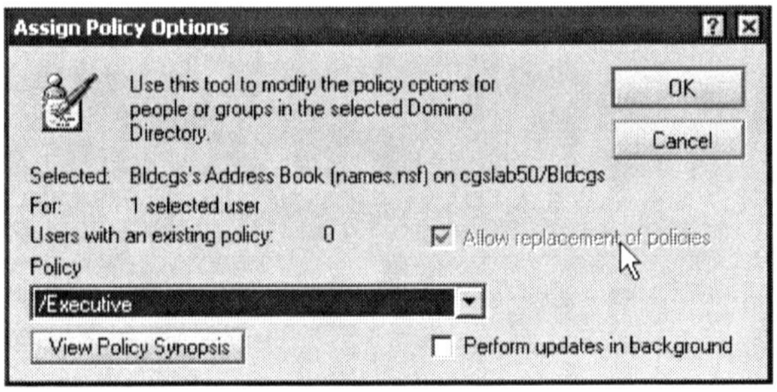

Figure B-4 Assign Policy Options dialog box.

1. Optionally, click the View Policy Synopsis button to see the new effective policy for the group or users to which you are assigning this policy.
2. In the Choose Organizational Policy dialog box, select the organizational policy you want to combine with the explicit policy to create the new effective policy.

After the Assign Policy tool populates the explicit policy field in each Person document in the Domino Directory, the inheritance tree for these users now looks like the structure shown in Table B-3.

Table B-3 Inheritance Tree for Users

User	Organizational Mail Policy
Sandy Beech	* */Acme */Sales/Bldcgs /Executives
Lee Smith	* */Bldcgs /Executives
Fran Jones	* */Bldcgs */Development/Bldcgs /Executives

View Policy

The View Policy tool lets you view your policy hierarchy and the relationships between policies and policy Settings documents. You can also view the effective policy for a selected user or group.

The View Policy tool has two views, which you access from the Configuration tab of the Domino Administrator. The By Hierarchy view shows the hierarchy of policies, and the settings associated with each one in the hierarchy. Use this view to display all policies, see details of a policy, or see which policies are assigned to a particular user or group. Figure B-5 shows an example of the By Hierarchy view.

The By Settings view shows policy Settings documents for each administrative area and the Policy documents that use each Settings document. Use this view to see where each setting is used and how a change to the setting affects other policies. Figure B-6 shows an example of the By Settings view.

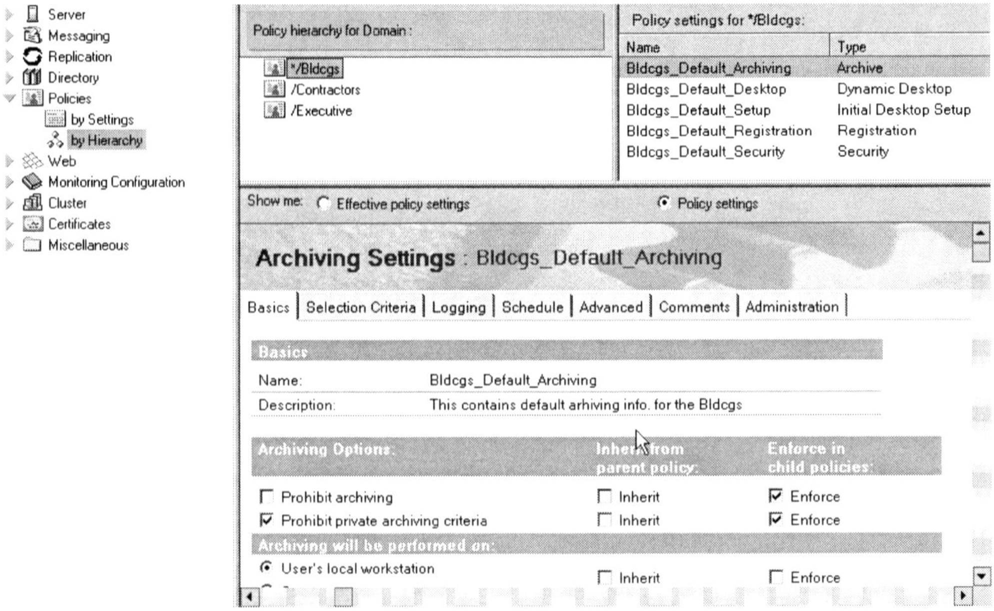

Figure B-5 "By Hierarchy" view using the View Policy tool.

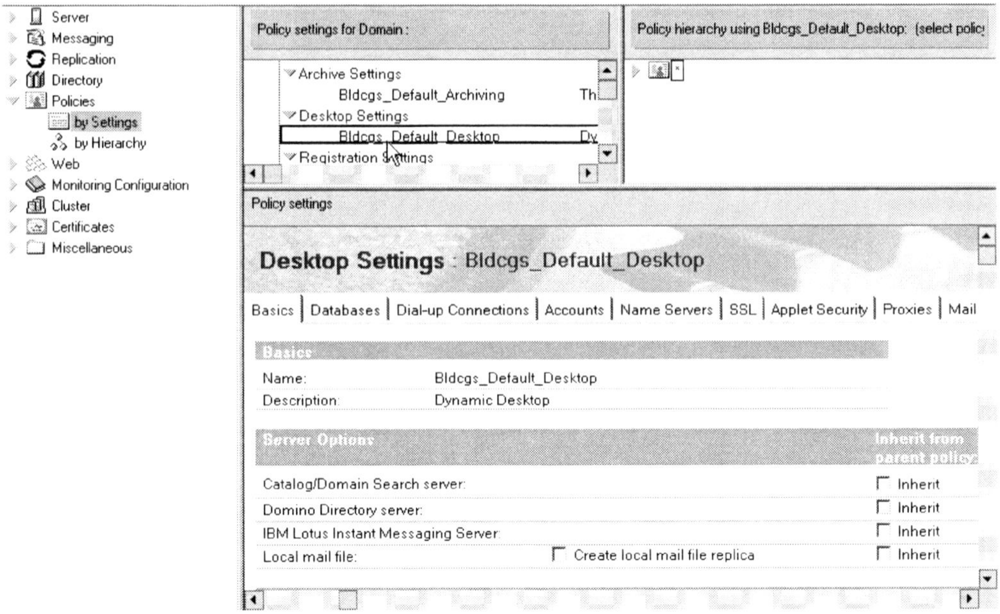

Figure B-6 "By Settings" view using the View Policy tool.

Policy Synopsis

With all this talk about effective policies with inherited and enforced settings, you may be asking yourself, "How can I keep track of where setting information is coming from, so I can be sure the right people receive the right settings?" We've anticipated this concern. Domino 6 includes a tool called Policy Synopsis that helps you figure out which settings fields come from which level in the hierarchy.

The Policy Synopsis tool determines the effective policy governing a selected individual and then creates a report that lists the policies and settings that apply to that user. Reports appear in a database called Policy Synopsis Results (polcysyn.nsf). You can create two types of reports, "Summary" and "Detailed." Summary reports display the policy hierarchy, listing both the organizational and explicit Policy documents used to derive the effective policy settings for the specified user. Detailed reports include everything in the Summary, plus actual settings in the effective policy.

To use the Policy Synopsis tool:

1. From the Domino Administrator, click the **People & Groups** tab.
2. Select the People view.
3. From the Tools pane, select **Policy Synopsis**.
4. Complete the Policy Synopsis dialog box.
5. Click **Results Database** to define the server location and file name of the Policy Synopsis Results database.
6. Click **OK**. When the Results database opens, double-click the report to read it.

Let's take a look at what a section from a sample Policy Synopsis for a person might look like.

Examples of Using Policies

Here are a few simple examples of policy-based system administration tasks that might have been difficult and/or tedious to perform in previous releases. By no means do these examples represent everything you can do with policy-based system administration—we're just whetting your appetite.

Creating a Corporate Policy

Imagine you're the Domino administrator for a corporation referred to as Bldcgs. Policy-based system administration greatly simplifies one of your major headaches—ensuring that all your employees are set up the right way, without having to rely on each user to select the exact options and preferences appropriate for them. So the first thing you do is create an organizational Policy document called for example */Bldcgs. This organizational policy applies to all users in your company unless you specifically choose otherwise. The */Bldcgs Policy document serves as a container for the Setting documents that control registration, setup, archiving, user desktops, and security. See Figure B-7 for an example.

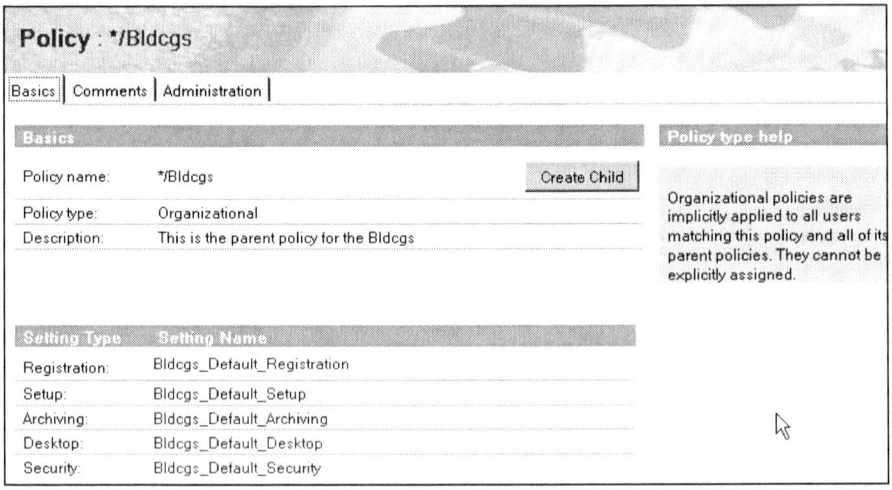

Figure B-7 Example of a Corporate Policy document.

Registration Settings

Bldcgs Corporation has numerous company policies. Among other things, to ensure customers can easily send email to your employees, all users must have the same Internet mail address format. Employees are to use Lotus Notes as their default mail system. And you want to enforce a standard password quality and certificate expiration date across the corporation. To do this, create a Registration Settings document (let's call it Bldcgs_Default_Registration) and enter all this information into it. You select the mail template to use and specify that certificates expire after 24 months. You can then choose Bldcgs_Default_Registration as the Registration Settings document for */Bldcgs. This means that by default, the information in Bldcgs_Default_Registration applies to all Bldcgs employees at registration time. See Figure B-8 for an example.

But as with most companies, Bldcgs has exceptions to every rule. For example, your on-the-go, multinational Sales force needs its own mail template. Further, Sales members are roaming users and require a different Internet mail address format. No problem—just create an organizational policy for your Sales OU called */Sales/Bldcgs. Then create a Registration Settings document called, for instance, Sales_Registration and enter the registration information appropriate to the Sales group. These settings will apply whenever you register a user with the Sales/Bldcgs certifier. See Figure B-9 as an example.

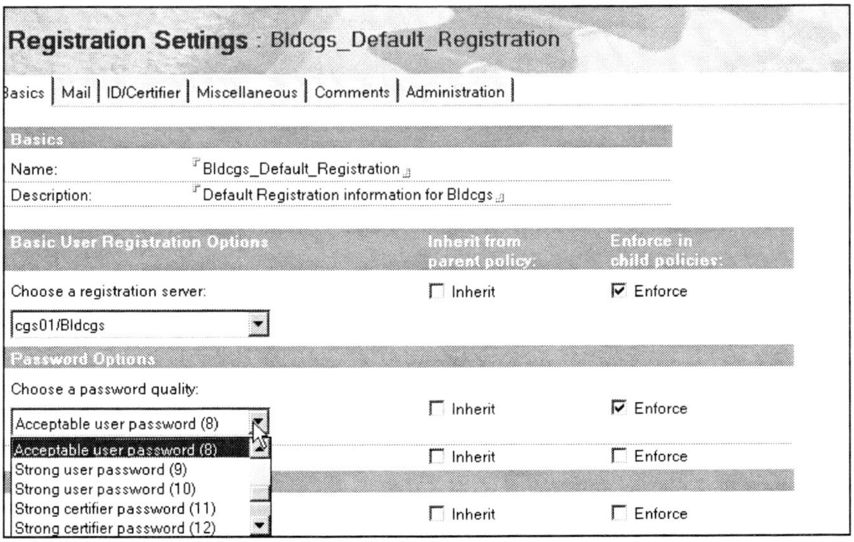

Figure B-8 Example of Registration Settings.

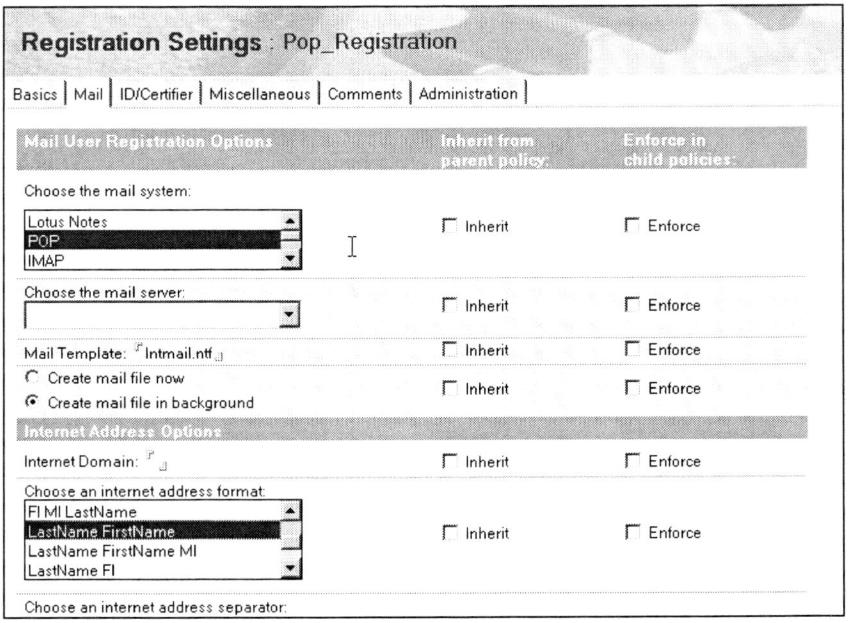

Figure B-9 Example of special Registration Settings.

So far so good. But here's a twist: Bldcgs corporate policy calls for certificates for all contractors to expire after six months, not 24; however, these contractors are spread throughout your

OUs. You can't create an organizational policy that covers all contractors. Instead, create an explicit policy, which you can name /Contractors (see Figure B-10).

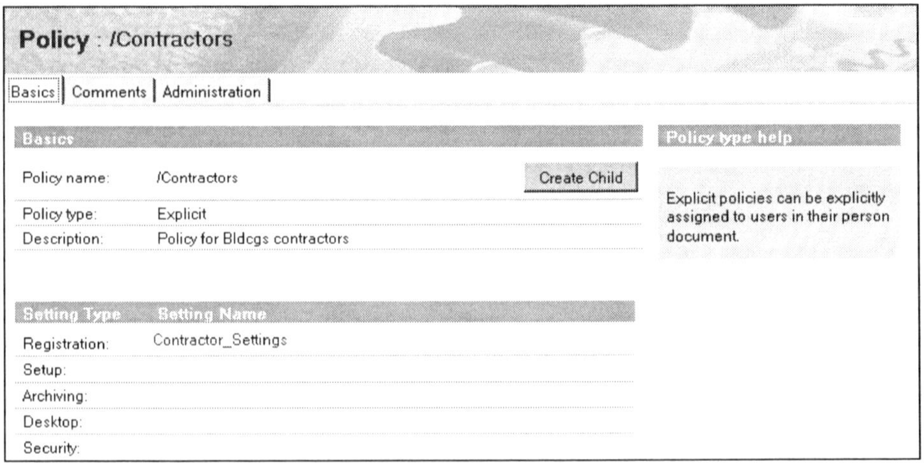

Figure B-10 Example of Explicity Policy for contractors.

Then create a new Registration Settings document for the /Contractors policy called, for example, Contractor_Settings. Enter the six-month certificate expiration setting and any other information that applies to your contractors (see Figure B-11).

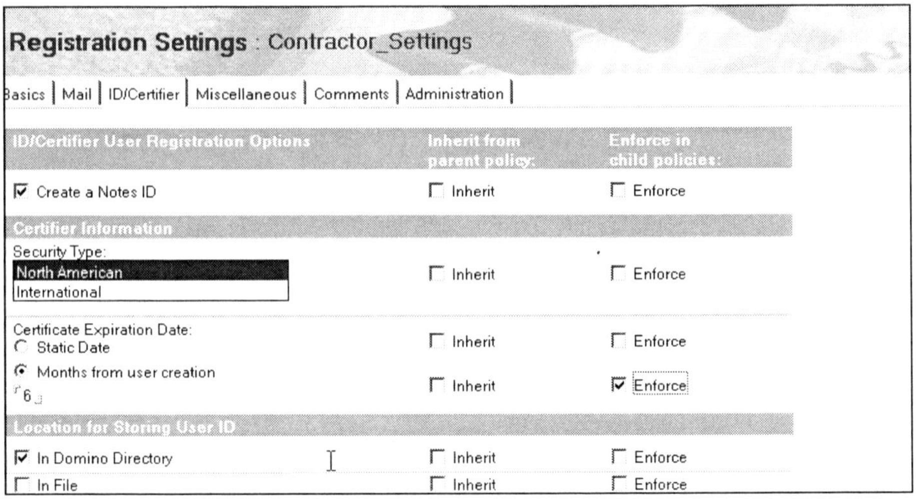

Figure B-11 Registration Settings for contractors.

To have these settings applied to contractors at registration time, open the contractor's Person document and type `/Contractors` in the Assigned Policy field.

Setup and Desktop Configuration

Now that you've gotten registration settings squared away, you can turn your attention to user setup, including enforcing a standard desktop for all your users. For example, you need to control user and location preferences, which up to now have been in the hands of each individual user—with sometimes unpredictable results. You also want the same initial set of databases and bookmarks for each user along with the same Welcome page. Perhaps most importantly, each user must have the same initial ECL settings for security.

Based on these examples, it is clear that policies make all this much easier to implement than in previous releases. No more trusting your users to select all the right preferences. No more days spent visiting one user at a time to verify that everything's set up correctly. Instead, just create Setup and Desktop settings documents for your */Bldcgs policy, and fill in this information. These settings will then apply to all users in your corporation. And as mentioned, you can handle exceptions to these rules using organizational and/or explicit Policy documents. See Figure B-12 for an example.

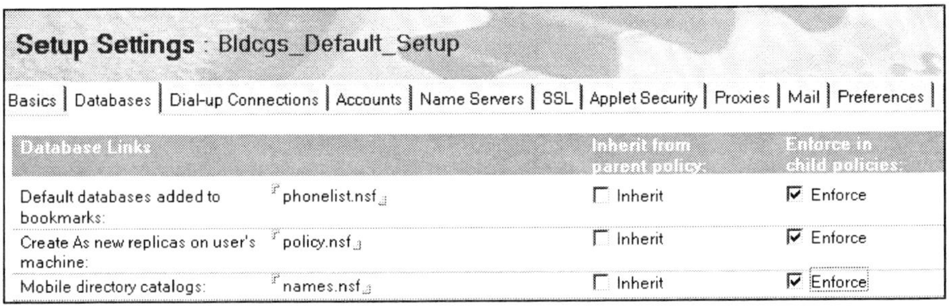

Figure B-12 Example of Setup Settings document.

Centrally Managing Mail File Archiving

Then there's the issue of mail archiving—the problem that prompted us to develop policy-based system administration in the first place. As Bldcgs's administrator, you've been struggling to deal with these mail file archiving issues:

- Lack of available space on your mail server
- Need for a central archive server
- Archiving must run off-hours
- Users aren't allowed to set their own archive settings
- Your Notes environment includes clients from different releases

To make your life easier, you create an Archive Settings document for the */Bldcgs policy. This lets you do the following:

- Centrally manage archive settings, with users prohibited from changing or creating their own
- Enable server-based archiving from a mail server to a designated archive server
- Designate a Domino 6 server as the archive server so policies can be enforced in a mixed environment
- Schedule archiving to occur during off hours for everybody

We've deliberately kept these examples simple. For instance, we haven't talked about parent-child policies, exceptions policies, or inherit/enforce. However, we have shown how policies can greatly simplify setting up users, even when certain groups need exceptions to the corporate rules.

Policy-Based System Administration: Making Your Job Easier

Policy-based system administration exemplifies the Lotus commitment to lowering total cost of ownership with centrally managed ways to set and enforce corporate-wide policies in an efficient and time-saving manner. This not only puts you more firmly in control of your environment, but also gives you a flexible way to manage a user down to the field level.

APPENDIX C

Detailed Steps for Building a Combined WebSphere and Domino System

"What we have to learn to do we learn by doing."

—Aristotle, Ethics

Preliminary Setup Details

This appendix gives detailed steps required to install WebSphere Application Server V5 and Lotus Domino 6.5. We first discuss the preliminary setup that will be needed and then walk through the installation steps for WAS V5 and Domino 6.5. We also discuss the installation of DB2 V8.1, which may be used with WAS. The installations presented are aimed at Windows platforms, with comments about Unix-based platforms included. Please refer to the WebSphere Info Center and Domino Administration Help documentation for further details.

Hardware and Software Prerequisite Details

In this section, we list the hardware and software requirements for the computer you want to install on, as well as the different product software levels required.

System hardware requirements:

- Pentium III or higher at 500MHz, 256MB memory at minimum—greater than 512MB recommended for both WAS and Domino.

- At least 1GB of free disk space. The approximate disk space requirements of the products after installation are Domino 500-600MB, IBM HTTP Server 30MB, WebSphere 500-800MB. At least 150MB must be available in the system temporary directory (/tmp). You must also provide space for the unzipped install files.
- CD-ROM drive.

The space requirement for Domino also can be reduced to a certain extent by choosing to install fewer components. For example, it is possible not to install the Help files. These files alone require 50MB. Similarly, the space for WebSphere specified here includes documentation and sample applications, which would not be needed in a production installation.

Software requirements:

- Microsoft Windows 2000 with Service Pack 4 or Windows Server 2003.
- Linux: Red Hat Enterprise Linux WA/ES/AS for Intel 2.1 or UnitedLinux 1.0 SP2a, UnitedLinux 1.0 for pSeries, or iSeries or zSeries (31-bit mode only for zSeries).
- TCP/IP networking with a fixed IP address for each machine.

Creating a User with Administration Rights

WAS, Domino, and DB2 must run under the permissions of a user or as system services. For testing purposes, it is more flexible to use a user ID with rights to run as an extension of the operating system rather than load the products as system services (however, WAS and DB2 must run as system services). For a production system, these products should run as services so that they will automatically load when the system is started without operator sign-on.

Here is how to create a Windows 2000 user ID with rights to run as an extension of the operating system. To do this, you must have the right to create an ID on your local machine (and in any Windows 2000 Domain Control server's user registry if your machine logs onto a network).

If you already have a user ID (no longer than eight characters if to be used for DB2) with the right permissions, you can skip this section.

A new user ID can be created in Windows 2000 using the Computer Management panel or the Users and Password wizard in the control panel. Here we describe the steps if using the Computer Management panel.

1. Start the Windows 2000 Computer Management panel by selecting **Start>Programs>Administrative Tools>Computer Management**.
2. In Computer Management select the left pane, expand the tree under Computer Management (Local), and then expand the node System Tools. Expand Local Users and Groups to display the Users and Groups folders.
3. Either open the Users folder or right-click on the Users folder and select **New**.

4. Fill out the user information. We specified the user name `db2admin`. Specify a password. Change the default setting for password expiration so only **Password Never Expires** is selected. Click the **Create** button to create your new user and then click the **Close** button to dismiss the dialog box.
5. Right-click the new user ID in the Computer Management window and select **Properties**. Once the properties are displayed, select the **Member of** tab on the top of the panel to show the groups the user belongs to. Click the **Add** button on the lower left of the panel. Highlight **Administrators** in the upper pane displayed and click the **Add** button in the middle of the panel. Click **OK** to confirm and return to the Computer Management window.
6. Close the Computer Management window.
7. Select **Start>Programs>Administrative Tools>Local Security Policy** from the Windows 2000 desktop to set the user rights. Expand Local Policies in the left pane and select **User Rights Assignment**. In the right pane you can see a list of system policies that users can be granted.
8. Right-click the entry **Act as part of the operating system** to get a pop-up menu and select **Security**. You will be presented with a panel that shows the current list of users with the right to act as part of the operating system (the list may be empty).
9. Click the **Add** button to select the new user ID (`db2admin` in our example) just created. Note that you can also select groups such as Administrators to grant this right.
10. Select the user and click the **Add** button to add the local security right to the user. Click **OK** to return to the Local Security Settings Policy window.
11. The new user also needs the following rights if DB2 is to run under the user ID just created: **Create a token object**, **Increase quotas**, and **Replace a process-level token**.

    You can go ahead and assign them to the user. However, the DB2 install process will add these rights if they are not set, provided the right **Act as part of the operating system** is set.
12. Finally, close the Local Security Policy Setting panel and log on with the new user ID just created.

    If you are unable to log on with this ID, you need to resolve the situation (perhaps with your administrator's help) before proceeding. It may be that you need to create a standalone server not associated with your existing Windows 2000 Domain for testing purposes. The user ID we just created only needs to be on the WebSphere computer. However, to make it easier to work with both computers, we recommend that you create the same user ID on both computers.

Installing WebSphere Application Server V5

Make sure that all HTTP servers or existing WAS servers on this computer are shut down before proceeding. Note that, by default, Windows 2000 server will be running the IIS HTTP service. Ensure that it is shut down and set to start manually in the services panel. The IIS HTTP service will be listed as World Wide Web Publishing Service. Ensure that it is stopped and its startup type is set to `Manual` by right-clicking the service, selecting **Properties**, and changing **Startup Type** in the General tab.

Switch to the administrative user (Windows) or `root` (Linux). Check that the `java` command points to a Java runtime greater than 1.1. (Issue: `java -fullversion` to see the Java version.) If not, adjust your default path or `JAVA_HOME` environment variable to use a current JDK.

You can install WebSphere from a product CD. The installation program should start automatically when you insert the CD. If not, run the setup.exe program on the CD (Windows) or change to the install directory /mnt/cdrom/... and run the LaunchPad.sh script (Linux). The LaunchPad application points to the documentation and WAS Installation Wizard. (To run the Wizard directly, change to the subdirectory containing the WAS install files and run the install.sh script.)

If your installation CD supports installation using different languages, you can select the language for the Wizard text. A panel similar to that shown in Figure C-1 is displayed.

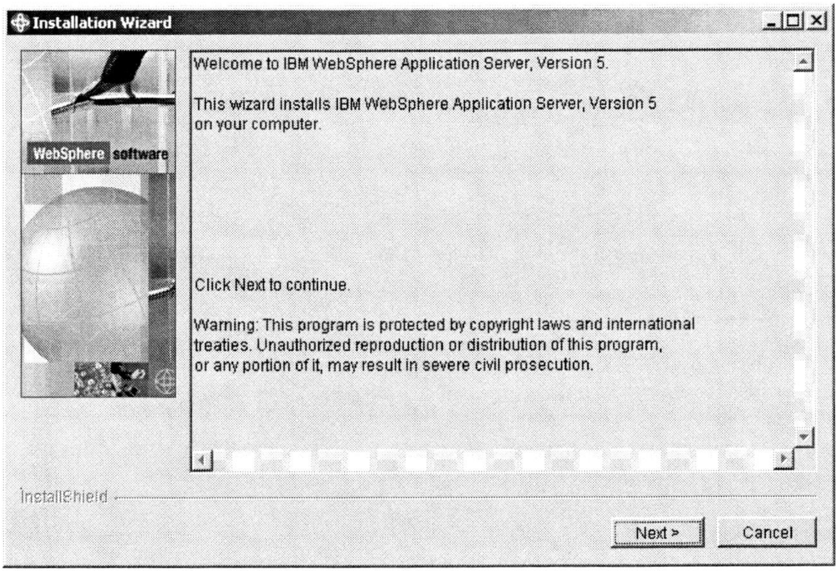

Figure C-1 WAS Installation Wizard (1).

Click **Next** to continue with the WebSphere installation. A panel is displayed indicating that the Wizard is checking the system prerequisites. If successful, the WebSphere license is displayed and the user is asked to accept the license.

A panel is displayed showing the default path in which WAS will be installed. You may accept this path or change it. If the Wizard detects another installation of WAS on the system, it will ask if this install is to update the existing installation(s) or create a new instance. (Note: if you choose to create a new instance, you will need to adjust the set of ports used by WAS to be different from the existing instances.)

The Wizard then verifies the operating system level. If running on a Linux system, which doesn't exactly match the IBM supported Linux distributions, an error message is reported, but the installation can be continued. Since WAS is Java-based, it will likely run on other distributions as long as the Java runtime is supported.

The Wizard then displays a panel allowing a choice between a full install or a selective install. This panel is shown in Figure C-2. Select the install type and click **Next**.

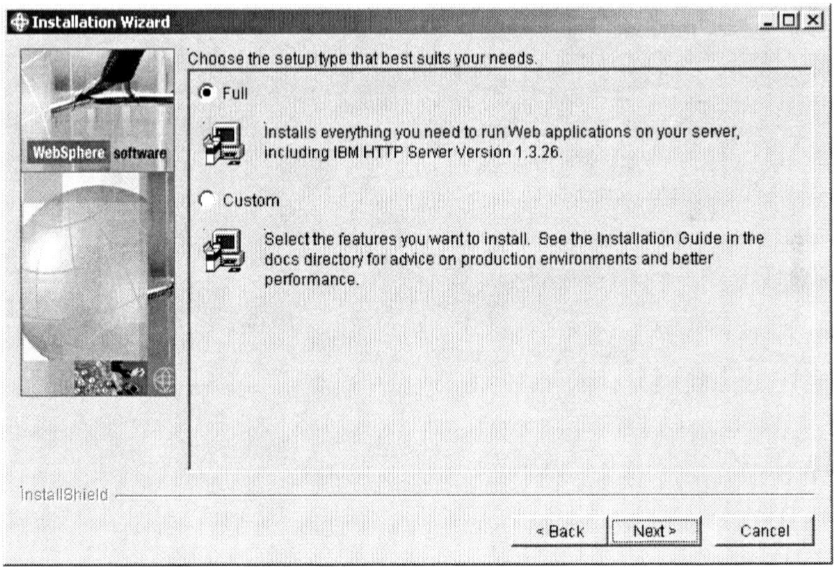

Figure C-2 WAS Install options.

If a custom install is selected, the Wizard displays a panel allowing the specific WAS components or features to be selected, as shown in Figure C-3. Select the components to install. If the Wizard detects HTTP servers present on the system, it will allow selection of WAS plug-ins to install and configure to the HTTP server. If no HTTP server is found, the IBM HTTP Server can be selected for installation and plug-in configuration.

When finished with the selection, click **Next** to continue. A panel is displayed asking for confirmation of the features selected. Click **Next** to begin the install.

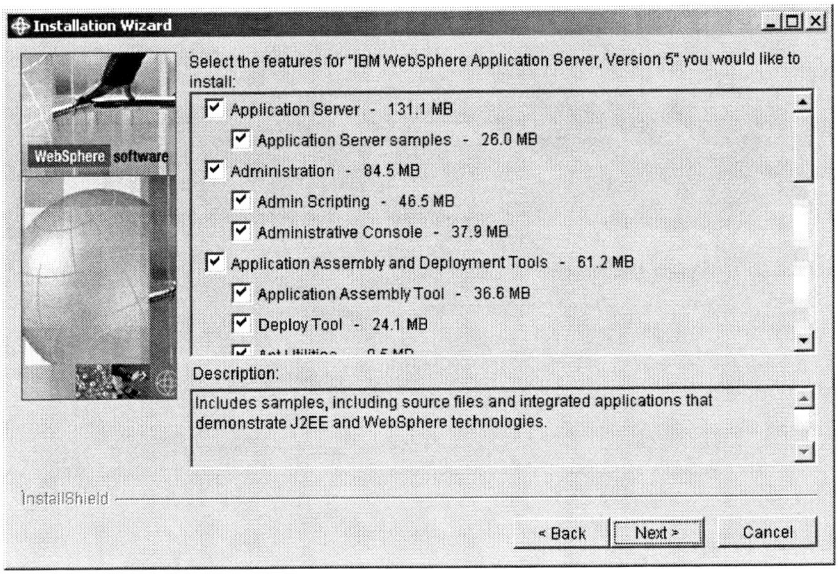

Figure C-3 WAS Install feature selection.

Verifying Installation of WebSphere V5

To verify the installation of WAS, you can run the Install Verification Tool (IVT). To run this program, change to the WAS bin directory under the install path (*install_root*), and issue the command: `ivt` (Windows) or `./ivt.sh` (Unix). The IVT program starts the server automatically if the server is not running. Once the server initializes, the IVT runs a series of verification tests and reports pass or fail status in a console window. It also logs results to the *install_root*\logs\ivt.log file.

Alternatively, you can start the application server and attempt to access the WAS administrative application via a browser. To start the server, issue the command: `startServer server1` (Windows) or `./startServer.sh server1` (Unix). Then, from a browser, attempt to open the URL: `http://`*`yourhost`*`:9090/admin`.

If the application server fails to start or there is a problem accessing the administrative application, check the WAS log files located in the *install_root*\logs\server1 directory. The application server process (JVM) writes its log messages to the files SystemOut.log and SystemErr.log.

Apply Any Relevant e-Fixes

If you have problems with your application in WebSphere and suspect you have found a WebSphere product problem, check the support download page for relevant e-fixes: `http://www-4.ibm.com/software/webservers/appserv/efix.html`.

E-fixes are individual fixes for critical problems. They have been individually tested, but not integration tested, and should only be applied if you have a critical problem without a valid workaround. They may be applied to both Standard and Advanced Editions, except where specifically noted. All e-fixes are rolled into the next scheduled FixPak. Each fix has a readme file with installation instructions.

This completes the installation of WebSphere and its FixPak. We are now ready to install Domino on a separate computer.

Installing and Configuring Domino 6

We now turn our attention to our Domino computer and describe how to install Lotus Domino 6.5 on a Windows system.

The Lotus Domino 6.5 Server family consists of Domino Utility Server, Domino Messaging Server, and Domino Enterprise Server. If you want to do more with Domino than just use its HTTP stack to integrate with WebSphere, you should install the Utility or the Enterprise server.

1. If you install Domino from a CD, the installation program should start automatically. If it does not, start the installation by running setup.exe from the installation CD. A welcome screen will be shown. Click **Next**.
2. Read the license agreement and click **Yes** to accept it.
3. Enter your name and company on the next panel if the installation program hasn't been able to pick up this information from the system. Click **Next**.
4. Specify where you want the Domino program and data files placed or accept the default locations. The default locations are C:\Lotus\Domino and C:\Lotus\Domino\Data. The installation program will create these directories if they don't already exist. Click **Next**.
5. Choose the type of server to install. We selected the Domino Enterprise Server, as shown in Figure C-4.

   You can use the Customize button to further refine the installation. If you click **Customize** you will get a list of server file sets to select or deselect. For example, you may not need the Help files because you will install the Help together with the administration client on another machine or in another directory later on. Deselecting Help files from the server install will save you around 40MB. Click **Next.**
6. Whether you picked the default installation or the customized one, you will be brought to the panel where you select which (Windows) Program Folder to which to add the Domino Server. Change the default if desired. Click **Next.**

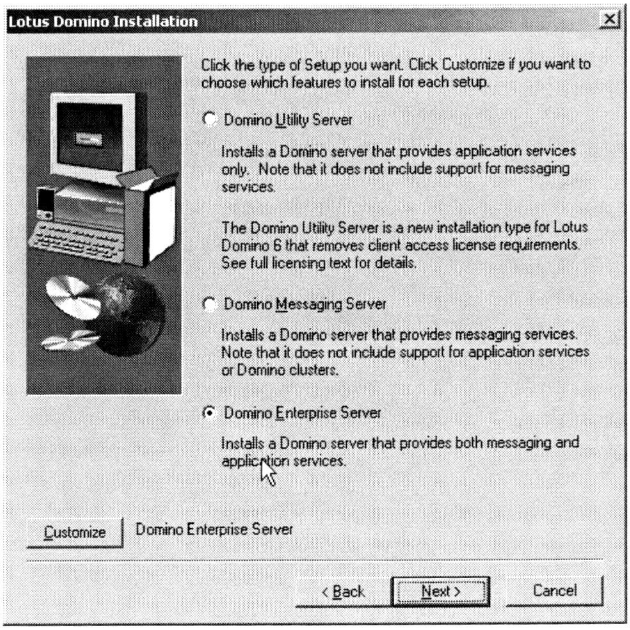

Figure C-4 Domino Server type selection.

At this point, the installation of the Domino server begins.

Once all the files are installed, the installation program shows a message confirming installation has been completed. You are instructed to start the server via the Program Folder you specified in Step 6. Click **Finish** to end the installation program.

Domino Server Configuration and Setup

After the Domino Server files are installed, the server must be configured before we can use it.

1. Start Domino by selecting **Start>Programs>Lotus Applications>Lotus Domino Server** from the Windows task bar. This will launch the Domino configuration program.
2. The initial configuration screen asks whether this is the first or an additional server in your Domino domain. We selected First Domino Server to set up a stand-alone test domain, as shown in Figure C-5. Click **Next**.
3. The next panel asks you to specify the name for the Domino server, as shown in Figure C-6. This name is used within the Domino domain to identify the server; it is not the TCP/IP host name. Specify a name and a descriptive title for the server and click **Next**.

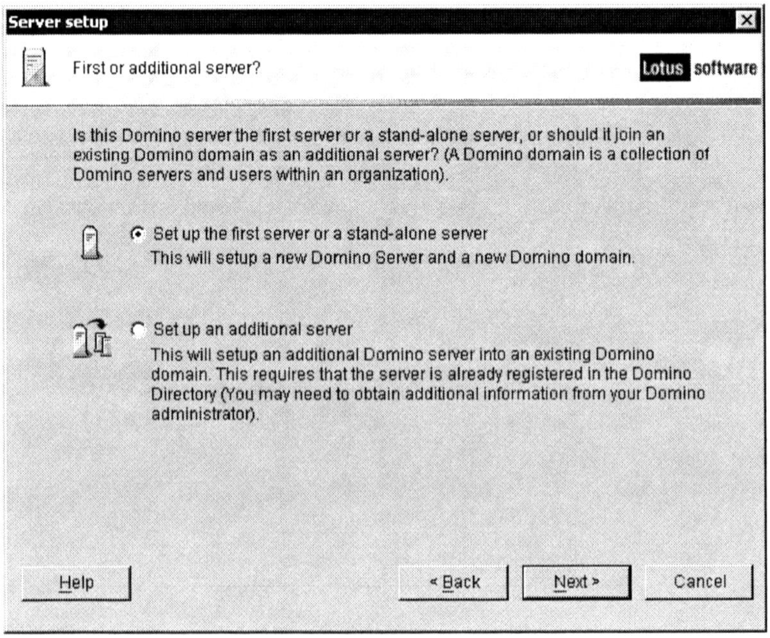

Figure C-5 Domino Server instance selection.

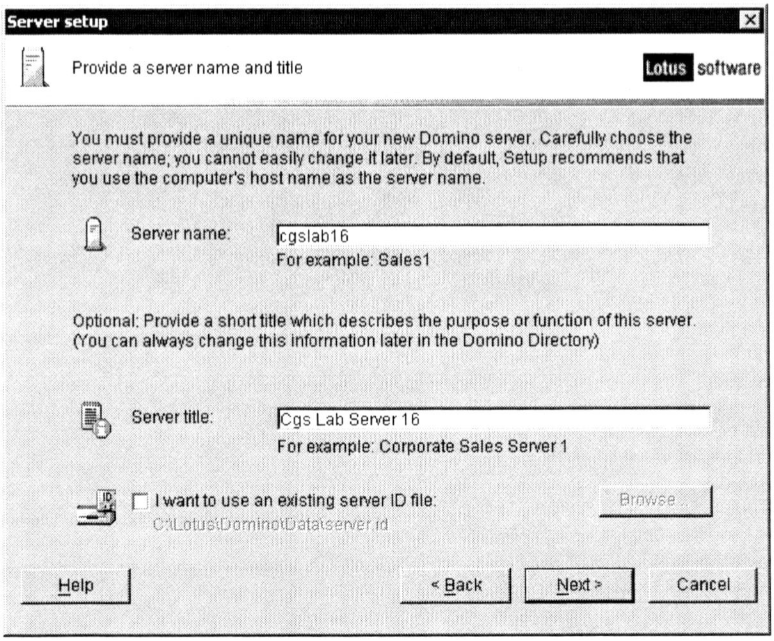

Figure C-6 Domino Server name selection.

4. The next panel asks you to specify the Organization name for the Domino server—that is, the top level qualifier of the Domino server name, as shown in Figure C-7. Specify an Organization name for the server and a password for the certifier (digital signature) used to sign user ids. You can add Organizational Unit names under this Organization name by selecting the Customize button. The Organizational Unit names are used to further qualify names in the Domino name space. Click **Next** when complete.

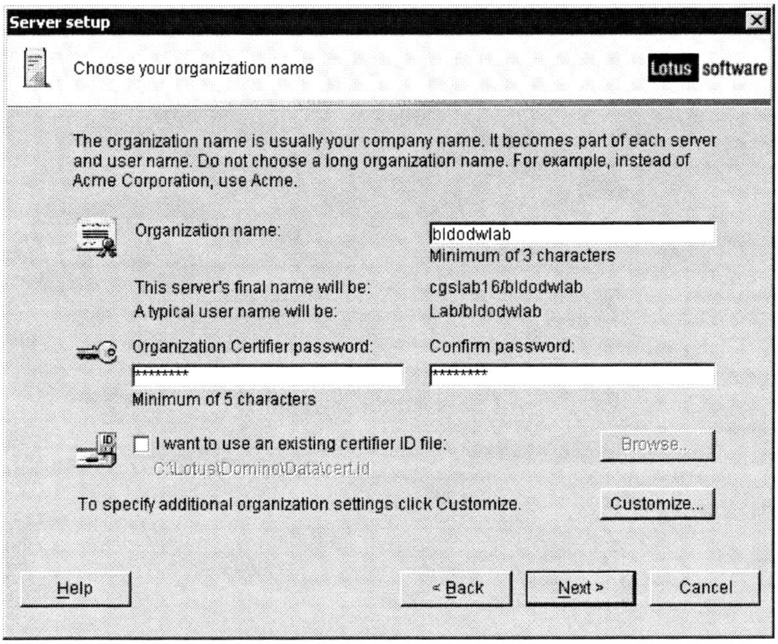

Figure C-7 Organization name specification.

5. The next panel asks you to specify the Domino domain name. This name represents a collection of servers. It is similar to a TCP/IP domain name and can be made the same. Click **Next**.

6. The next panel asks you to specify an administrator name for the server. A Notes ID file and entry in the Domino directory is created with this name. A password must be specified for the Notes ID as shown in Figure C-8. Click **Next**.

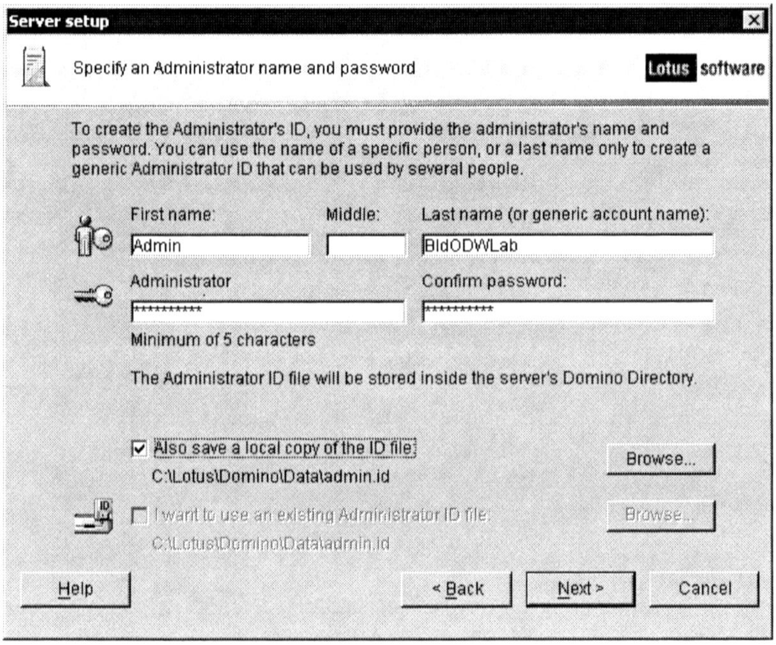

Figure C-8 Administrator name specification.

7. The next panel allows selection of the Internet services, which the Domino server will provide. Three basic sets of services, HTTP, mail, or LDAP can be selected from this panel, see Figure C-9. The Customize button brings up a second panel from which a more specific selection can be made, as shown in Figure C-10.

   We selected the basic HTTP and mail services and then refined this selection to include only the SMTP service. We selected the DIIOP service since it is required for remote access from Domino Java Classes. Nonessential services such as calendaring or DOLS can be turned off if they are not going to be used. If you are going to use the Domino directory as your user registry for WAS security, you will need to enable the LDAP service since this provides the LDAP access to the Domino directory. If you are going to use a stand-alone LDAP directory (and Directory Assistance to access it from the Domino directory), you do not need to enable the Domino LDAP service.

   Any of these service selections can be changed via the Domino server configuration document once the server is running.

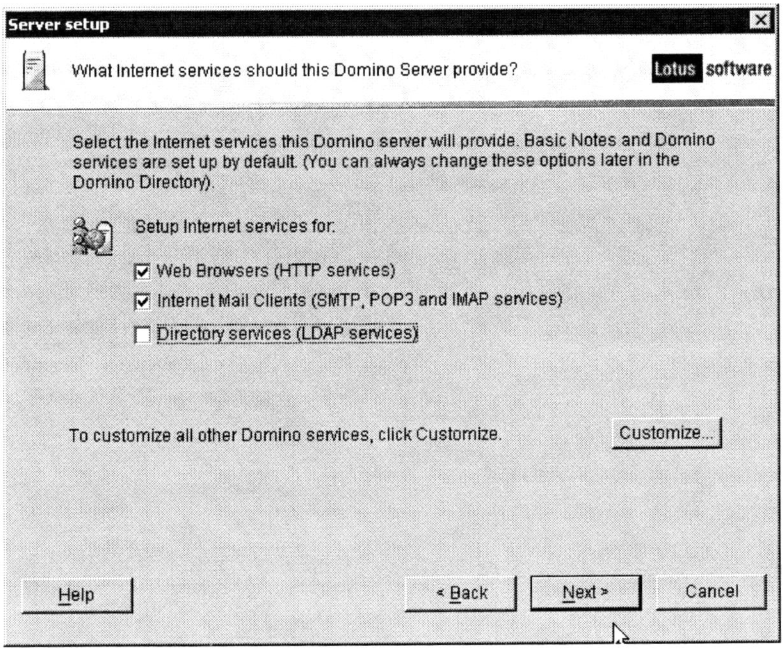

Figure C-9 Internet Services selection.

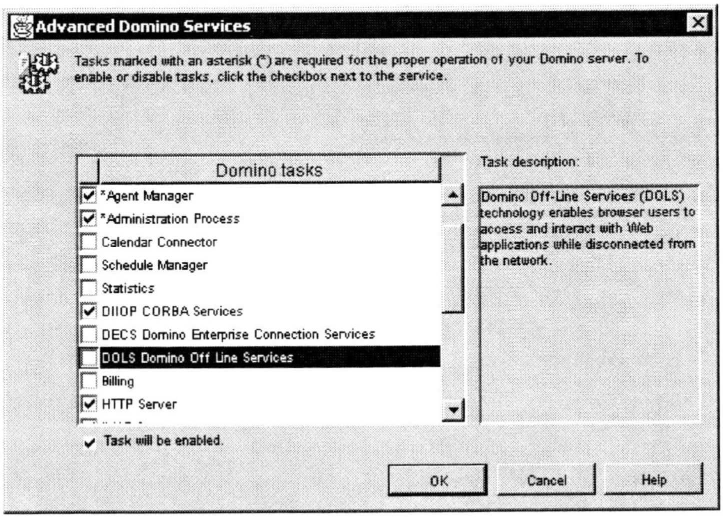

Figure C-10 Advanced Services selection.

8. The next panel allows configuration of the network settings to be used for the Domino server, primarily the network ports and TCP/IP host name. The default settings detected by the configuration program can be changed via the Customize button. The host name specified should be the fully qualified TCP/IP host/domain name for the server. See Figure C-11.

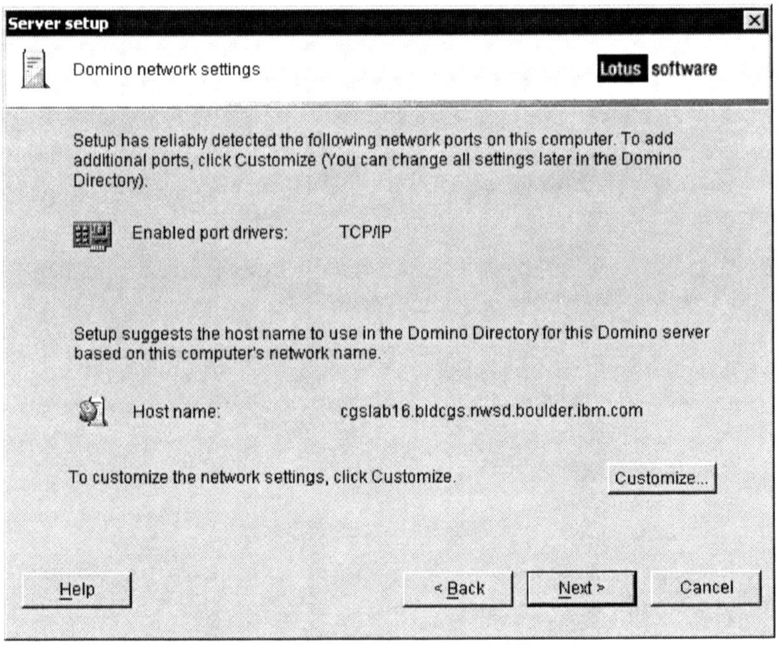

Figure C-11 Network Settings configuration.

9. The next panel allows two security related options to be disabled, but both should be left enabled. They are: **Prohibit Anonymous access to databases and templates** and **Add LocalDomainAdmins group to databases and templates**. Click **Next**.
10. A summary panel is displayed next, and the user is asked to confirm the selections made. Click **Finish** to have the configuration program complete the server configuration.

Verifying the Domino Server Configuration

We now check that the Domino HTTP server loads correctly and that it can be reached from a Web browser.

1. Start the Domino server by selecting **Start>Programs>Lotus Applications>Lotus Domino Server** from the Windows task bar.

2. Ensure that the HTTP task is running by issuing a `show tasks` command from the console or by observing the console log when the server starts. Note: The Domino HTTP task will fail to start if another HTTP task is running using the default HTTP port 80. For example, if you have IIS installed on your server, it should be stopped from the Windows Services panel and set to be started manually.
3. Next, HTTP access to the Domino server can be checked by using a browser. Start your Web browser and type in the host name of your Domino server as the URL. (If you are running on the same machine, you can just type **`localhost`** as the host name in the URL.)

Installing the Domino Administration Client

You need to install the Domino administration client on your server or another workstation (IBM recommends using a separate workstation for administration). This will allow you to change the server's settings, especially those in the Domino directory.

We will not describe the installation of Domino administrator in detail. Be sure that the Administration Client is selected for installation; you can accept all other default values during the installation. One of the first things you can do after installing Domino Administrator is allow your users to run Java programs on the Domino server.

This completes the installation and configuration of Domino 6.5.

Configuring Domino to Use the WebSphere Plug-In

The plug-in itself is provided as a shared library (DLL for Windows, .a for AIX, .so for Solaris; and as of this writing no plug-in exists for Domino on Linux) and a default configuration file: `plugin-cfg.xml`. The plug-in files are packaged with both the Domino and WAS products. The plug-in files can be found under the Domino install path: /Domino/Data/domino/plugins/was5.

There are two essential steps to installing the plug-in: 1) specifying the plug-in library file to the Domino HTTP server and 2) specifying the location of the plug-in configuration file to the plug-in itself. The first step is straightforward and simply requires a change to the Domino server document in the Domino directory. You specify the plug-in library file as a DSAPI filter file on the Internet Protocols/HTTP page of the Server document.

The second step is not as straightforward and is operating system-dependent. The plug-in library itself looks for the plug-in configuration file via an operating system environment variable (registry key on Windows). Unfortunately, these variable names are not well advertised in the WebSphere or Domino documentation. For Windows systems, a registry key value needs to be set as shown in Figure C-13 (you can use the `regedit` utility program to create the required key path if it is not already defined). For Unix systems, the plug-in looks for the configuration file in the config sub-directory under the path defined by the `WAS_HOME` environment variable, that is, `$WAS_HOME/config`. For IBM iSeries systems, the configuration file location is specified in the Domino notes.ini file.

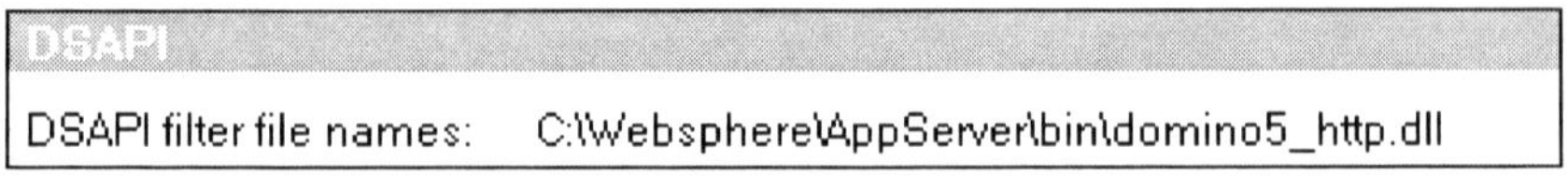

Figure C-12 Specifying WAS plug-in to Domino.

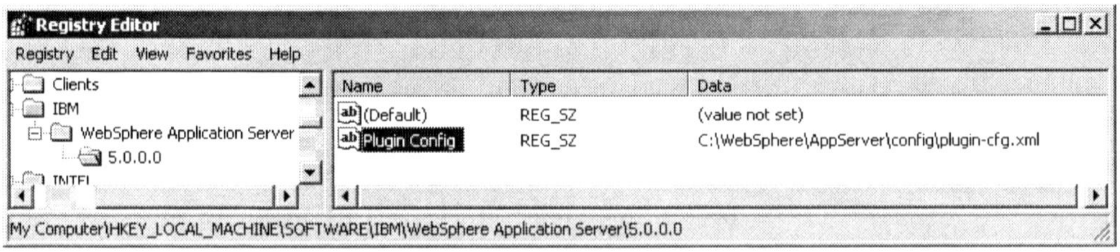

Figure C-13 Registry Setting for WAS plug-in.

Once these settings have been made, you can restart the Domino HTTP server task to have it load the WebSphere plug-in as a DSAPI filter. If the settings are correct, you will see the following messages at the Domino console:

```
HTTP Server: Java Virtual Machine loaded
WebSphere HTTP DSAPI filter loaded
HTTP Server: DSAPI WebSphere HTTP DSAPI Filter Loaded successfully
HTTP Server: Started
```

Next, let's take a closer look at the plug-in configuration file, plugin-cfg.xml, to see what it controls and how it can be tailored. This configuration file specifies three key aspects of how the plug-in operates: which URL hostnames and which URL types get routed to WAS and to which WAS servers they get routed. It also allows specification of the log file written by the plug-in and the level of logging. The following shows the configuration file that comes with Domino:

```
<Config>
   <!-- The LogLevel controls the amount of information that gets
   ➥written to the plugin log file. Possible values are Error,
   ➥Warn, and Trace. -->
   <Log Name="C:/WebSphere/AppServer/logs/native.log"
   ➥LogLevel="Error"/>

   <!-- Server groups provide a mechanism of grouping servers
   ➥together. -->
   <ServerGroup Name="default_group">
      <Server Name="default_server">
         <!-- The transport defines the hostname and port value
         ➥that the web server plugin will use to communicate with
         ➥the application server. -->
```

```
      <Transport Hostname="localhost" Port="9080"
      ➥Protocol="http"/>
    </Server>
  </ServerGroup>

  <!-- Virtual host groups provide a mechanism of grouping
  ➥virtual hosts together. -->
  <VirtualHostGroup Name="default_host">
    <VirtualHost Name="*:*"/>
  </VirtualHostGroup>

  <!-- URI groups provide a mechanism of grouping URIs together.
  ➥Only the context root of a web application needs to be
  ➥specified unless you want to restrict the request URIs that
  ➥get passed to the application server.  -->

  <UriGroup Name="default_host_URIs">
    <Uri Name="/servlet/*"/>
    <Uri Name="/webapp/examples/*"/>
    <Uri Name="*.jsp"/>
    <Uri Name="*.jsv"/>
    <Uri Name="*.jsw"/>
    <Uri Name="/ErrorReporter"/>
    <Uri Name="/j_security_check"/>
    <Uri Name="/tradetheme/*"/>
    <Uri Name="/theme/*"/>
    <Uri Name="/WebSphereSamples/*"/>
    <Uri Name="/estore/*"/>
  </UriGroup>

  <!-- A route ties together each of the above components. -->
  <Route ServerGroup="default_group" UriGroup="default_host_URIs"
VirtualHostGroup="default_host"/>
</Config>
```

As with all WAS configuration files, the plug-in configuration file has XML syntax and thus can be edited with any text editor. The `Log` element specifies the file name of the log file written by the plug-in and the level of logging desired. The `ServerGroup` element specifies the WAS server(s) to which URLs are routed and the protocol used. The default is to route to the WAS server on the same host at port 9080 (the embedded WAS HTTP server) over HTTP. Obviously, you must change the `Hostname` attribute to point to the WAS server host name if WAS is running on another host. Additional servers can be specified here, as well as HTTPS instead of HTTP. If more than one server is specified within a server group, the plug-in routes to the servers in a round-robin fashion. If HTTPS is used, additional system environment variables must be set pointing to the SSL libraries and the certificate and encryption key files required by the plug-in for HTTPS.

One or more `VirtualHostGroup` elements can be specified to define the hostname patterns, which the URLs must match to be considered part of that host group. Similarly, one or more `UriGroup` elements can be specified to define the types of URLs belonging to the URI group. Finally, the `Route` element specifies a single routing definition in terms of the virtual host and URI group elements. That is, a Route element defines that URLs found to belong to the specified VirtualHostGroup and UriGroup should be routed to the specified ServerGroup.

Enabling Tracing in the Domino WebSphere DSAPI Plug-In

By default, very little information is captured during the running of the plug-in. You may wish to log more information to understand the underlying processing or to diagnose error situations.

Stop the Domino HTTP server by issuing the console command: `tell http quit`. Modify the plug-in XML file, changing `LogLevel` attribute of the `Log` element to `Trace`. Adjust the `Name` attribute to point to the log file to contain the trace messages. Restart the Domino HTTP task.

Installing Domino and WebSphere on the Same Computer

Simply follow the installation steps described in the previous sections but install Domino on the same server used by the WebSphere server; then go to the previous section "Configuring Domino to use the WebSphere plug-in." There is no need to configure the plug-in configuration file as described previously.

You should ensure that the IBM HTTP server and/or the Microsoft Internet Information Server do not start automatically in the Windows Services control panel; otherwise, the Domino HTTP server will not be able to bind to port 80. Instead, it will exit with a message:

```
HTTP Socket Bind Error, hostname/ip <hostname>
HTTP Server: Could not bind port 80. Port may be in use
```

If for some reason you do want to have both Domino and the IBM HTTP server (or another Web server) both run on the same computer, you can do this by changing the port number (for example, to 8080) that the Domino Server uses for its HTTP server in the Domino Directory (Server document, Ports, Internet Ports, Web, or TCP/IP Port Number). If you also want both HTTP servers to access the WebSphere application server, you have to add a host entry under the default_host, specifying the new port number to be used (8080). If you do not do this, the plug-in will not be able to resolve the host entry and will reject the request. The error message would be returned by the Domino server (404—Not Found). If you have other virtual hosts defined that you want to access, you also would have to do the same for them.

Again, you can increase the detail traced in the WebSphere plug-in log by editing the plug-in configuration file as described in the previous section, "Enabling Tracing in the Domino WebSphere DSAPI Plug-In."

Installation of DB2 UDB Release 8.1

This section describes the installation of DB2 on the WebSphere computer. The supported level of DB2 for WAS use is 6.1 plus Fix Pack 4 or higher. We show the installation for DB2 UDB Version 8.1.

Make sure you are logged on to Windows with the user ID you created earlier or another user ID with similar rights. Although this is not strictly necessary to run the install program, you will need to be logged on with this user ID when accessing DB2 after installation.

DB2 UDB can be installed either from a zip file from the IBM software Web site or from a product CD. If the installation program on the CD doesn't start automatically (or if you are installing from an unpacked zip file), you must run the setup.exe program manually.

1. First, the Welcome screen is displayed. Click on the **Install** tab in the left column of the screen to begin the installation.
2. The product selection screen is displayed next. You must select **DB2 Enterprise Edition** and should select **DB2 Administration Client** if you want to have WebSphere connect to DB2 on a remote computer in the future.
3. You are presented with a selection of a Typical, Compact, or Custom installation. Select **Typical** and press **Next**.
4. On the next screen, you can select where to install DB2 (**Choose Destination Location**).
5. The installation directory can be changed to a disk with adequate space if the default drive does not have enough space. Click the **Browse** button to specify a new location for DB2.
6. You are then prompted to supply a user ID and password for the DB2 administration server to run under.

   If necessary, replace the defaults provided with the user ID set up earlier. This user ID will also become the default user ID in DB2.

   Ensure that the option, **Use the same values for the remaining DB2 Username and Password settings**, is selected.

Click **Next** to continue. A warning will be shown if you did not assign all necessary local security rights to your user, as described earlier in this chapter under, "Creating a User with Administration Rights." However, as long as you have assigned your user the right to act as part of the operating system, the DB2 installation can assign the other rights, and you can continue with the installation.

7. The next screen displays the installation options chosen. You can review these and, if necessary, click **Back** to change your selections. Click **Next** to confirm your choices and start the installation of the program files.

   DB2 will install and configure itself. At the end of the configuration process you will be presented with the option to install the OLAP starter kit. Select **Do not install the OLAP starter kit** and press **Continue**. You will be presented with a confirmation screen. Click **Finish** to complete the installation.

   You may be prompted to reboot your machine. If so, do so.

By default, DB2 will display the First Steps program, which you can use to create a sample database if you wish. Creating a sample database is not necessary to continue, but it is an easy way to verify that DB2 has been installed successfully.

Summary

In this appendix, we have covered the steps necessary to install WAS V5 and Domino R6.5 on different Windows machines. We have also discussed how to verify correct installation.

Appendix D

What's New in WAS 6?

> **"There are many enhancements with WebSphere 6.0, but the most important is support for the J2EE 1.4 standard, which brings numerous improvements related to Web services."**
>
> **—www.midrangeserver.com, January 2004**

WebSphere Application Server 6 was not yet available when this book was written. So this Appendix describes many of the new features expected to be in this version of WAS. Most of the early information available was about WebSphere Application Server Technology for Developers. The J2EE 1.4 compliant WebSphere Application Server Technology for Developers V6 will help us understand the expected new features in WAS 6. The J2EE 1.4 specification includes a number of improvements to the existing technology:

- Web Services for J2EE, V 1.1
- Java Servlet 2.4
- JavaServer Pages (JSP) 2.0
- Enterprise Java Beans (EJB) 2.1
- Java Message Service (JMS) 1.1
- JDBC 3.0
- Java API for XML Processing (JAXP) 1.2

The features that IBM has announced for WebSphere 6.0 suggest a renewed commitment to the Web services model and open standards generally. Among the features that WebSphere developers could look at early are support for Java Connector Architecture (JCA) 1.5 and JSR 109. JCA is a set of mechanisms for application integration at the level of data, allowing for bidirectional communication between J2EE servers and external applications by way of resource adaptors. The development of JCA runs alongside the absorption of specialist Enterprise Application Integration (EAI) tools into the WebSphere portfolio. Both trends suggest that back-end integration, implemented by way of open and industry-standard mechanisms, plays a major role in IBM's On Demand Business road map—and in the future of WebSphere.

Support for JSR 109 (the number refers to the underlying Java Specification Request) is an even clearer signal of where IBM is heading with WebSphere. JSR 109 defines the programming model and runtime architecture for implementing Web services in Java. By mapping J2EE concepts to their Web services equivalents, extending existing frameworks, and proposing new ones, JSR 109 offers the prospect of a transparent implementation of Web services within the J2EE framework—a call to a Web service would be no more elaborate than a standard remote procedure call.

WebSphere Application Server on J2EE 1.4 helps you improve integration with customers, suppliers, and partners with complete Web Services support, including

- JAX-RPC 1.1 API, which supports service endpoints based on servlets and Enterprise Beans and provides interoperability with Web Services based on the WSDL-based SOAP bindings.
- Web Services for J2EE specification (JSR 109), which defines deployment requirements for Web services and utilizes the JAX-RPC programming model.
- Platform Web Services interoperability via the WS-I Basic Profile 1.0.
- Leverage existing assets with new functions using Java Connector Architecture (JCA) 1.5.
- Resource adapters that provide bi-directional connectivity (synchronous and asynchronous) between EISs (Enterprise Information Systems) and J2EE-compliant applications.
- Benefits from a highly scalable and efficient embedded JMS messaging provider built in to the application server.
- The ability to take advantage of server enhancements that enable faster, easier development with new Java Servlet and Java Server Pages (JSP) technologies.
- Servlets that now support request listeners and enhanced filters.
- A simplified JSP technology page and extension development models that make it easier than ever for developers to build JSP-enabled pages.

The J2EE 1.4 Standard

Here are the significant changes incorporated in the J2EE1.4 specification when compared to the previous technology.

Java Servlet 2.4

The Java Servlet 2.4 represents a relatively small upgrade with small enhancements and clarifications to the prior version 2.3. Here are the changes:

- The deployment descriptor for the servlet 2.4 now uses the XML schema instead of the DTD.
- New attributes are added in the RequestDispatcher for the forward method.
- SingleThreadModel interface is deprecated.
- New methods are added to the ServletRequest interface and the ServletRequestWrapper.
- The HttpSession.logout method is removed.
- Servlets now require HTTP/1.1 and J2SE (Java 2 Platform, Standard Edition) 1.3 or higher and can work with J2EE 1.4.
- New APIs are added to the Request Listener.

JavaServer Pages (JSP) 2.0

Java Server Pages 2.0 introduced the most significant changes to the JSP technology for the ease-of-development with the introduction of the new Expression Language (EL) into the JSP container, defining reusable custom actions using JSP technology, and the use of the Java Server Pages Tag Library (JSTL).

Expression Language was initially defined by the Java Server Pages Standard Tag Library (JSTL) 1.0 specification, but now it is incorporated in the JSP 2.0 specification. The Expression Language uses familiar syntax with `$` and `{}` such as:

```
${expr}
```

EL is available for use in attribute values in standard and custom tags. In both cases, it is invoked using the construct `${expr}`. Expression Language is a simple language based on relational, logical, and arithmetic operations, a set of implicit objects, extensible functions mapping into static methods in a Java class, and available PageContext Attributes.

An expression language makes it possible to easily access application data stored in JavaBeans components.

Example 1:

For nested calls like `<%=accountBean.getAccount().getBalance()%>`, it is written in EL as `${accountBean.account.balance}`.

Example 2:

```
<c:if test="${bean.number < 3}">
```

The customer number `<c:out value="${bean.customerNumber}"/>` has 2 accounts.

```
......
</c:if>
```

Same code using JSP 2.0 can be written as

```
<c:if test="${bean.number < 3}">
```

The customer number `${bean.customerNumber}` has 2 accounts.

The JSP expression language EL allows you to define a function that can be invoked in an expression. Functions are defined using the same mechanisms as custom tags.

To define a function, in a public class the `testMethod` should be a public static method. JSP also supports a method of inserting dynamic content using custom tags. The JSP pages can generate dynamic content using simple tag syntax instead of Java code. Custom tags are incredibly useful, because they provide further separation of responsibilities between developers (who create the custom tags) and page authors (who use them).

Creating a custom tag requires several steps to connect your JSP code to the business logic of your Java components as compared to simple scriptlets in the JSP. However, custom tags are easier to distribute and reuse.

A custom action is invoked by using a custom tag in a JSP page. A tag library is a collection of custom tags. A tag library descriptor (TLD) file is an XML document that describes the library.

JDBC 3.0

Java Database Connectivity (JDBC) adds features to savepoint interface support, connection pooling, retrieval of auto-generated keys, and cursor support. JDBC has emerged as part of Java 2 Platform, Standard Edition (J2SE), and part of Java Enterprise Edition (J2EE). JDBC 3.0 API has two packages java.sql and javax.sql, both of these are included in J2SE and J2EE.

JDBC can be used to access one or more data sources in a distributed heterogeneous environment. JDBC 3.0 has some more new features such as ability to have multiple open result set objects, passing parameters to `CallableStatement` objects by name, and retrieval of parameter metadata.

Enterprise Java Beans (EJB) 2.1

The new features of EJB 2.1 include the following:

- Enhancements to EJB query language (EJB-QL), many new functions are added to this language: ORDER BY, AVG, MIN, MAX, SUM, COUNT, and MOD.
- Support of Message-Driven Beans (MDBs) for J2EE Connector 1.5: Now accept messages from sources other than JMS.

- Support for linking of Message Destinations (the same idea as EJB references, resource references).
- Support for Web services usages within EJB: Stateless session beans can be invoked over SOAP/HTTP. EJB can easily access a Web service using the new service reference.
- A container managed timer service: A new event-based mechanism for invoking EJBs at specific times.
- EJB containers must now support the newest specifications: J2SE 1.4, JMS 1.1, JavaMail 1.3, JAF 1.0, JAXP 1.2, JAXR 1.0, and JAX-RPC 1.1.
- The ejb-jar.xml standard deployment descriptor is now specified with XML schema, rather than DTD.

Java Connector Architecture (JCA) 1.5

The Java Connector Architecture defines the standards that allow J2EE application servers to interact with a back end, which is called an Enterprise Information System (EIS). An EIS generically designates a system that provides some sort of service to a number of clients. Examples of such systems include legacy infrastructures—such as relational databases or transaction processing monitors—and Enterprise Resource Planning (ERP) applications. JCA defines a significantly large number of aspects that regulate the interactions between J2EE applications and such systems. Those aspects include system contracts, for use of the Application Server, and client interfaces, for use with the J2EE application developers who intend to create applications that interact with an EIS.

A key idea in the Connector Architecture is the concept of Resource Adapter. The resource adapter is responsible for implementing the system contracts and the client interfaces—and it is structured so that it can be plugged into any J2EE-compliant application server. Resource adapters are specific to the EIS they provide an interface for—but any J2EE application server could use a certain resource adapter to interact with a certain EIS. Resource adapters therefore eliminate the need to write code that is specific to an application.

Java Message Service (JMS) 1.1

For this Technology for Developers release, the embedded messaging provider of WebSphere Application Server Version 5 has been removed and is replaced by a default messaging provider that complies with J2EE 1.4. The default messaging provider is also intended as the messaging foundation for a future Web services integration bus technology.

For the default messaging provider, the core messaging functionality is provided by a messaging engine. The messaging engine manages messaging resources and, through destinations associated with the messaging engine, provides a connection point to which applications connect to access messaging resources. A default messaging engine is created automatically at installation time.

Web Services Enhancements in J2EE 1.4

One of the primary goals of the J2EE 1.4 release is to support Web services that are based on SOAP and HTTP. With this release, J2EE technology can be integrated with Web services in a variety of ways. J2EE components (Java Beans and EJBs) can be exposed as Web services. These services can be accessed by clients that are written in the Java programming language or by existing Web service clients that don't take advantage of Java technology. Also J2EE components can act as Web service clients. The new Web Service APIs in J2EE 1.4 are JAX-RPC 1.1, SAAJ 1.2, JAXR 1.0, and Web Services for J2EE 1.1 (JSR 109).

JAX-RPC (Java API for XML-Based RPC)

The objective of the JAX-RPC (Java API for XML-based RPC) specification is to support the development of simple Web services and Web service clients in such a way as to shield developers from the underlying complexity of SOAP communication.

Basically, JAX-RPC is a specification for building Web service providers and requesters. A requester is a client of a Web service. On the surface, JAX-RPC looks like another instantiation of RMI. Essentially, JAX-RPC allows clients to access a Web service as if the Web service was a local object mapped into the client's address space even though the Web service provider could be on the other side of the world.

However, instead of IIOP or JRMP—as in the RMI world—in the JAX-RPC world, a remote method invocation is done via the XML-based protocol SOAP 1.1, which typically rides on top of HTTP.

The JAX-RPC Web service can be created either from a Java Bean or from a Session Bean. The Web service provider/developer can specify the remote procedures by defining remote methods in a Java interface. The Web service provider/developer only needs to code one or more classes that implement those methods. The rest of the classes and other artifacts can be generated by the Web service vendor's tools.

SAAJ (SOAP with Attachments API for Java)

SOAP with Attachments API for Java (SAAJ) is used mainly for the SOAP messaging that goes on behind the scenes in JAX-RPC and JAXR implementations. Secondarily, it is an API that developers can use when they choose to write SOAP messaging applications directly rather than using JAX-RPC. The SAAJ API allows you to do XML messaging from the Java platform: By simply making method calls using the SAAJ API, you can create, send, and consume XML messages over the Internet.

JAXR (Java API for XML Registries)

The Java API for XML Registries (JAXR) provides a convenient way to access standard business registries over the Internet. Business registries are often described as electronic yellow pages because they contain listings of businesses and the products or services the businesses offer.

JAXR gives developers writing applications in the Java programming language a uniform way to use business registries that are based on open standards (such as ebXML) or industry consortium-led specifications (such as UDDI).

The high-level architecture of JAXR consists of the following parts:

- A JAXR client, which uses the JAXR API to access a registry via a JAXR provider.
- A JAXR provider, which implements the RegistryService interface and various interfaces in order to allow a client to access registries.

APPENDIX E

The WebSphere Portal

> "Some are born great, some achieve greatness, and some have greatness thrust upon 'em." The line is Shakespeare's—it appears in Twelfth Night—but it applies surprisingly well to the greats of IT. The Apple Macintosh was born "insanely great" (in the typically modest words of Steve Jobs) and has stayed that way ever since, at least in the eyes of the Mac community. Windows, by contrast, achieved greatness after several years of development, including a period of joint development with IBM. And the story of e-business provides the names of many vendors who had greatness thrust upon them, all too briefly in many cases.
>
> A new entrant in this third category is the portal server. Both WebSphere and its main rival, BEA's WebLogic, provided new portal server offerings during 2002. IBM also made a commitment to expanding the portal market by introducing WebSphere Portal-Express: an affordable low-end product that, we're told, can be installed with a few mouse clicks."
>
> —Phil Edwards, a freelance writer from the U.K. in a January 2003 editorial in *e-Pro Magazine*

This appendix gives information on the IBM WebSphere Portal. In December 2003, the IBM WebSphere Portal was augmented with Lotus Collaboration and J2EE Office tools. The Collaboration Center comprises several portlets that are integrated with Lotus instant messaging, team workspaces, and directories. In general, the WebSphere Portal is focused on delivering capabilities that allow users to interact in highly collaborative settings. IBM's goal in this is to help companies build portals that extend collaboration in a wide variety of settings, including employee intranets within organizations and extranets that extend outside organizations to partners, suppliers, and others. The first version of the Collaboration Center included a team workplaces portlet that integrates with Lotus Team Workplace (formerly QuickPlace); a Web conferencing portlet that brings Lotus Web Conferencing (formerly Sametime) e-meeting features right into the portlet; and a people-finder portlet that lets users search a corporate directory by last name, first name, or other criteria.

IBM will release other portlets as they become available through a download site called the Portlet Catalog. This will allow users to add portlets incrementally as they need them. Also IBM will incorporate J2EE-based Office productivity applications—including rich-text editing and spreadsheet functionality—into WebSphere Portal. Actually, IBM is targeting portal customers who may want to avoid the expense of Microsoft Office and don't need or use many of its features. WebSphere Portal will sport a relatively basic J2EE rich-text editor and spreadsheet. The idea is that as people build and deploy portal-based applications, there will be some who need rich-text editing; however, the rich-text tools now are often either too feature-rich or not enough. The browser-based apps aren't rich enough, but the Office apps are overkill, and many users only tap into a small bit of their functions. The WebSphere Portal will provide function-rich capabilities in a much more cost-effective way.

Enterprise portals that present Web-enabled applications alongside collaborative workspaces, search facilities, e-mail, and news are a familiar concept. In 2004, portal technology started to become very popular. It's taken awhile for people to realize what a great idea they are. Portals were discredited by the early implementations, which gobbled up screen area with not that much benefit. Later, less flashy implementations delivered more tangible benefits: always-on access to key applications, e-mail, chat, and a browser. Also, the portal is the logical next step for both IBM and BEA. After all, both companies have already provided a J2EE application server, in-depth support for XML, and application integration tools. What's more, these offerings themselves sharpen the need for portal services. Server-side Java means assembling applications from reusable interoperable objects, XML and Web services making it possible to combine remote and local components, and legacy integration promising to open a whole new range of data sources. Applications built under these conditions will be significantly more open to extension and customization than the ones most of us are used to. Portals mirror this flexibility on the client side, offering users a personalized combination of application components and data sources.

Portals are built on the same open standards used by application servers. Using WebSphere Portal Server, you can either plug in an application with a Web service interface as a remote portlet or bind it to an existing portlet (for example, to return information from an Internet search to a local application).

Benefits of the Portal Model

IBM's overall software strategy favors portal development. The company has grouped its software under four key middleware brands: WebSphere, DB2, Tivoli, and Lotus. IBM now intends to integrate the four more closely. The portal model demonstrates the benefits of this approach, tying the WebSphere platform to Tivoli system management tools and a range of Lotus collaboration and knowledge management technologies, including QuickPlace, Sametime, and Lotus Extended Search (formerly Domino Extended Search).

The WebSphere Portal

WebSphere Portal is software for building and deploying enterprise portals. It provides an extensible framework for delivering access to enterprise applications and information. As portals enter their third generation of maturity, WebSphere Portal has introduced new concepts of delegated administration, cascading page layouts, portal federation through Web services, advanced portlet application concepts, business process integration, and advanced personalization.

WebSphere Portal is available in three different sizes (also known as editions): Enable, Extend, and Experience. These are cumulative. For example, Extend includes all functionality available in Enable—plus additional functionality.

WebSphere Portal Server

WebSphere Portal Server (WPS) provides common services such as application connectivity, integration, administration, and presentation that are required across portal environments. It is installed as another application within WebSphere Application Server, with its own Web and Enterprise Java Bean (EJB) container. Since it is part of the Application Server, it provides the same benefits of management, cloning, and scaling. The main components running in the Portal Server's Web container are portlets and portlet applications.

Installing the WebSphere Portal Server

The WPS Server is installed on the WebSphere Application Server (WAS). However, if you want to install the WPS and you don't have a WAS server, then the installation procedure can still be quite straightforward since you can choose to have WPS, WAS, and IBM HTTP Server (IHS) all installed together, as is indicated in Figure E-1.

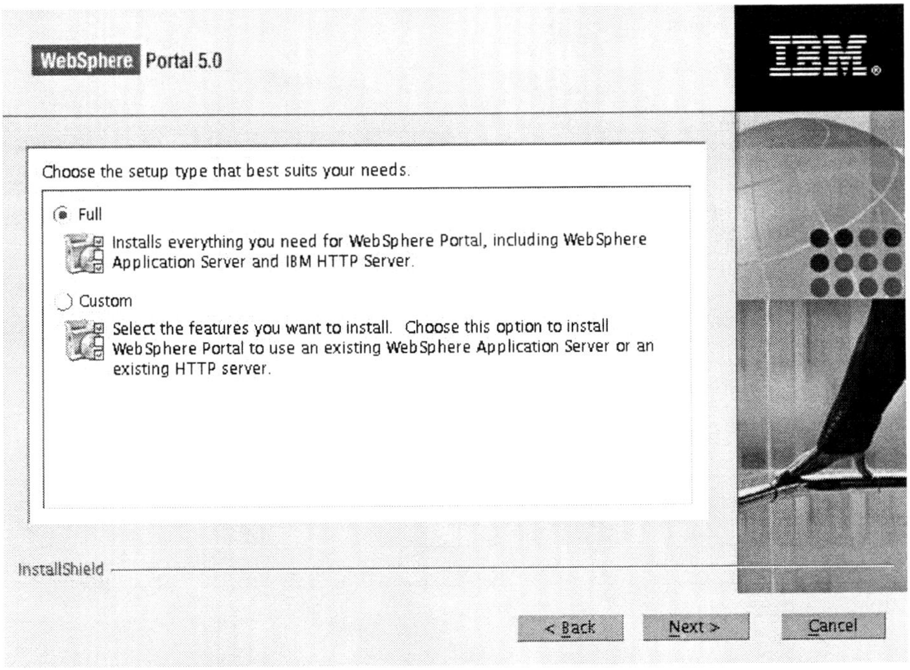

Figure E-1 Installing the WebSphere Portal Server.

Portlets

A portlet is a small portal application that is displayed in a rectangular area on a Web page. This rectangular area is akin to a miniature browsing area. Portlets are the heart of a portal. They are independent, reusable components that provide access to applications, Web-based content, and other resources. Web pages, services, applications, and syndicated-content feeds can be accessed through portlets.

Portlets are written in Java utilizing a portlet API. They are developed, deployed, managed, and displayed independent of other portlets. Portlets are becoming prevalent in the IT world. Enterprises can create their own portlets or select from a portlet catalog available from IBM and other third-party vendors. Until the portlet API gets adopted as a standard, however, a portlet written for one portal server cannot be used in another vendor's portal server without modifications.

The IBM portlet catalog is available at `http://www.ibm.com/websphere/portal/portlet/catalog`. There are well over 250 portlets available for download. Most of these are free, but some are written by third-party software vendors and require a usage/license fee. Each portlet is design for a certain version of WebSphere Portal Server—so keep that in mind when you are downloading.

Portlet Applications

Portlets are actually complete applications. A portlet application can contain one or more portlets and follow the model-view-controller (MVC) design. Portlet applications are packaged as Web Archive (.war) files, per the J2EE specifications.

Like servlets, portlets run inside a Web container that is part of the portal server. As in the case of servlets, the Web container provides a runtime environment in which portlets are instantiated, used, and finally destroyed. Portlets rely on the portal infrastructure to access user profile information, participate in window and action events, communicate with other portlets, access remote content, look up credentials, and to store persistent data. Portlets, however, are administered more dynamically than servlets. For example, portlet applications consisting of several portlets can be installed or removed while the server is running. An existing portlet can even be dynamically updated with a newer version.

The Portlet API

Portlets are assembled into a larger portal page, with multiple instances of the same portlet displaying different data for each user. Portlets rely on the portal infrastructure for their functionality. Since they are very similar to servlets, it makes sense that portlets are a special subclass of HttpServlet, with properties that allow them to easily plug into and run in the portal server. The portlet Application Programming Interface (API) provides standard interfaces for the portlets' functionality. The portlet API defines a common base class and interfaces for portlets in order to cleanly separate a portlet from the portal infrastructure. In most respects, the portlet API is an extension of the servlet API, except that it restricts certain functions to a subset. This makes sense for portlets running in the context of a portal. For example, unlike servlets, portlets may not send errors or redirects as a response. This is only done by the portal itself, which controls the overall response page.

Normally, one or more portlets are invoked in the course of handling a single request, each one contributing its content to the overall page display. Some portlets can be rendered in parallel so that the portal servers assemble all the markup fragments when all the portlets finish or timeout. Portlets that are not considered thread-safe will be rendered sequentially. The markup fragments that portlets produce may contain links, actions, and other content. The portlet API defines URL-rewriting methods that allow portlets to transparently create links without needing to know how URLs are structured in the particular portal.

WebSphere Portal Architecture

The WebSphere Portal is a third-generation portal server with a lot of components, offering one of the most comprehensive portal solutions. This is possible because WebSphere Portal server mirrors the proven architecture of the WebSphere Application Server. With its own portlet API and the standardized J2EE API, WebSphere Portal not only integrates well with existing backend components but also is easily extendable. Enterprises can use exiting security systems by using *Trust Association Interceptors* to achieve portal authentication. If enterprises already have content-management systems, WebSphere Portal toolkits are available to interface with them. WebSphere Portal gets installed as a Web application in WAS. Like every other application, it is comprised of Enterprise Applications. The four major Enterprise Applications in WebSphere Portal server are as follows:

1. WebSphere Member Subsystem (WMS)
2. WebSphere Portal Server (WPS) Enterprise Application
3. WebSphere Content Manager (WCM) Publish Webapp
4. WebSphere Proxy

In addition to these, approximately 26 portlets get installed as Enterprise Applications. When you view the WebSphere Application Server Administrative Console, you will see numerous Enterprise Applications, all running on one or more nodes of the WebSphere Application Server. Keep in mind that companies can start with the WebSphere Portal Extend Edition and then add components in the future to move up to the Enterprise architecture.

The Portal Blueprint

Here are some of the concepts, definitions, standards, and architecture that form the basis on which all Java-based portal server software is built. Many of these concepts, definitions, standards, and architectures hold equally true for both WebSphere Application Server and WebSphere Portal server.

J2EE Architecture in WebSphere Portal

Since WebSphere Portal is based on Java, its architecture closely follows the J2EE architecture. As such, the terminology and concepts will seem familiar to those who already know the Java language. J2EE (Java 2 Platform, Enterprise Edition) defines the standard for developing multitier enterprise applications. J2EE simplifies enterprise applications by basing them on standardized, modular components; by providing a complete set of services to those components; and by handling many details of application behavior automatically, without complex programming. These modular components are servlets, Java Server Pages (JSPs), and Enterprise Java Beans (EJBs). The set of services is provided by the containers in which these components operate, such as the Web server or the EJB container. As such, the architecture lets a client indirectly access back-end enterprise applications while protecting enterprise-sensitive data. J2EE comes with the standard application model and platform. It not only comes with a reference implementation but also a compatibility test suite for products wanting to achieve J2EE certification.

Customizing the Portal

Once you've installed the portal, the fun part is usually customizing and personalizing the portal so that users feel welcome. Usually as a first step, customers want to give the portal a pleasing and consistent corporate look. In WebSphere Portal terminology, this is called "customizing the portal."

Default Portal Look

WebSphere Portal has the concept of pages that exist within places. The user's access rights determine the places that get displayed. Each place can contain one or more pages, and only users who have adequate viewing rights can see those pages. Pages can contain column containers, row containers, and portlets. You can populate row or column containers with portlets or other containers.

Self-Registration Feature

If available, new users can click on the Sign Up icon and "register" themselves with the portal.

Personalizing the Portal

Personalization support in WebSphere Portal is provided by WebSphere Personalization Server and Macromedia's Likeminds Recommendation Engine. However, we will focus primarily on WebSphere Personalization Server.

Overview of WebSphere Personalization

Personalization is about targeting Web content and applications to specific users based on certain information that is gathered and stored about them. This information is analyzed, and the user is presented with content that would be interesting to him or her. Personalization techniques can be used by a site to customize news feeds, provide specialized advice, target e-mail and advertising, and promote products.

The Portal Gatekeeper

This section has to do with security and how security is set up for the WebSphere Portal Server. Some questions are: Who controls the Access Control Lists (ACLs), how often do users have to sign in, and how do you use Secure Certificates and integrate with third-party authentication products?

Users and Groups

WebSphere Portal offers a centralized point of access control over portal resources (which can be augmented with a security manager external to the product, such as Tivoli Access Manager). It also offers a centralized point of administration for users and groups if a customer does not already have an administration tool for its user registry. These tools allow customers to better define portal users and manage user's access rights.

WebSphere Portal Components

Most of the WebSphere Portal Components were or still are stand-alone products but are nicely integrated with the WebSphere Portal Server.

Personalization

Optimizing each user's experience with the portal is one of the key goals of WebSphere Portal. It provides end-user and administrative interfaces for customizing the content of portal pages, as well as the look and layout of the pages. With these tools, users can customize their own pages by selecting portlets and customizing the settings of each one. WebSphere Portal also comes with the Personalization Server. The purpose of this server is to select content for users based on information in their profiles and on business logic. It provides facilities that allow subject-matter experts to select content suited to the unique needs and interests of each portal visitor. The Personalization Server and portal server share a common user profile and a common content model. This model is based on the WebSphere resource framework interfaces classes. This means that personalization rules can easily be added to portlets to select portal content and target it to the portal's registered users.

Search

WebSphere Portal provides integrated text-search capabilities, including a search portlet, a crawler, and a document indexer. The search service can search the portal's document repository, as well as Internet content. Customer's seeking support for large document collections or support of searching a wide range of document types and data sources should consider using IBM's Lotus Extended Search (included with the WebSphere Portal Extend offering) or IBM's Enterprise Information Portal. WebSphere Portal has specific portlets that work with each kind of search engine.

Site Analyzer

WebSphere Site Analyzer is a software component that comes with the Extend and Experience Editions of WebSphere Portal. Both technical and business users can use Site Analyzer (SA) to monitor and maintain a very effective portal site. Portal administrators and site developers can obtain overall usage statistics, such as logins and logouts, enrollments, and error conditions. Site developers can get portlet and page usage statistics, including portlet actions, number of views, and modifications. The business sponsors can measure the effectiveness of the business rules and campaigns and use visitors' profiles to market products more effectively via the portal.

Collaborative Components

Lotus Domino, Lotus Sametime, Lotus QuickPlace, and Lotus Discovery are the building blocks for the Collaborative Components in WebSphere Portal. Lotus Software Collaborative Components are listed in Table E-1. They provide Java API methods and tags for Java Server Pages (JSPs), extending the functionality of Lotus's advanced collaboration in portlets. Using the Collaborative Services not only hides the complexities of integrating with these components but also takes care of managing the security context.

Table E-1 WebSphere Portal Collaborative Services

Collaborative Service	Brief Description
People Service	Provides online awareness capability so that portlets can determine who is online and what each user's status is.
Domino Data Service	Provides access to Domino data, including views, documents, and directory-related information. This service is used to locate Domino servers, create and edit documents, and work with views.
Discovery Service	Provides methods to find people, documents, knowledge maps, and profiles.
QuickPlace Service	Provides methods for creating a new Lotus QuickPlace.
Menu Service	Displays a pop-up menu in the portal, used with the People Service to create chat, e-mail, and search menus.

All of the Collaborative Components, except the People and Menu Services, are UI-neutral. That is, the portlet developer can design the user interface for the collaborative features they are implementing. See the more expanded section on WebSphere Portal Collaborative Components later in this Appendix.

Content Management

When companies deploy portals, they are primarily focused on content aggregation, especially the delivery of personalized content. WebSphere Portal meets the portal's content-delivery needs by supporting syndicated content, by integrating with Web content-management systems, and by providing built-in content-organizing portlets. Content syndication is about delivering fresh, personalized, and filtered content from multiple sources to subscribers. Typically, the content is related to news, finance, and entertainment. Companies are embracing syndication concepts and standards to automate the publishing of electronic catalogs and other internal information and to make this information available to workers through enterprise portals. Web content-management deals with creating, approving, and publishing Web content from content creators to Web servers. This process involves defining content types, roles, publication options, destination specifications, and workflow processes.

Portal Content Organizer (PCO) is a built-in content-management application that also can serve as an interface to integrate content received from outside the portal. This can be content published from Web content-management systems, allowing you to build organization-wide informational content sites. PCO enables you to integrate content of various formats from these systems into the portal. It also serves as a content-management integration framework and helps users organize content that they have already viewed, plan to view, or would like to share with other users.

Example of IBM's Portal for IBM Employees

IBM makes use of the WebSphere Portal Server for the Intranet Web site used by all IBM employees. Every employee can easily customize the "Work" and "Career and Life" views indicated in the upper left-hand corner of Figure E-2 showing the employee "home" view. The "edit settings" link (also shown in the upper-left corner of Figure E-2) provides an easy way for each employee to customize his portal—including the addition of portlets. IBM uses the "Career and Life" portal view as a way to help support (e.g., offload work) IBM's Human Resources (HR) group.

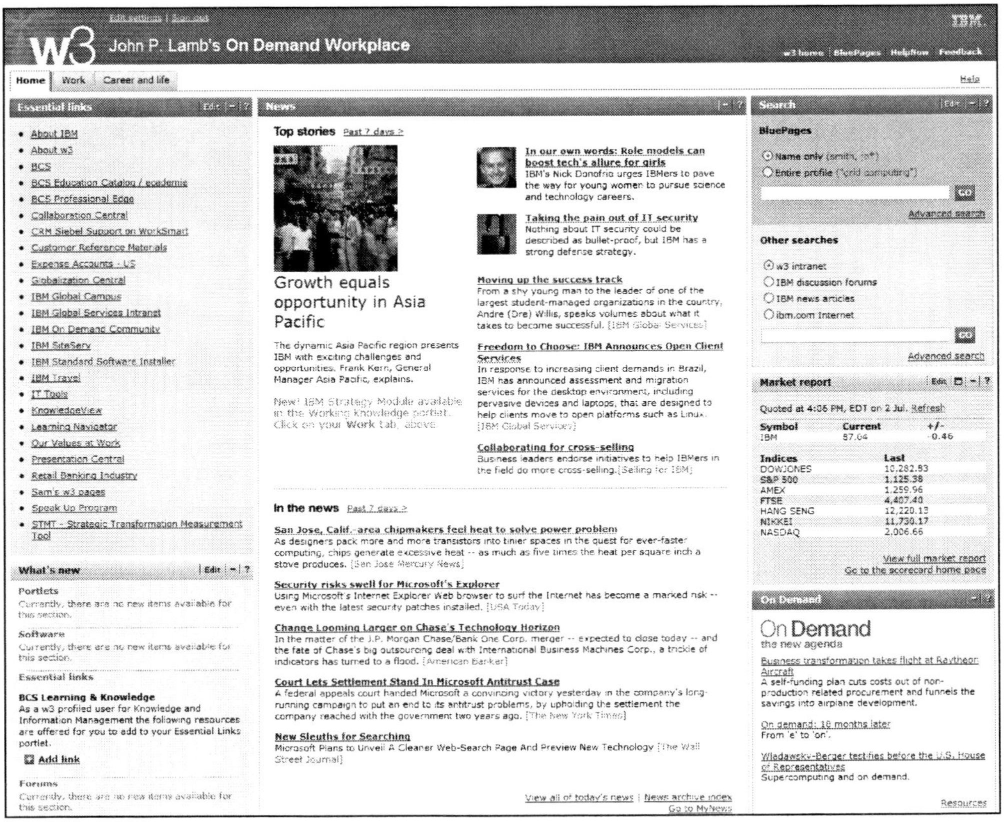

Figure E-2 IBM Employee Portal.

One of the ideas behind the IBM Employee portal (and almost any corporate portal) is to save corporate expenses. The "HelpNow" link shown in the upper right-hand side of Figure E-2 is a simple way for the IBM employee to get on-line help in many areas, usually without speaking to someone at an IBM Help Desk—thereby reducing corporate costs. Figure E-3 shows the screen displayed when an IBM employee clicks on the "HelpNow" link from his employee portal.

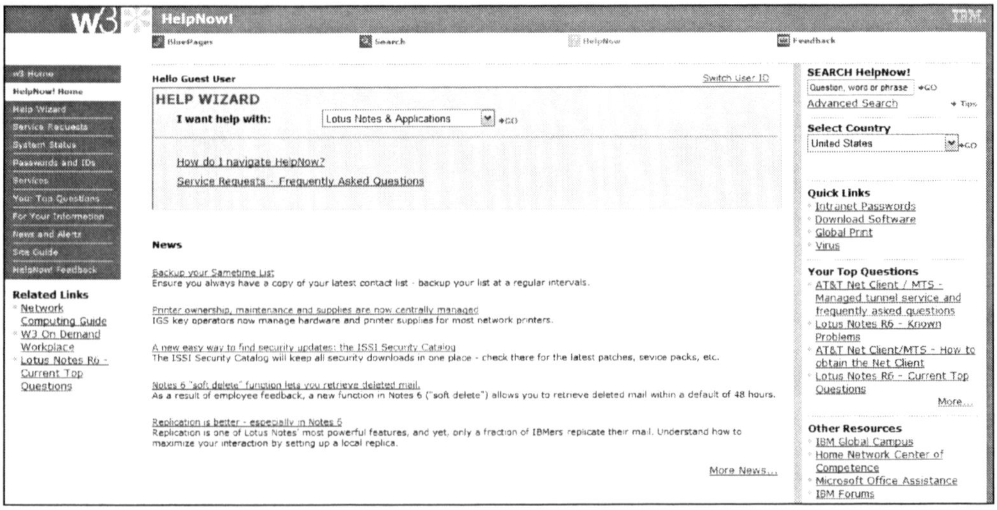

Figure E-3 IBM "HelpNow" link from the Employee Portal.

Another link from the IBM Employee Portal is the "Collaboration Central" link indicated in the list of links on the left-hand side of the portal screenshot shown in Figure E-2. The IBM Collaboration Central Web site is shown in Figure E-4. This IBM Web page lets employees find out about collaboration tools used within IBM such as e-meetings (Lotus Web Conferencing), Instant Messaging, QuickPlace (Lotus Team Workplace), Lotus TeamRoom Plus, and Forums (IBM's name for Web-based discussion databases).

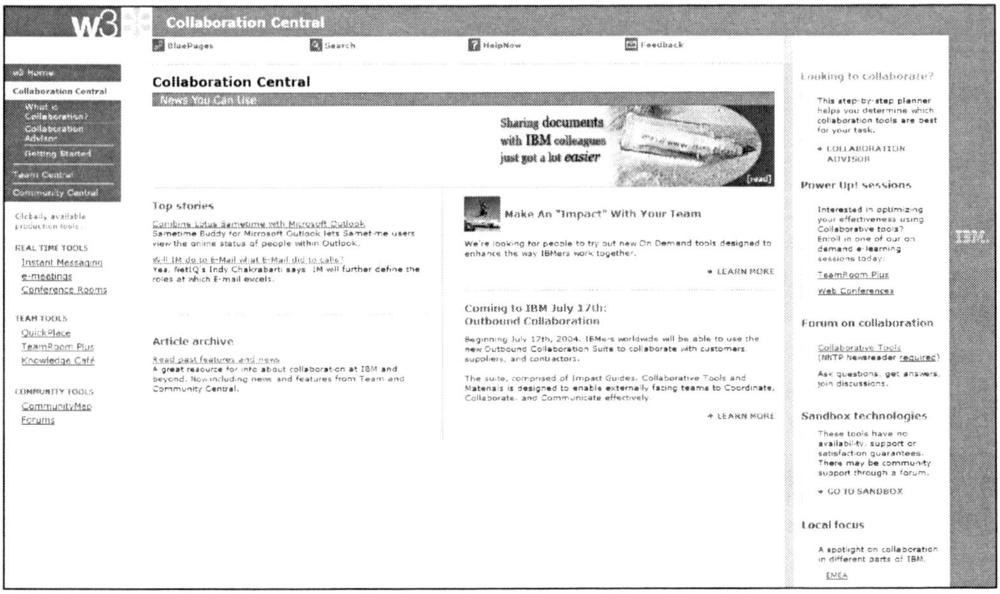

Figure E-4 IBM "Collaboration Central" link from the Employee Portal.

So, IBM is "eating its own cooking" when it comes to the use of the WebSphere Portal Server and Lotus Collaboration tools.

WebSphere Portal Collaborative Components

Since this book is about WebSphere and Lotus Collaboration, the WebSphere Portal Collaborative Components take on a very significant role. This section gives additional details on this significant subject. The WebSphere Portal Collaboration Components allow you, for example, to find out who is using the portal; have a quick chat with suppliers or business partners while they are all accessing the enterprise portal—thereby decreasing the need for e-mail; or have an e-meeting to discuss a press release before it is published.

Collaborative Places

Collaborative Places in WebSphere Portal are community areas in the portal where common pages and applications are organized. Members use these places to work together on common projects, communicating with each other to make effective decisions. A collaborative place can be *personal* or *shared.* A personal place provides a private workspace for individual users to manage daily work needs. This includes portlets that access e-mail, calendar portlet, portlets for corporate resources, and project database portlets tailored to that user. A shared place can be set up for teams working on a project or for enterprise-wide communities that need access to the same resources or a place to store and share information and discussions. A shared place can be restricted to certain members of the community or made public. Access to a collaborative place is controlled through a membership list. Members can be designated as managers, designers, or participants in the community. As the names suggest, managers can add new members to the community, designers can change the layout and content, and participants are only allowed to visit and use the workplace.

Collaborative Portlets

The collaborative portlets for Lotus Advanced Collaboration software are included in the WebSphere Portal Extend and WebSphere Portal Experience offerings. They have built-in features that allow portal users to take actions on documents or user names that appear in a portlet. Some of these portlets are enhanced versions of the portlets provided in WebSphere Portal Enable. From within the portlet, a portal user can see if other users are on-line and then select from a menu of options to interact with those users. Based on the context of the portlet, users can take actions directly from the portlet—with or without launching the entire application. For example, Lotus Notes e-mail portlets make any e-mail author's name live, with on-line awareness and menu options without altering the underlying e-mail application.

Collaborative portlets are available for all Lotus Advanced Collaboration technologies, including Lotus Notes/Domino, iNotes, Sametime, QuickPlace, Extended Search, and Discovery Server. Table E-2 lists the collaborative portlets that come with WebSphere Portal Extend Edition. In addition to the collaborative portlets provided out-of-the-box with WebSphere Portal

Extend, many other portlets for Lotus collaborative technologies can be downloaded from IBM's portlet catalog, including these:

- Lotus Discovery Mini-Search Portlet
- Lotus Discovery Search Results Portlet
- Lotus Discovery K-map Portlet
- IBM Lotus Extended Search Portlets
- MyiNotes Portlets, which auto-detect users' mail file settings
- Configurable iNotes, which is a single portlet that is configurable for the various iNotes views.

The IBM WebSphere Portlet catalog can be accessed from the URL: `http://www-3.bim.com/software/webservers/portal/portlet/catalog`.

The collaborative portlets do not support caching in their current form. Caching was disabled because Domino users are used to receiving dynamic information. If you are deploying the Domino portlets in a large enterprise, you might want to consider enabling caching on the portlets by adding the caching tag to the portlet.xml file within the Notes portlet's Web application (notes.war).

Table E-2 Collaborative Portlets in WebSphere Portal Extend

Name	Brief Description
iNotesMail Portlet iNotesCalendar Portlet iNotesContacts Portlet iNotesNotebook Portlet iNotesToDo Portlet	Provide iFrame-based access to iNotes-enabled Domino servers and mail files. This includes access to the Welcome, Mail, Calendar, To-Do List, Contacts, and Notebook iNotes functions.
MyNotesMail Portlet MyNotesCalendar Portlet MyNotes ToDo Portlet	Display users' mail, calendar, and to-do lists from their traditional Lotus Domino mail files. The "my" name indicates that these portlets auto-detect each user's mail file settings and thus do not need to be configured individually for each user.
NotesMail Portlet	Displays any view in a traditional notes mail database.
NotesDiscussion Portlet	Displays a Notes database built with the Notes Discussion database template.
NotesTeamroom Portlet	Displays a Notes database built with the Notes Teamroom database template.
NotesView Portlet	Displays any view in any Notes database.
QuickPlace Portlet	Launches a designated QuickPlace in a separate browser window. This portlet is also included in WebSphere Portal Enable.

continues

Table E-2 Collaborative Portlets in WebSphere Portal Extend (Continued)

Name	Brief Description
Sametime Portlet	Launches the Sametime Connect Java client. This portlet is also included in WebSphere Portal Enable.
Webpage Portlet	Displays any Web page inside an iFrame. This is not quite a "collaborative" technology, but is included with WebSphere Portal Extend, as it is based on the same technology as the QuickPlace and iNotes iFrame portlets.

The Technology of Collaborative Components

Collaborative portlets are really JavaBeans with a set of methods that expose collaborative functionality to portlet developers. They are designed to access Lotus software products like Domino, QuickPlace, Sametime, and Discovery Server. These interfaces are not a replacement of the existing APIs. They are useful for building custom portlets or adding functionality to portlets, such as menus or on-line awareness. There are two categories of collaborative components:

1. **Java Classes and Methods in cs.jar.** This package contains all the Java implementations of the Collaborative Components. These are classes and methods for leveraging Domino, QuickPlace, Sametime, and Discovery Server.
2. **JavaScript Tag Libraries like people.tld and menu.tld.** These provide Sametime awareness and contextual menus for JSPs.

Deploying Collaborative Components

You can deploy the Collaborative Components samples available in the Enterprise Application (cs.ear) by following these high-level steps:

1. Ensure that SSO is enabled on the server where the collaborative components enterprise application (cs.ear) is installed—either the Application Server or the Portal Server.
2. Open the WebSphere Advanced Administrative Console.
3. Use the Wizard to install an Enterprise Application, selecting the **cs.ear** file from its default location (<WAS_HOME>/InstallableApps). See Figure E-5.
4. Click **Next** in the series of screens that follow. When the screen that prompts you to select a server appears, select **Default Server** and then click **Finish.**

Start the Enterprise Application.

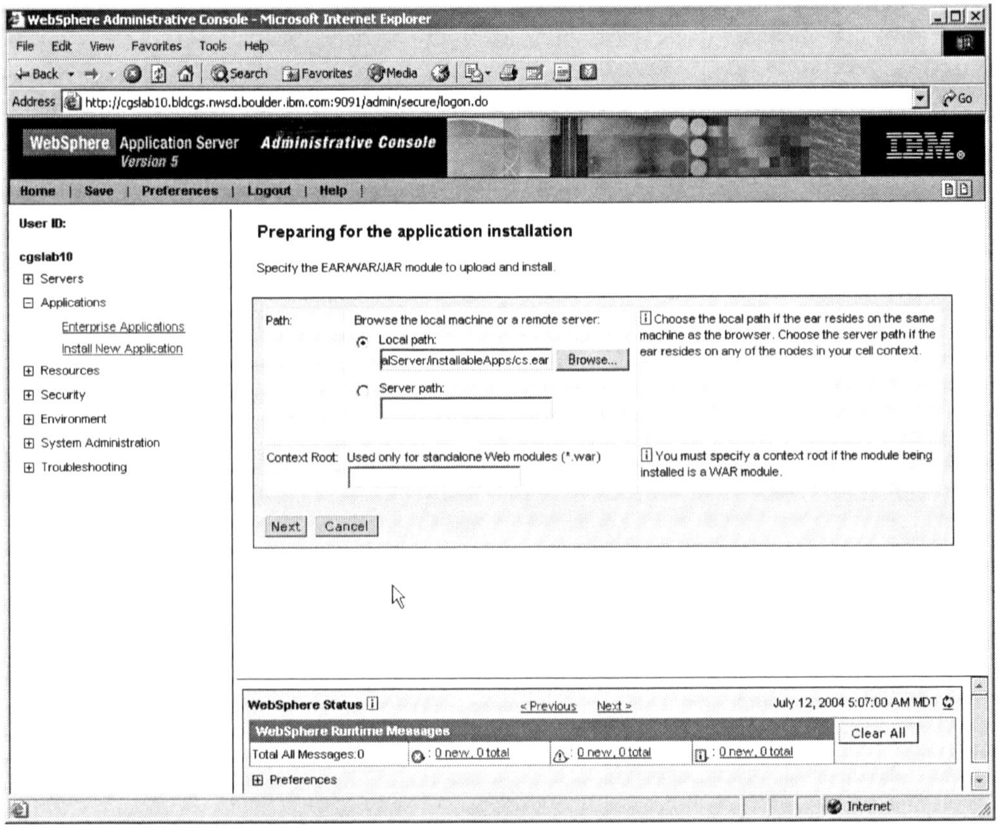

Figure E-5 Deploying Collaborative Components.

Resources to Extend WebSphere Portal Server for Collaboration

Setup Manager (SUM) installs the following files on top of WebSphere Portal server to extend Portal Server for collaboration.

The Collaborative Components enterprise application, cs.ear, contains the following:

- Collaborative Components Web application, cs.war.
- Collaborative Components java archive files: cs.jar, KDSW.jar, kdsapi.jar, NCSOW.jar, stcommsrvrk.jar, and commres.jar. The default location is <WPS_HOME>\app\wps.ear\wps.war\WEB-INF\lib.
- A directory of files related to People Awareness. The default location is <WPS_HOME>\app\wps.ear\wps.war\peopleawareness.
- A People Awareness tag-language descriptor file, people.tld. The default location is <WPS_HOME>\app\wps.ear\wps.war\web-inf\tld.

- A directory of files related to menus. The default location is <WPS_HOME>\app\wps.ear\wps.war\menu.
- A menu tag-language descriptor file, menu.tld. The default location is <WPS_HOME>\app\wps.ear\wps.war\web-inf\tld.
- A set of CSResources properties files. The default location is <WPS_HOME>\lib\app\nls.

The Collaborative Components API contains the following.

- Javadoc HTML files. The default location is <WPS_HOME>\app\wps.ear\wps.war\doc\Javadoc\CollabServices.
- JSP code samples in cs.war.

After you install Lotus Collaborative Places and Components on the WebSphere Portal Server, your operating system registry will show certain entries. Table E-3 lists the entries for each OS and indicates where they will be found.

Table E-3 OS Registry Entries

Operating System	Location and Entries
Windows	Windows Registry: HKEY_LOCAL_MACHINE\Software\IBM\WebSphere Portal Server\4.1: "Extend" "4.2.0.0"
AIX	AIX ODM: Fileset: wpsExtend Level: 4.2.0.0
Solaris	Solaris PK Registry: Name: wpsExtend Abbreviation: wpsExtend 4.2.0.0.DSP=4.2.0.0 Package Instance: IBMWPSExt
Linux	No platform registry entry. AppServer/extendVersion.txt will show "wpsExtend.version=4.1.2.0."

Collaborative Components Samples

The Collaborative Components enterprise application (cs.ear) that is installed at <WAS_HOME>/installableApps/cs.ear contains samples for Collaborative Components. You may deploy the standalone samples (JSPs) as follows:

1. Bring up the WebSphere Advanced Administrative Console.

2. Ensure that Single Sign-On (SSO) is enabled on the server where the Collaborative Components enterprise application is installed—either the Application Server or the Portal Server. Chapter 12, "Security and Single Sign-On," gives details on setting up SSO for both WAS and Domino.
3. Click on **Console>Wizards >Install Enterprise Application**.
4. Click **Browse** and select **cs.ear** from <WAS_HOME>/installableApps directory.
5. Click **Next** in the series of screens that follow.
6. When the screen that prompts you to select a server appears, select Default Server and then click **Finish**.

Once the samples are deployed, you can launch them from WebSphere Portal using URLs such as `http://<HOST_NAME>:9080/cs/qpsamp`.

The samples that are provided in the Collaborative Services enterprise application include the following:

- DominoService Samples
  - dbpicsamp.jsp (Domino directory sample: show selection lists for servers, databases, and views)
  - viewsamp.jsp (Notes view sample: add collaborative features to Notes database views, launch columns, and people links)
  - calsamp.jsp (Notes calendar sample: add collaborative features to a Notes calendar view)
  - dbdocsamp.jsp (Notes document sample: add a document to a Notes database)
  - dbsamp.jsp (Domino database sample: create a Domino database)
- QuickPlaceService sample, qpsamp.jsp (create a QuickPlace)
- DiscoveryServerService sample, dssamp.jsp (show Discovery Server functionality, including profile and K-Map)
- PeopleService sample, psamp.jsp (add on-line awareness and person menus to create people links for named people)
- Menu tag sample, msamp.jsp

Configuring Collaborative Component Samples

Figure E-6 shows the configuration screen for the Lotus Collaboration Component Samples at an IBM Boulder lab. The screenshot indicates that we had not completed the configuration. The installation of WebSphere Portal 5.01 at IBM Boulder, Colorado, was tested on a Redhat Linux Advanced Server 2.1. The testing was done by Charles Lin.

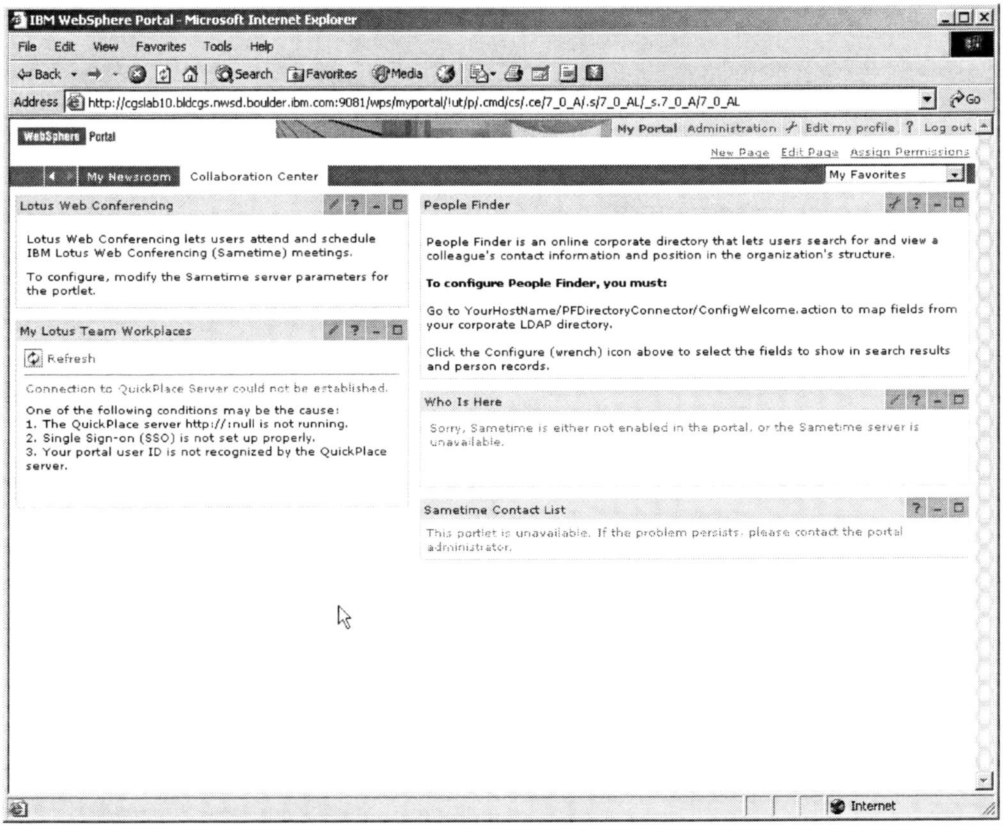

Figure E-6 Configuring Collaborative Component samples.

Lotus Domino Integration

A Domino server is needed to make all features of the collaborative portlets and Collaborative Components available to portal users. The Domino Directory Server or the Domino LDAP Server is required to look up various user values and attributes, like the `mailfile`. The Domino server is used as the data source, and the Domino Data Source servers must have HTTP, LDAP, and Domino IIOP enabled.

Portal Settings to Support Domino

You should enable the Domino Directory Server during WebSphere Portal server installation. If you install the Domino Directory after the Portal server is installed, you will need to modify the collaborative environment properties file on the Portal Server called CSEnvironment.properties, found in the <WAS_HOME>/lib/app/config directory.

The following entries pertain to Domino integration. They are required to enable the Domino Directory for collaboration:

```
CS_SERVER_DOMINO_DIRECTORY.enabled=true
CS_SERVER_DOMINO_DIRECTORY_1.hostname=ashoknote.encinitas.ibm.com
```

After modifying the CSEnvironment.properties file, remember to restart the portal server application via the WebSphere Administrative Console.

Domino Server Settings

Next you need to configure the Domino Server to support WebSphere Portal server. The Domino server, Domino LDAP server, and Single Sign-On have to be configured. Please reference Chapter 6, Appendix C, and Chapter 12 on SSO for Domino and WebSphere.

- First, make sure the Domino server is running (Start>Programs>Lotus Applications>Lotus Domino Server). Then bring up the Domino Server Administrator (**Start>Programs>Lotus Applications>Lotus Domino Administrator**).
- On the Basics tab of the Server document, make sure the fully-qualified host name is entered for the server name.
- On the Ports tab and its Notes network Ports sub-tab, ensure that the net address of the TCP/IP port is set to either the IP address of the server or its fully-qualified host name.
- On the Internet Protocols tab and its HTTP sub-tab, check that the host name is also specified as the fully-qualified server name.
- On the Internet Protocols tab and its HTTP sub-tab, make sure "Allow HTTP Clients to Browse Databases" is set to yes, to allow the selection of databases from within the Notes portlets.

Domino LDAP Settings

Whether or not you use Domino LDAP for WebSphere Portal authentication and authorization, you must have Domino LDAP enabled to take full advantage of WebSphere Portal Extend's collaboration capabilities.

Domino Directory Access Control List

At installation time, two users (wpsadmin and wpsbind) and the wpsadmins group should exist in names.nsf. They should be designated as managers. They should also have the Group Creator, Group Modifier, User Creator, and User Modifier roles set and have the ability to delete documents.

LDAP Field Settings

In the names.nsf database, make sure the MailFile and MailServer person fields and the HTTP hostname field are all available via LDAP (i.e., they are visible).

Single Sign-On Settings

Enabling SSO (Single Sign-On) between Domino and WebSphere Portal is key for collaboration. A user logging on to the portal should be able to use Domino-based collaboration portlets without being challenged a second time. Using LTPA (Lightweight Third-Party Access) tokens, you can easily do that. The high-level steps are as follows:

1. Create an IBM LTPA token in WebSphere Application Server via the WebSphere Administrative Console's Security Center.
2. Import the LTPA token generated by WebSphere Application Server into Domino, which is then used for all Domino servers within the Domino domain.
3. The detailed steps for creating and importing LTPA tokens are given in Appendix C.

Domino IIOP Settings

Internet Inter-Orb protocol (IIOP) settings need to be set in Domino because some of the collaborative portlets and services use IIOP to access the Domino server and get information from the Domino database. You can either load the IIOP task in your Domino Server Console or add it to the NOTES.INI file and restart the Domino server. If you are using IBM HTTP Server as the Web server in the WebSphere Portal environment, the collaborative portlets and components use the values in the diiop_ior.txt file to open IIOP sessions with the collaboration servers. You need to copy the diiop_ior.txt file from its original location at C:\Lotus\Domino\Data\domino\html to the IBM HTTP Server \htdocs directory.

Adding People Awareness to a Portlet

In order to add people awareness to a portlet, include the following lines in your JSP, and then use the person tag

```
<%@page language="java" import="com.lotus.cs.*,java.util.*" %>
<%@taglib uri="/WEB-INF/tld/people.tld" prefix="sametime" %>
```

Using Lotus Workplace 1.1 to Install WebSphere Portal Server

The installation program for Lotus Workplace 1.1 includes the installation of WebSphere Application Server (WAS), WebSphere Portal Server (WPS), and all necessary Fix Packs. With the Lotus Workplace installation, the installation of WAS and WPS have been canned and totally scripted. Thus, in order to have a noncustomized installation of WebSphere Portal Server, installation of Lotus Workplace 1.1 is a simple way to go. See Appendix F for details on Lotus Workplace 1.1.

Internationalization Support

WebSphere Portal customers in non-English speaking countries and multinational corporations need to have content in their native language. WebSphere Portal server supports multiple languages by trying to match the user with a language preference. If the language preference was not specified, the portal tries to make a match based on the language of the browser. This capability also extends to any portlets that have been enabled for translation. Since the default language is English, on a given page, any portlets that have the ability to show translated content will be shown in the language of choice. Otherwise, they will be displayed in English.

Supported Languages

Currently, WebSphere Portal server supports the following languages:

- English
- French
- German
- Italian
- Spanish
- Czech
- Hebrew
- Polish
- Turkish
- Brazilian Portuguese
- Japanese
- Korean
- Simplified Chinese
- Traditional Chinese

Following are the steps needed to change the language properties of WebSphere Portal server into a language that is not on the supported list.

Selecting and Changing the Language

There are various places and ways to change the language supported by the portal. Here are ways to select or change a language.

Installation

The installation program uses the default locale of the operating system to display the panels in the language supported by that locale. Once installation is done, this default language is stored in the LocalizerService.properties file in the <WAS_HOME>/lib/app/config/services directory. This file has three primary properties, as shown in Listing E-1.

Listing E-1 The LocalizerService.properties File

```
# Licensed Materials - Property of IBM, 5724-B88, ( C ) Copyright IBM
  ➥Corp. 2001, 2002 - All Rights reserved.
# ------------------------------------ #
# Properties of the Localizer Service #
# ------------------------------------ #

# Language, country, and variant settings for the default locale
#
# Default: <empty> [ = Locale.getDefault() ]

locale.default.language =en
locale.default.country =
locale.default.variant =

#
# List of BiDirectional languages separated by ',' or ';'
# Add ar,ur,yi,… when there are resources for those languages.
#

locales.bidi = he
```

To change the default language and locale supported, the locale.default.language and locale.default.country parameters need to be changed. Sometimes, there is a locale variant. For example, in order to change to American English, the parameters should be set as follows:

```
locale.default.language = en
locale.default.country = US
```

A portlet can support one or more locales. All portlets must have their own default language specified in the deployment descriptor; otherwise, the portlet cannot be installed.

Changing Titles and Descriptions for Places

In the manage Places and Pages tab, use Work with Pages to change the title for places and pages. This portlet allows you to change the title and description for each available place and page. The following instructions outline the steps needed to change the title for places.

1. In the Work with Pages place, go to **Manage Places and Pages** tab.
2. Select Home as the place and click Set locale-specific-titles.
3. Select the radio button corresponding to the required locale and then click Set title for selected locale.
4. Enter the new title for the place, and click OK. The place will be displayed with its new "localized" name.

Changing Titles and Descriptions for a Page

To change the title for a page, follow these steps.

1. Highlight the page from the list of available pages and click Manage Pages.
2. Select the page with the title you want to change and click Manage Page Properties. From here, the steps are similar to what you did for changing the title of a place.

Language Selection by the User

A user can select a personal language preference for rendering portal content during the enrollment process. If the user later decides to change the locale, he can use the Edit My Profile portlet to specify the preferred language form the drop-down list. The list of languages matches what is specified in the <WAS_HOME>/lib/app/config/language.properties file.

Language Selection by the Portal

The portal goes through the following sequence to determine the language in which to render content.

1. If the user has logged in, then use the language specified by the user. (For anonymous users, the portal skips this step.)
2. If the user did not specify a language preference, use the language specified in the browser.
3. If the browser's language cannot be determined, use the default locale of the portal.
4. If no supported language is found by the first three steps, then use the default language of the portlet.

The language then applies to the entire portal. If the portal or any component does not find the appropriate resource, a similar language is used. The language is determined by the file names for the resource bundles and the hierarchy of the directory structure for the JSPs. This search sequence applies to all portal components, down to the portlets. That is how a portal can display individual portlets in different languages.

Depending on how a portlet is written, a portlet can support multiple locales. A portlet needs to specify the default language in its deployment description; otherwise, the portlet couldn't be installed.

Changing the Character Set for a Language

The character set for a language is stored in a database. The default encoding is UTF-8. The values in use for the various languages are shown in Table E-4.

Table E-4 Language Encoding in Use

Language	Encoding in Use
Japanese ja	Shift-JIS
Simplified Chinese zh	GBK
Traditional Chinese zh_TW, HTML	Big5
Traditional Chinese zh_TW, WML	UTF-8
Korean ko	KSC5601
All Others	UTF-8

Follow these steps to change a character set for a language:

1. Log in to the portal as the Portal Administrator (**wpsadmin**). Under Portal Administration, navigate to the **Portal Settings** tab.
2. Select Manage Markups.
3. Highlight the markup you want to modify. Click **Edit selected markup**.
4. The next screen lists the MIME type and current encoding. Click **Set locale specific setting**.
5. On the next screen, you can change the character set for the selected language in the selected markup. Click **OK**.

Adding Support for a New Language

To support a new language, you need to add that language to the file language.properties. You do this by inserting resource bundles and adding JSPs for the new language where applicable. Some JSPs use resource bundles, while others, such as help JSPs, are translated directly.

Resource bundles are located in <WPS_HOME>/app/web/WEB-INF/classes/nls. They store text displayed in JSPs or in Java code. The naming convention for resource bundles is *[bundle]_[language]_[country]_[variant].properties.*

The ISO standard ISO-639 is used for the language codes of most languages, and ISO-3166 is used for the country codes. (Hebrew uses the old language code, iw.)

WebSphere Portal uses the properties file called by the Java class java.util.ResourceBundle to store text rendered in JSPs. The resource bundles are searched in the following order:

1. *[bundle]_[language]_[country]_[variant].properties.*
2. *[bundle]_[language]_[country].properties.*
3. *[bundle]_[language].properties.*
4. *[bundle].properties.*

The default bundles, [bundle].properties, are in English. To add a resource bundle, follow these steps:

1. Copy an existing resource bundle and translate it.
2. Name the new bundle according to the naming convention for resource bundles.
3. Convert it into unicode with the native-to-ASCII converter, native2ascii. It comes as part of the JDK and is found in <JAVA_HOME>/bin.

Here is a sample command to convert a Chinese text file, using UTF-16 encoding:

```
Native2ascii -encoding UTF-16 helloWorld_zh.txt
➥helloWorld_zh.properties
```

JSPs that contain mostly static text are translated directly, which means the text is contained in the JSP instead of in the resource bundle. You have to copy and translate the existing JSP and store it in the appropriate location. The location of JSPs can be, for example, jsp/[mime_type]/[language]/[country]/[variant]/sample.jsp. The exiting Help JSPs in WebSphere Portal are already translated and placed in the relevant [language] and [country] subdirectories. Rather than create and use multiple JSP files, it is recommended that you create a single JSP file that contains the tags to load multilingual strings from the resource bundles. It is easier to maintain one JSP file than to keep multiple files in sync.

APPENDIX F

Lotus Workplace

> "IBM Lotus Workplace is a new innovation for collaboration and human interaction that uses a single, open platform to integrate people with business processes. Users can access collaborative tools such as messaging, e-meetings, and calendaring and scheduling in the context of the work they're engaged in, rather than as separate applications. The results can include improved responsiveness and business productivity. In this tutorial, we explore how you can install a stand-alone system with the IBM Lotus Workplace V1.1. This environment is a good starting point to get a feel for what version 1.1 offers. The installation can be used for a proof of concept."
>
> —Rohit Saharabudhe, Lotus Workplace 1.1 Tutorial Author from IBM Austin, Texas, February 2004

This Appendix gives information on Lotus Workplace, with details on Workplace 1.1. Lotus Workplace has been described as a "Web-based integrated solution for collaboration." The first section of this appendix describes the Lotus Workplace concept, while the second section of the appendix consists of a tutorial developed by Rohit Saharabudhe of IBM Austin, Texas.

The Lotus Workplace Concept

IBM Lotus Workplace is a new concept that allows companies to easily and quickly set up collaboration tools on a single, open platform. The idea behind any set of corporate collaboration tools should be to help in the efficient execution of business processes. Lotus Workplace users can access collaborative tools such as messaging, e-meetings, and calendaring and scheduling in the context of the work they're engaged in, rather than as separate applications. The results can include improved responsiveness and business productivity. The supported operating systems are Windows NT (4.0 with Service Pack 6a), Windows 2000 Server (or Window 2000 Advanced Server with Service Pack 2), UNIX Solaris (Version 7 or Version 8), UNIX AIX (Version 4.3.3 or Version 5.1), Linux Red Hat, and Linux (Other) (SuSE Linux Enterprise Server, Version 7). Other requirements are Lotus Domino or IBM Websphere application server software, or both, and IBM DB2. WebSphere Portal provides the framework for bringing together the modular components of the Lotus Workplace platform. It provides both Lotus Domino and WebSphere developer communities with a modular architecture for building integrated collaborative solutions.

IBM Lotus Workplace Team Collaboration 1.1 features integrated instant messaging, Web conferencing, and team workspaces. Executives stressed the simplicity of setting up an e-meeting, likening it to a "browser-based, broadcast kind of approach." Using the roles-based IBM Lotus Workplace Web Content Management 1.1, nontechnical users can publish material within the Workplace environment, on a portal, or on an application server. IBM Lotus Workplace Collaborative Learning 1.1 offers self-managed employee training and supports scheduling for both classroom-based and e-learning programs. Each of the modules can easily be embedded within others. For example, users can see which colleagues are on-line from e-mail, use a "people finder" to access information about colleagues from the corporate directory, send an instant message, or enter a collaborative workspace to trade documents and search for learning materials. Target users for this product include contact center workers who often need to solve customer problems by collaborating with colleagues.

The Workplace Version 2.0 release will include collaborative document management, a connected rich e-mail client, and new development tools for business users. In order to gather "hands-on" experience for this book, one of our IBM colleagues, Charles Lin of IBM Boulder, installed and tested both WebSphere Portal Server 5.01 and Lotus Workplace 1.1 at the IBM Boulder lab. Charles did point out that since the Lotus Workplace 1.1 installation included the configuration for the LDAP directory and DB2, that installation was greatly simplified when compared to his installation of WebSphere Portal Server 5.01. Even the later WebSphere Portal Server 5.02 extended version still requires separate instillations of Sametime and QuickPlace servers. Lotus Workplace 1.1, on the other hand, does not need to connect to a separate Sametime server for Web conferencing or instant messaging. Lotus Workplace 1.1 also includes a component called Teamspaces, which is similar to but has fewer features than QuickPlace. The Lotus Workplace 1.1 components are all J2EE-based applications running on top of the WebSphere server. Here are some screenshots from the Boulder lab installation of LWP1.1. Figure F-1 shows the Lotus Workplace Mail Portlet while Figure F-2 shows the Lotus Workplace Web Conferencing Portlet.

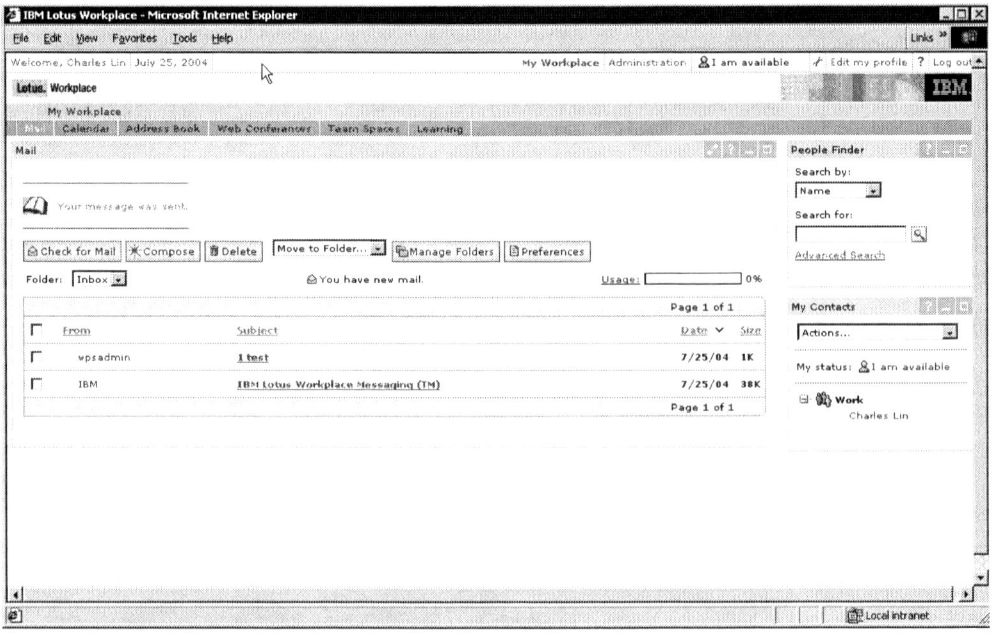

Figure F-1 Lotus Workplace mail portlet.

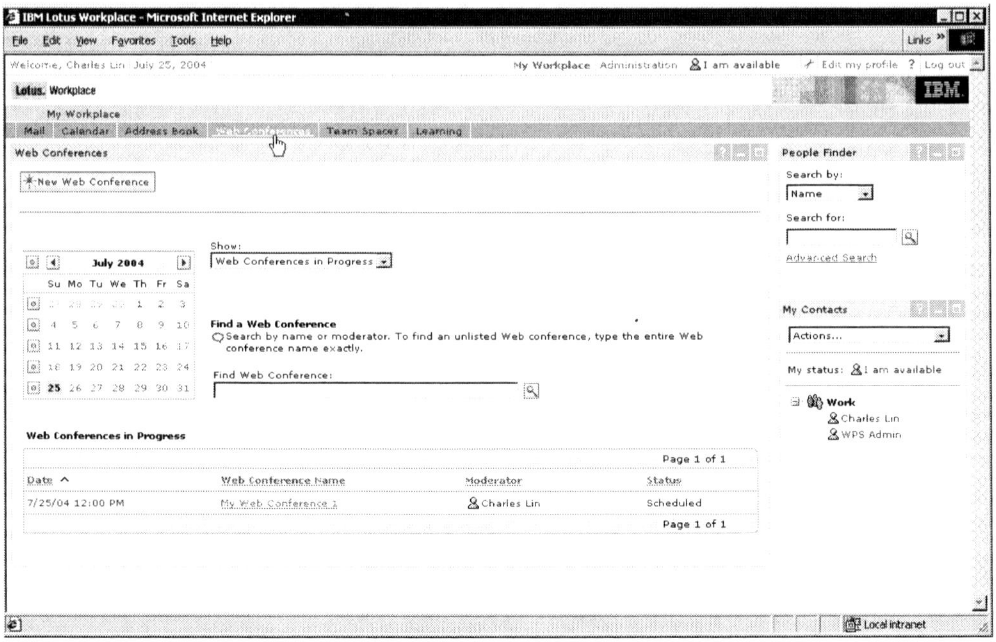

Figure F-2 Lotus Workplace Web Conferencing portlet.

Installing the IBM Lotus Workplace Tutorial

To look at this tutorial, access the URL `http://www-136.ibm.com/developerworks/lotus/` and then choose "Tutorials" in the Learning Resources section of that portal. The tutorial described here is the one titled, "Installing IBM Lotus Workplace 1.1." In this tutorial, Rohit Saharabudhe of IBM Austin helps you explore how you can install a stand-alone system with the IBM Lotus Workplace V1.1. This environment is a good starting point to get a feel for what version 1.1 offers. The installation can be used for a proof of concept. This tutorial is based on the "Getting started with IBM Lotus Workplace 1.1" workshop, held at Solution Partnership Centers around the world. The workshop introduces the Lotus Workplace (LWP) environment and teaches attendees how to install and configure version 1.1. In this tutorial, you can explore how you can install a stand-alone system with the IBM Lotus Workplace V1.1 code.

In order to follow this tutorial for a stand-alone, "sandbox" style controlled environment, you may want to set up a Loopback adapter on your system. Even as part of the workshops, attendees use an image created with a Loopback adapter. This adapter will allow you to set up a virtual network on a Windows environment. The underlying middleware components used in IBM Lotus Workplace require a fully qualified domain name. For you to follow this tutorial, you either need to have your own static IP address and domain name registered or utilize the Loopback adapter and the HOSTS file to make the system think that a particular hostname and IP being used is the true IP.

For IBM Lotus Workplace V1.1, it is recommended that you start with a clean Windows 2000 Server environment with Service Pack 4. The following instructions will help you set up a Loopback adapter on this system:

1. Choose Start>Settings>Control Panel>Add/Remove Hardware.
2. Click **Next**.
3. Select **Add/Troubleshoot a device** and click **Next**.
4. Select **Add New Device** and click **Next**.
5. Select **No, I want to select the hardware from a list** and click **Next**.
6. Scroll down and select **Network adapters** as the hardware type and click **Next**.
7. Scroll down and select **Microsoft** as the manufacturer and **Microsoft Loopback Adapter** as the network adapter and click **Next**.
8. Click **Next** and click **Finish**.
9. From the Windows desktop, right-click My Computer, select **Properties**.
10. Select the **Network Identification** tab and then click the **Properties** button.
11. Click **More** and add the suffix: `techpreview.ibm.com` in the text box. Check the box underneath (change primary DNS suffix) as selected.
12. Click **OK** all the way through, and you will probably be asked to reboot.

Once you've rebooted, set your IP address as follows:

1. Choose Start>Settings>Network and Dial-up Connections.
2. Right-click the Microsoft Loopback Adapter.
3. Click **Properties**.
4. Scroll down and double-click **Internet Protocol** (TCP/IP).
5. Enter an IP address `(10.0.0.1)` and Subnet Mask `(255.0.0.1)`.
6. Reboot.

Add this fully qualified hostname (assumed by this tutorial: `lwp11.techpreview.ibm.com` the IP address `(10.0.0.1)` to your `c:\winnt\system32\drivers\etc\hosts` file.

Check your network settings. Start a command-line window, and execute the command:

```
ping lwp11.techpreview.ibm.com
```

The IP address should be 10.0.0.1 instead of 127.0.0.1.

The tutorial assumes that the DB2 Administrator and the Windows Administrator user is the following:

```
Userid: db2admin
Password: db81admn
```

> **TIP**
> **Installation**—Be patient. Some of the scripts take time to execute. And keep the Task Manager in view to get a clear picture of the installation process.

Installing IBM DB2 UDB 8.1 Enterprise Server

1. Create a local db2admin account, and assign it a password (this tutorial assumes: `db81admn`) if one doesn't already exist and make this account a member of the Administrators group.
2. If you haven't already, log in as `db2admin` and make sure you have the following Local Security Policies set for your db2admin user. Assign these policies by choosing Start>Programs>Administrative Tools>Local Security Policy, Select Local Policies> User Rights Assignments:
   - Act as part of the operating system
   - Create a token object
   - Log on as a service
   - Replace a process level token

3. Verify that the db2admin account is a member of the local Administrators group. If not, add the user in that group.
4. From the DB2 Enterprise Edition CD (CD 5-1), run either Setup.exe or Launchp.exe.
5. On the installer screen, click **Install Products**.
6. Click **Next** (DB2 UDB Enterprise Server Edition is selected).
7. From Welcome to the DB2 Setup Wizard, click **Next**.
8. Review the license agreement. If you accept the terms, click **I accept the terms of the agreement**, and click **Next**.
9. In the Setup the installation type window, click **Typical** and click **Next**.
10. The default screen does not have **Data Warehousing and Satellite Administration capability** selected. Do not change the defaults. Click **Next**.
11. A warning screen appears regarding APPC. Click **OK**.
12. Select the installation action, accept the default, Install DB2, Enterprise Server Edition on this computer. Do not select **Save your settings in a response file**. Click **Next**.
13. At Select installation folder for DB2, select the installation folder. This tutorial assumes the folder to be C:\SQLLIB.
14. Set user information for the DB2 Administration Server as follows, and click **Next** when you are done.
    - For user name, enter `db2admin`, and be sure all the rights are selected for this user.
    - Specify the password created in Step 1.
    - Leave **Use the same user name and password for the remaining DB2 services** checked.
15. At Set up the administration contact list, accept the default (Local), and click **Next**.
16. When you see the warning screen Notification SMTP server, click **OK**.
17. In the Configure DB2 instances screen, click **Protocols** and configure the TCP/IP protocol. Keep the default settings (Service name, db2c_DB2 and Port number, 50000), click **OK**, then **Next**.
18. In the Prepare the DB2 tools catalog window, click **Prepare the DB2 tool catalog in a local database** and click **Next**.
19. In the Specify a local database to store the DB2 tools catalog window, accept the defaults and click **Next**.
20. In the Specify a contact for health monitor notification window, click **Defer the task after installation is complete** and click **Next**.
21. In the Start copying files window, review the list and click **Install**.
22. Click **Finish** when installation completes.
23. Reboot the machine, and then follow the instructions to Apply DB2 UDB 8.1 Fix Pack 2.

Apply DB2 UDB 8.1 Fix Pack 2

You must apply at least Fix Pack 2 to DB2 UDB 8.1.

1. Log on as `db2admin`.
2. Go to Control Panel/Services and stop all the DB2 services (that are started automatically), except for DB2 DB2-0.
3. From the DB2 Fix Pack 2 CD (CD 5-10), run update.exe. It will automatically find the previous installation of DB2 8.1. If prompted to stop DB2 services, click **YES**.
4. Click **OK**, and **UPDATE** when prompted. Wait until this process is complete.
5. If you see a screen stating DB2 is currently running and locked by the following processes, click **YES** to shutdown those processes and continue.
6. Click **Finish** at completion.
7. Reboot your machine.

Next, you'll need to install IBM Directory Server 5.1.

Installing IBM Directory Server 5.1

Probably the most important component of a successful LWP deployment is the LDAP server schema. It is imperative that the LWP administrator collaborate with the LDAP administrator. This tutorial assumes a clean install of IBM Directory Server 5.1 and import of provided LDIF. If you are using an existing LDAP server, you can skip the instructions here and go to Lotus Workplace installation.

Follow these steps to install IBM Directory Server 5.1 on the same system as Lotus Workplace.

1. Log on to the local system as `db2admin`.
2. From the IBM Directory Server 5.1 CD (CD 3-1), change to the ids_ismp directory and run Setup.exe.
3. Select **English** as the language to be used, and click **OK**.
4. Click **Next** on the Welcome to Installshield Wizard.
5. Review the license agreement. If you accept, accept the terms in the license agreement and click **Next**.
6. Click **Next** at the verification that DB2 is already installed.
7. Change the directory name to `C:\LDAP` and click **Next**.
8. Select **English** as the language for IBM Directory Server, and click **Next**.
9. It is *critical* to select **CUSTOM** as the setup type, and click **Next**.
10. Because IBM Directory Server 5.1 uses WebSphere Application Server to deploy the graphical administration client, you must uncheck **Web Administration and IBM WebSphere Application Server**. Therefore, only select **Client SDK**, **Server 5.1**, and **GSKit** as indicated in Figure F-3, and then click **Next**.

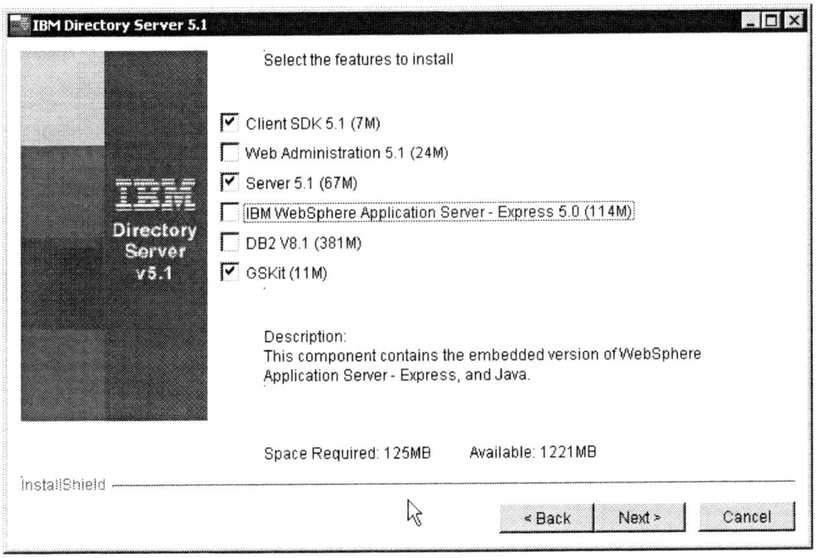

Figure F-3 Installing IBM Directory Server 5.1.

11. Review the installation summary, and click **Next**.
12. Click **OK** to begin GSKit installation.
13. Click **Next** (possibly twice) at the completion of installation.
14. Click **Next** to restart your server.
15. Click **Finish**.
16. Reboot your system.

IBM Directory Server 5.1 Configuration

Upon reboot, there are a few configuration steps required using the IBM Directory Server Configuration Tool:

- Click **Administrator/DN password**.
  1. For the Administrator DN, use the default, `cn=root`.
  2. For the Administrator password, use `password`.
  3. Confirm password, use `password`.
  4. Click **OK**.
  5. Click **OK**.
- Click **Configure Database**.
  1. Check **Create New Database**, and click **Next**.

2. Enter your DB2 Administrator user ID, `db2admin`.
3. Enter the DB2 Administrator password, `db81admn`.
4. Click **Next**.
5. Enter a database name, **`db2ldap`**, and click **Next**.
6. Check **Create a universal DB2 database (UTF-8/UCS-2)**, and click Next.
7. Use drive C for the database location, and click **Next**.
8. Verify the settings, and click **Finish**.
9. Upon completion, click **Close**.

- Click **Configure Suffixes**.
  1. Under Suffix DN, enter **`dc=ibm, dc=com`** and click **Add**.
  2. Click **OK**.
- Click **Import LDIF File**.
  1. Click **Browse** and navigate to select the sample LDIF file, provided in `Resources`.
  2. Choose **Standard Import**.
  3. At the bottom of the page, click **Import**.
  4. Upon completion, click **Close**.
- Close IBM Directory Server Configuration by choosing File>Close.

Now start IBM Directory Server via Windows Services IBM Directory Server 5.1.

Lotus Workplace Installation

1. Make sure you are logged in to Windows as `db2admin`.
2. Run install.bat from the Lotus Workplace Setup CD.
3. In the Select a language to be used for this wizard window, select **English**, and click **OK**.
4. In the IBM Lotus Workplace Installer window, click **Install**.
5. In the Welcome to the IBM Lotus Workplace Installation program window, click **Next**.
6. In the Results of system check…window, make sure you see a supported IBM DB2 Enterprise Server is installed on your server. Click **Next**.
7. In the The system check detected a DB2 Enterprise Server on this computer window, select the option, **This computer will host Lotus Workplace data and products**, then click **Next**.
8. In the Before continuing Lotus Workplace installation, WebSphere Portal must be installed window, click **Next**. You will see the screen Installing WebSphere Portal. *Do not* close this window. You can't click this installer.

9. The Lotus Workplace Version 1.1 Installer opens to install Portal. In the Select a language to be used for this wizard window, choose **English** (the default), and click **OK**.
10. In the Welcome to the Installshield Wizard window, click **Next**.
11. Review the license agreement. If you accept the terms, click **I accept the terms**, then click **Next**. If you see a box asking you to disable any firewall products running on this machine prior to installation, click OK after disabling them manually.
12. In the Choose the setup type that best suits your needs window, select **Full**, and click **Next**.
13. In the WebSphere Application Server will be installed in the following directory window, accept the default, `C:\WebSphere\AppServer`, and click **Next**. Make sure this directory path contains no spaces.
14. In the IBM HTTP Server will be installed in the following directory window, accept the default, `C:\IBMHttpServer`, and click **Next**.
15. In the You can use Windows Services to run the following WebSphere Application Server features window, just select the Run IBM HTTP Server as a service option. Enter your Windows administrator username (`db2admin`) and password (`db81admn`), and click **Next**.
16. In the Enter a node name for this instance of WebSphere Application Server window and in the Enter the hostname for this installation of WebSphere Application Server window, enter a node name (**`lwp11`**), and enter the fully qualified host name (**`lwp11.techpreview.ibm.com`**); then click **Next**.
17. In the WebSphere Portal will be installed in the following directory window, accept the default, `C:\WebSphere\PortalServer`, then click **Next**.
18. In the Enter the Portal administrative user and password window, enter **`wpsadmin`** for the Portal administrator username and **`wpsadmin`** for the password, confirm the password, and then click **Next**.
19. In the window Lotus Workplace is ready to install. The following components will be installed; click **Next**. You will see multiple screens as installation progresses. Installation takes approximately one hour. You will use CD1-1, CD1-6, and CD2 for this part of the installation. (This tutorial assumes Portal Administrator of `wpsadmin` with password of `wpsadmin`.)
20. In the Installation is successful window, which lists all the components that have been installed, uncheck **Launch First Steps** and click **Finish** to launch the Lotus Workplace Installer.
21. In the Results of system check window, click **Next**.
22. Make sure that the prerequisite product installation has been completed:
    - Windows 2000 Server with Service Pack 4

- WebSphere Portal Server V5.0
- DB2 V8.1.2

Lotus Workplace Products Installation

1. In the Click Next to install Lotus Workplace Products window, click **Next**. You will use the Workplace CD for this part of the installation.
2. In the IBM Lotus Workplace Installation window, click **Next**.
3. In the Select the Lotus Workplace products that you have purchased licenses for window, select all the products, and then click **Next**, as shown in Figure F-4.

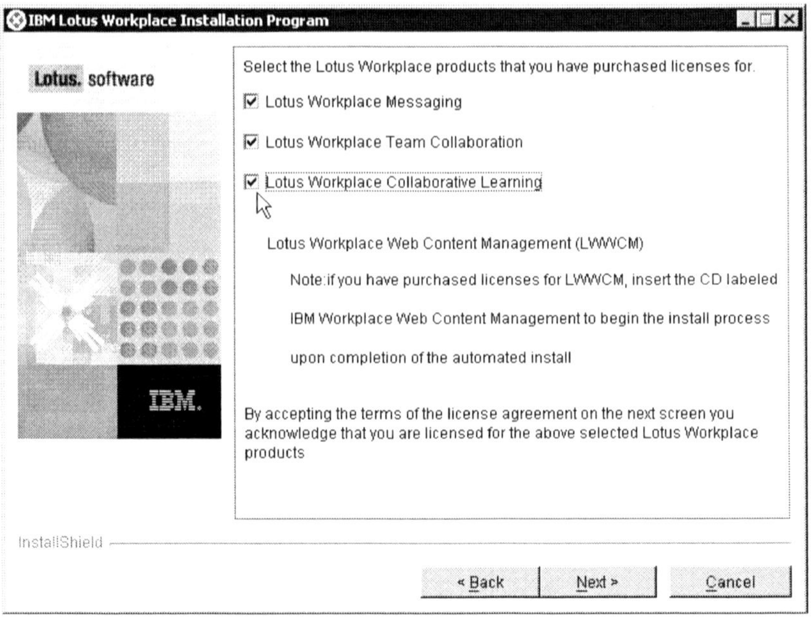

Figure F-4 Selecting the Lotus Workplace products.

4. Review the license agreement. If you accept the terms, click **I accept the terms** and click **Next**.
5. At the next screen, which describes the install options, select **Install and set configuration properties** and click **Next**, as shown in Figure F-5.
6. In the Click Next to install IBM Lotus Workplace to this directory window, accept the default, `C:\WebSphere\WorkplaceServer`, and click **Next**.
7. In the window, The following WebSphere Application Servers were found on this computer, the system should display `C:\WebSphere\AppServer`. Click **Next**.

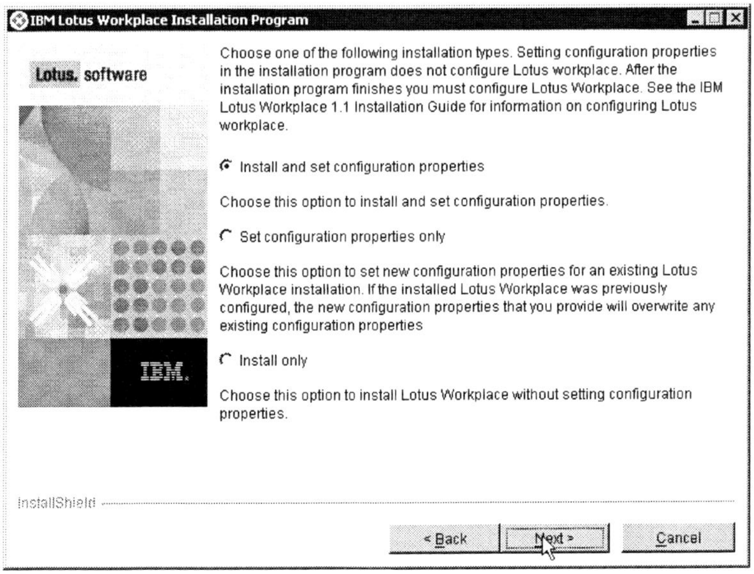

Figure F-5 Select the install and set configuration properties.

8. Based on the products you selected on the previous panel, these features are available for configuration window, leave all the defaults selected, and then click **Next**, as shown in Figure F-6.

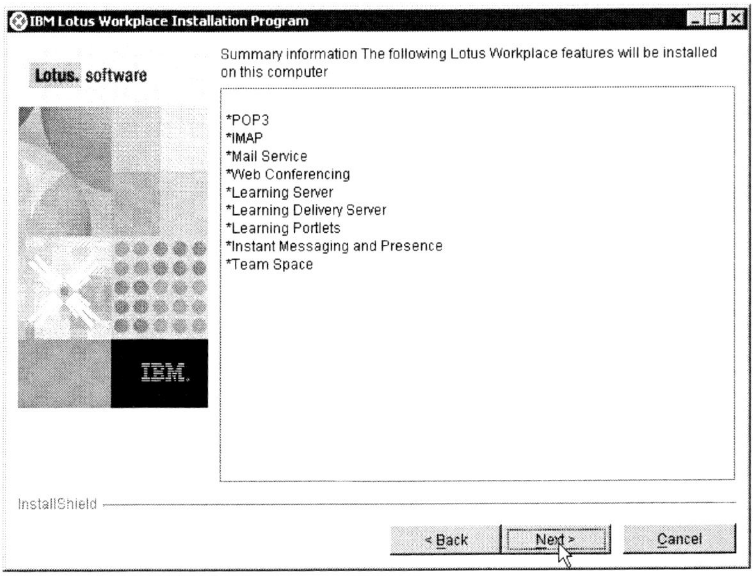

Figure F-6 Select the features to configure.

Granting Access

1. In the window Lotus Workplace will install a WebSphere Application Server on this computer, Lotus Workplace requires administrator access to this server. Enter **wpsbind** as the administrator name and **password** as the password. Then click **Next**.
2. In the window Lotus Workplace will install a WebSphere Portal on this computer. Lotus Workplace requires administrator access WebSphere Portal. Enter **wpsadmin** as the administrator name and **wpsadmin** as the password. Then click **Next**.
3. In the To create its databases, Lotus Workplace requires window, enter the local DB2 program directory, **C:\SQLLIB**. Then enter **db2admin** for the DB2 administrator name, and **db81admn** as the password. Click **Next**.
4. In the window Lotus Workplace sends failure notifications and other automatically generated administrative messages from a postmaster account, enter the mail service queue directory as **c:\qfilestore** and a valid e-mail address for the postmaster account. Click **Next**.

Learning Management Server Settings

1. In the Learning Management Server Settings window, enter **http://lwp11.tech-preview.ibm.com**, and enter **ldsadmin** as the username and **password** as the password. The username does not have to exist in a directory.
2. In the Additional Learning Management Server Settings window, do the following (see also the figure below), and then click **Next**:
   - For Learning Management Server administrator LDAP user or group name, enter **wpsadmin**.
   - For Learning Management Server administrator name type, keep the default selection, User.
   - For Path for Learning Management Server application log files, enter **c:\lms_logs**.
   - For Path for imported course packages, enter **c:\lms_packages**.
3. At the second Additional Learning Management Server Settings screen, enter **C:\juruindex** for the Juru index path and **English** for the Juru index language. Click **Next**.

Learning Delivery Server Settings

1. In the Learning Delivery Server Settings window, enter **`http://lwp11.techpreview.ibm.com`**, **`ldsadmin`** as the username, and **`password`** as the password. The delivery server will use this information to authenticate with the Learning Management Server. The username does not have to exist in a directory. Click **Next**.
2. In the Additional Learning Delivery Server Settings window, enter the following, and click **Next**:
   - For Server ID, enter **`DS1`** for the Learning Delivery Server.
   - For Description, enter **`Learning Delivery Server 1`**.
   - For Path for Learning Delivery Server application log files, enter **`c:\lms_logs`**.
3. At the second Additional Learning Delivery Server Settings window, enter the following, and click **Next**:
   - For E-mail address to receive system notifications sent by the Learning Delivery Server, enter a valid e-mail address for the administrator, **`wpsadmin@lwp11.techpreview.ibm.com`**.
   - For E-mail address from which Learning Delivery Server system notifications are sent, enter an e-mail address to display for Learning Delivery Server notifications, **`wpsadmin@lwp11.techpreview.ibm.com`**.

Learning Delivery Server Content Deployment Settings

1. In the Learning Delivery Server Content Deployment Settings window, enter **`http://lwp11.techpreview.ibm.com`**, and click **New** to specify a server to provide course content.
2. In the Learning Delivery Server Content Server Settings window, enter the following, and then click **Next**.
   - For Server name, in the Content Deployment Settings panel, enter a name to identify this Content server, **`Lwp11`**.
   - For the Deployment type, select **File system**.
   - For Location for deployed content, enter **`C:\IBMHttpServer\htdocs\en_US\content`**, and click **OK**.
3. Click **Next** with the Learning Delivery Deployment Settings defined.

Lotus Workplace Learning Settings

1. In the Lotus Workplace Learning Settings window, enter **`http://lwp11.techpreview.ibm.com`** as the URL for the Learning Management Server, and click **Next**.

2. In the window Lotus Workplace can create a single database for all installed products, or separate…, select **Create multiple databases**, and click **Next**.
3. In the window To create its databases, Lotus Workplace requires the location of the DB2 databases. Specify the following (see also Figure F-7) and then click **Next**.
   - For DB2 program directory, enter **`C:\SQLLIB`**, which should be the default selection.
   - For DB2 administrator name, enter **`db2admin`**.
   - For DB2 administration password, enter **`db81admn`**.
4. At the Warning, to create new database screen, click **Next**, and answer **Yes** to the subsequent warning specifying data will be lost.
5. In the DB2 data directory, specify the DB2 data directory window, accept the defaults `C:\lwpdata` and `C:\db2backups` as the directory to hold the backup files. Click **Next** and wait. This step may take a minute to complete. *Do not* click **Next** again.

Summary and Installation Wrap Up

1. In the Summary information window (see Figure F-7), click **Next** and wait. This step may take a minute to complete. *Do not* click **Next** again.

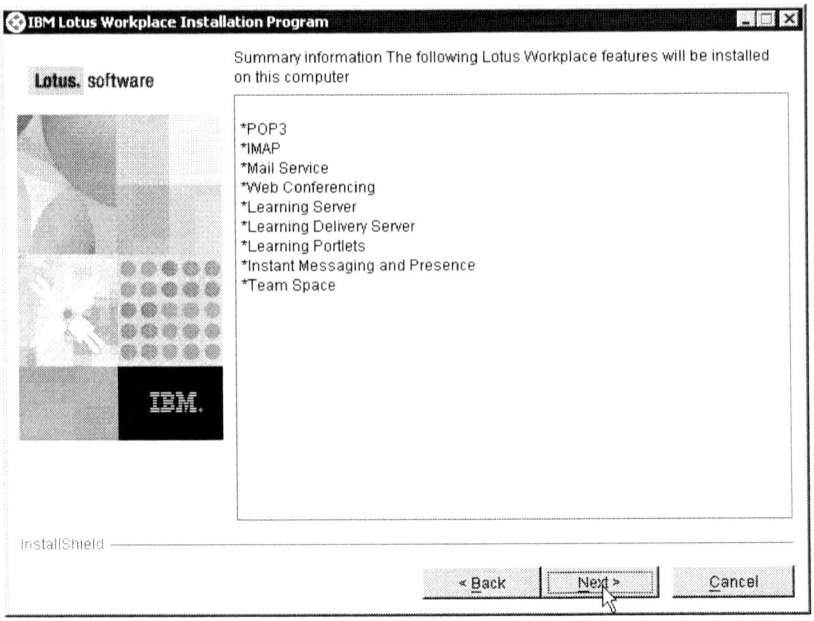

Figure F-7 Summary of Lotus Workplace features to be installed.

2. In the Please read the information below window, click **Finish**, and click **Exit** on the IBM Lotus Workplace Installer screen.
3. Do not start Lotus Workplace yet. First, configure Lotus Workplace as described in the next section.

Configuring the IBM Lotus Workplace Tutorial

Create Local Databases

After you have installed the WebSphere Application Server, WebSphere Portal Server, and Lotus Workplace products, follow these steps to configure the software. We will be capturing the output in a log file to help debug any problems that occur.

1. To create the necessary local databases:
   - Log on to your Windows system as db2admin.

     Ensure the default application server, server1, is started: `C:\websphere\appserver\bin\serverstatus all`. (Do not be concerned that WebSphere_Portal or LotusWorkplace_Server are not listed as application servers. They will be added in a subsequent steps.)
   - If server1 is not running, start it: `c:\websphere\appserver\bin\startserver server1`.
2. At the DOS prompt, go to `C:\WebSphere\WorkplaceServer\config`, and enter the following command, as shown in Figure F-8.
   - **`LwpDBconfig lwp-create-db > createdb.log`** (processing takes about 20 minutes).

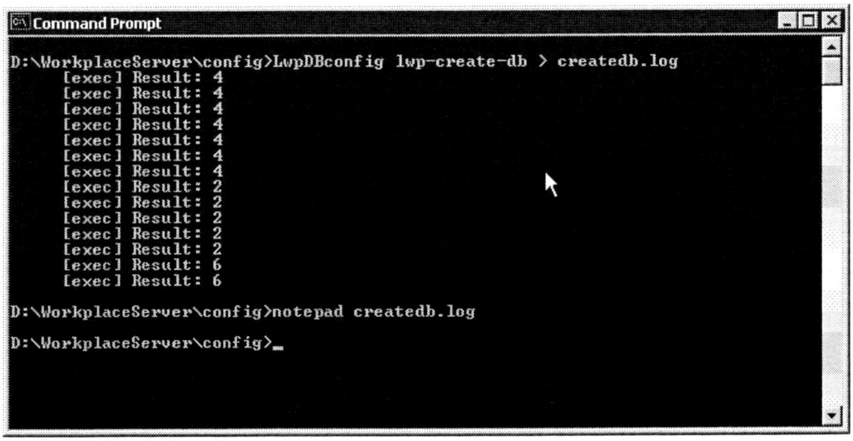

Figure F-8 DOS prompt to create local databases.

- Verify that you see `BUILD SUCCESSFUL` in the log file indicating that the databases are created.

3. Verify the list of DB2 databases, as shown in Figures F-9 and F-10.
   - At the command prompt, type the following:

```
db2cmd
db2 list database directory
```

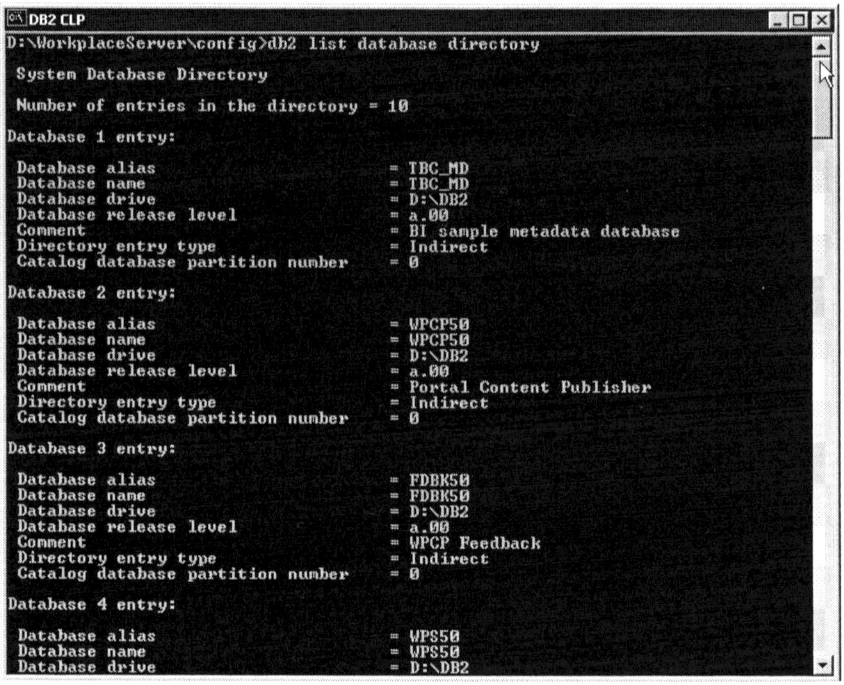

Figure F-9 Verify list of DB2 databases (screen 1 of 2).

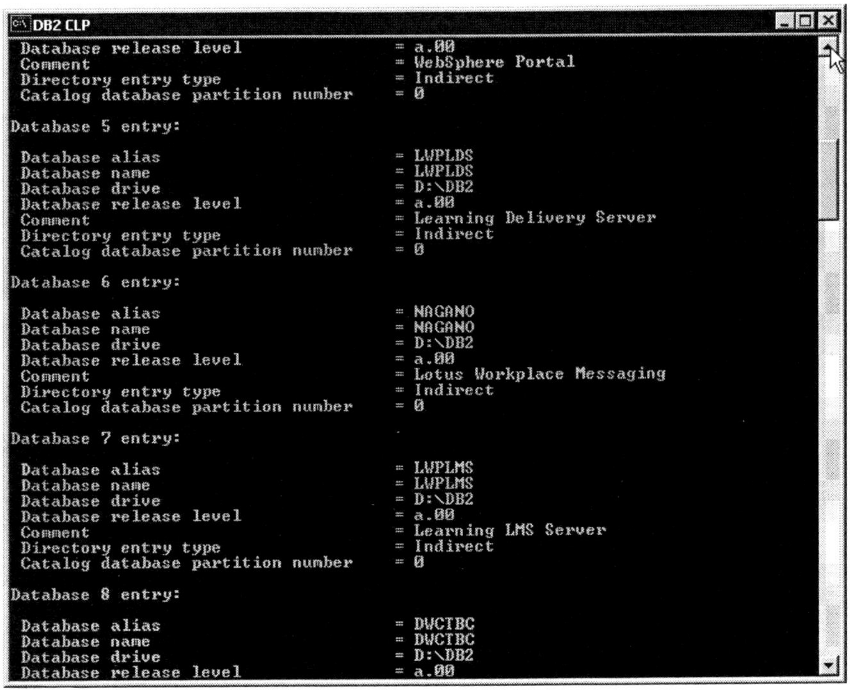

Figure F-10 Verify list of DB2 databases (screen 2 of 2).

Configure Lotus Workplace Servers

1. At the DOS prompt, go to `C:\WebSphere\WorkplaceServer\config\lwp-props.properties` and edit/configure the following property:
   - LWPHostName is set to `lwp11.techpreview.ibm.com`, as shown in Figure F-11.

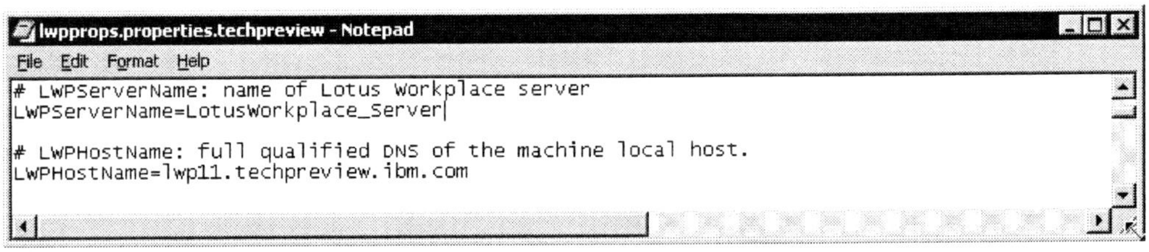

Figure F-11 Set LWPHostName.

   - `PortalSafeMode` is set to False, as shown in Figure F-12.

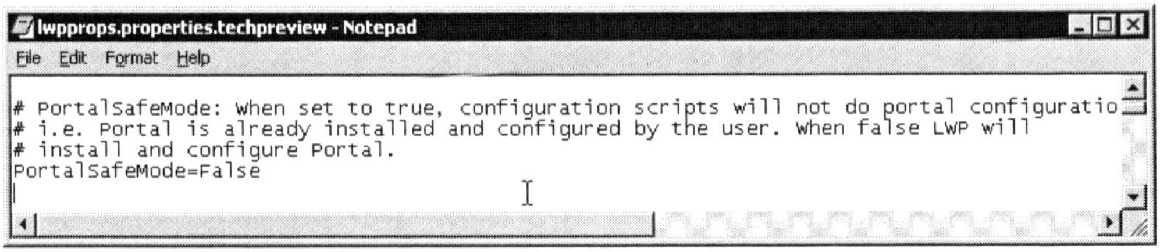

Figure F-12 Set PortalSafeMode=False.

2. Go to `C:\WebSphere\WorkplaceServer\config` and enter the following command, as shown in the Figure F-13.
   - **`lwpconfig lwp-base-setup - basesetup.log`** (this process takes about 90 minutes)

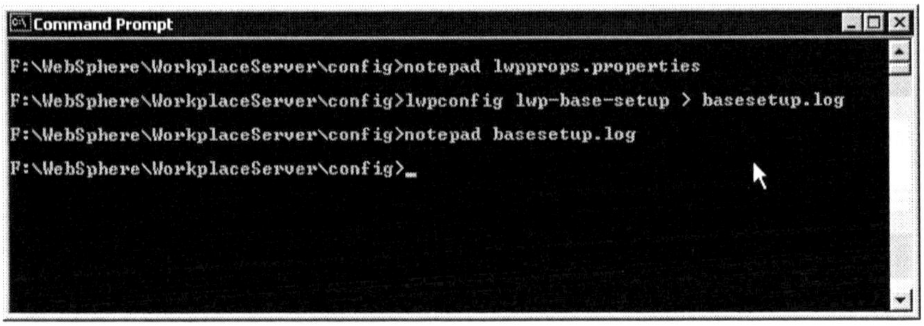

Figure F-13 Entering the lwpconfig command.

   - Verify that you see `BUILD SUCCESSFUL` in the log file indicating that the base-setup is complete.
3. Enter the following URL in your browser to verify that all servers are running: **`http://lwp11.techpreview.ibm.com:9081/lwp/workplace`**

You should see a plain login screen. *Do not* log in to the Workplace yet.

Update the Web Server Plug-In File

1. Enter the following URL in your browser: **`http://lwp11.techpreview.ibm.com:9091/admin`**
2. Log in as any user name as indicated in Figure F-14. The name is not verified, as we don't have security enabled yet.
3. Select Environment>Update web server plug-in. Click **OK** to update (see Figure F-15).

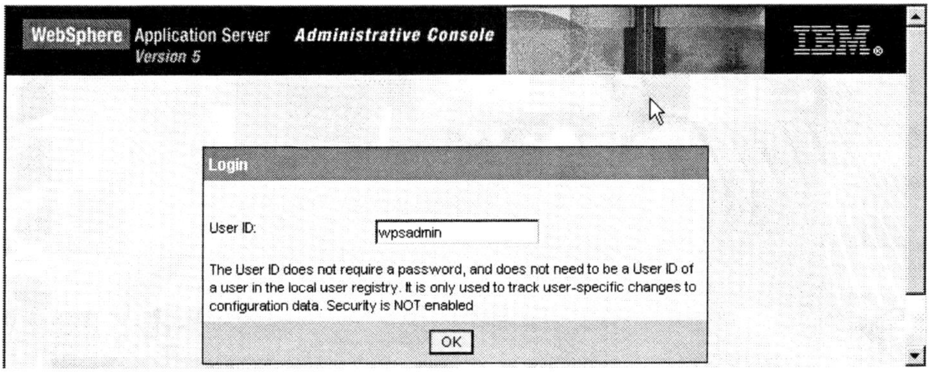

Figure F-14 Log in to WebSphere Administration.

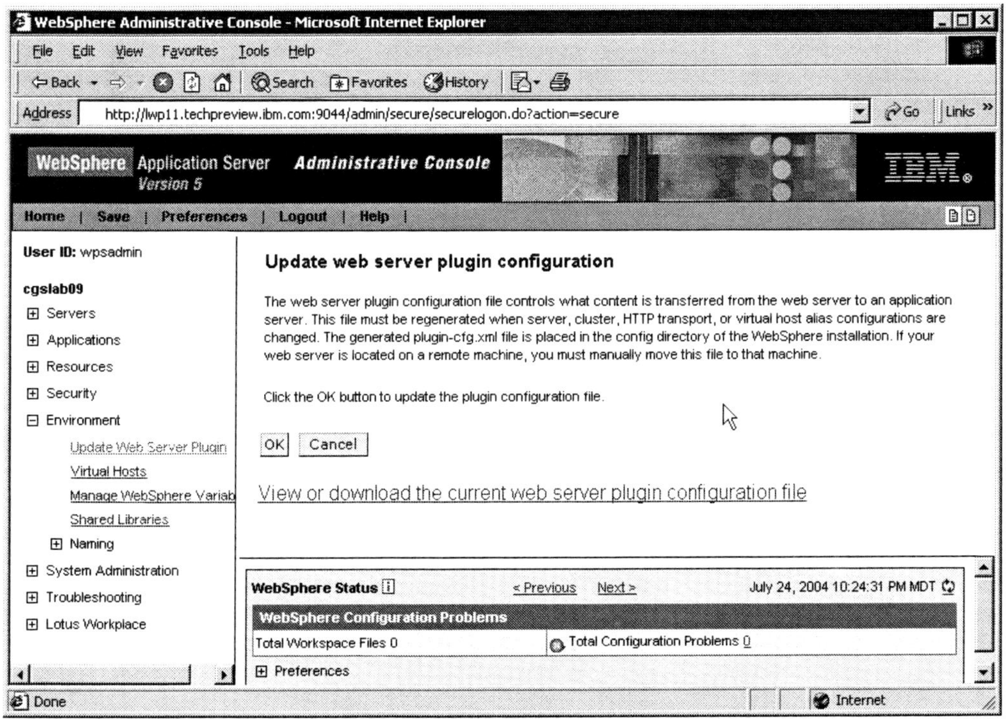

Figure F-15 Update the Web server plug-in file.

Enable Security

1. Make sure the following entries exist in the LDAP directory. Create them if necessary before continuing to the next step.
   - **A user entry for the WebSphere administrator.** The entry must have write access to the directory. The tutorial set up assumes it to be `wpsadmin` or `uid=wpsadmin,cn=users,dc=ibm,dc=com` as found in the LDAP.
   - **A user entry for the WebSphere Portal administrator.** The entry must have write access to the directory. Do not create a separate entry. Instead use the entry for the WebSphere administrator.
   - **A user entry for the LDAP administrator.** The entry must have write access to the directory. Do not create a separate entry. Instead use the entry for the WebSphere administrator

     Note: You can use one entry to represent the WebSphere administrator, WebSphere Portal administrator, and LDAP administrators, rather than a separate entry for each. This tutorial assumes one entry to represent all three administrators.
   - **A user entry with read access or greater used for WebSphere Portal searches of the directory.** This entry can be the same one used for the LDAP administrator. The tutorial set up assumes it to be `wpsbind` or `uid=wpsbind,cn=users, dc=ibm,dc=com` as found in the LDAP.
   - **A Portal administrator group entry with write access to the directory.** This group must include the WebSphere administrator, Portal administrator, and LDAP administrator names. The tutorial set up assumes it to be: `wpsadmins` or `uid=wpsadmins,cn=groups,dc=ibm,dc=com` as found in the LDAP.
2. At the DOS prompt, go to `C:\WebSphere\AppServer\bin` and enter the following command to create a backup configuration:
   - `backupconfig.bat`
3. Go to `C:\WebSphere\AppServer\bin`, and enter the following command to restart the server (see also the Figure F-16).

```
serverStatus -all
startServer server1
```

4. **Note:** Included in the Resources, you will find a completed lwpprops.properties.tech-preview file. You can copy it to `c:\websphere\workplace\config` and rename it to lwpprops.properties or use it as a working template.

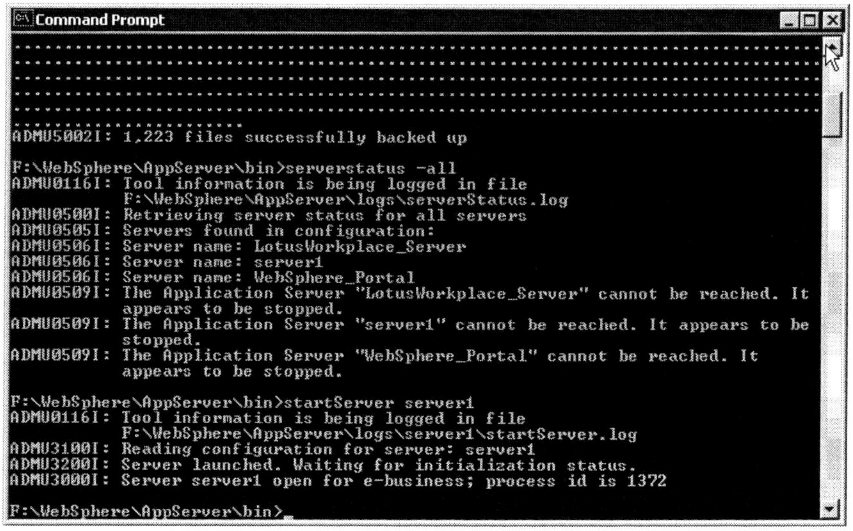

Figure F-16 Restart the server.

5. Go to `C:\WebSphere\WorkplaceServer\config`.
   - Edit lwpprops.properties, and specify the following LDAP server properties, all of which are required, as listed here in Table F-1.

Table F-1 LDAP Server Properties.

Property	Description
Wps.LDAPWasUserid=<DN>	Fully distinguished name that corresponds to the WebSphere administrator entry in the LDAP directory. This name must have write access to the directory. This tutorial assumes `uid=wpsadmin, cn=users,dc=ibm,dc=com`.
WasAdmin=<non-qualified name>	Non-qualified name of the WebSphere administrator entry. This tutorial assumes `wpsadmin`.
WasPassword=<password>	Password for the WebSphere administrator entry. This tutorial assumes `wpsadmin`.
Wps.PortalAdminIdLong<DN>	Fully distinguished name that corresponds to the Portal administrator entry in the LDAP directory. This name must have write access. The name can be the same as the one used for the WebSphere administrator. This tutorial assumes `uid= wpsadmin,cn=users,dc=ibm,dc=com`.

Property	Description
PortalAdminId=<non-qualified name>	Non-qualified name of the Portal administrator entry. This tutorial assumes `wpsadmin`.
PortalAdminPwd=<password>	Password for the Portal administrator. This tutorial assumes `wpsadmin`.
Wps.LDAPAdminUId=<DN>	Fully distinguished name that corresponds to the LDAP administrator entry in the LDAP directory. The name must have write access to the directory. The name can be the same as the one for the WebSphere administrator or Portal administrator. This tutorial assumes `cn=root`.
Wps.LDAPAdminPwd=<password>	Password for the LDAP administrator. This tutorial assumes `password`.
Wps.LDAPBindID=<DN>	Fully distinguished name that Portal uses when searching the directory. The name requires only read access to the directory. Can be the same as the LDAP administrator name. This tutorial assumes `uid=wpsbind,cn=users,dc=ibm,dc=com`.
Wps.LDAPBindPassword=<password>	Password for the name used for searches. This tutorial assumes `password`.
Wps.PortalAdminGroupId=<DN>	Fully distinguished name of the Portal administr-tor group entry in the LDAP directory. The group must have write access to the directory and include as members the WebSphere administrator, Portal administrator, and LDAP administrator names. This tutorial assumes `cn=wpsadmins,cn=groups,dc=ibm,dc=com`.
Wps.PortalAdminGroupIdShort=<non-qualified name>	Non-qualified name of the WebSphere Portal administrator group. This tutorial assumes `wpsadmins`.
WpsHostName=<hostname>	Fully qualified host name of the Portal server, for our set up: `lwp11.techpreview.ibm.com`.
Wps.SSOEnabled=<true or false>	True to use single sign-on (SSO) with Lotus Workplace products, false otherwise. SSO is configured through WebSphere Portal. Enter `true`.
Wps.SSODomainName=<SSO domain>	If using SSO, SSO domain of the Portal server, for example, .acme.com. This tutorial assumes `.techpreview.ibm.com`. ***Note****: There is a leading "." before the SSO domain name.*

continues

Table F-1 LDAP Server Properties.

Property	Description
Wps.LDAPHostName=<LDAP host name>	Fully qualified name of the LDAP server, `lwp11.techpreview.ibm.com`.
Wps.LDAPType=<type of LDAP directory server>	Type of LDAP directory. This tutorial assumes `IBM_DIRECTORY_SERVER`
Wps.LDAPSSLEnabled=<true/false>	True to use SSL for connections to the LDAP directory server, or false otherwise. Enter `false`.
Wps.LDAPSuffix=<suffix>	Suffix for the LDAP directory server. This tutorial assumes `dc=ibm,dc=com`.
Wps.LookAside=<true/false>	True to use WebSphere Portal Member Manager (recommended), false otherwise. Enter `true`.
LWMLocalDomainNames	The domains that will be considered local for receiving mail. Only users with addresses that match this will be able to use Workplace Messaging. This tutorial assumes `techpreview.ibm.com`.
LWMFullyQualifiedDomainName	Your full DNS server name. This tutorial assumes `lwp11.techpreview.ibm.com`.
LWMDnsNames	A DNS local to you. On a stand-alone system, you may need to specify a dummy name. This tutorial assumes `lwp11.techpreview.ibm.com`.

6. Go to `C:\WebSphere\WorkplaceServer\config`, and enter the following command (see also Figure F-17):

   `lwpconfig lwp-enable-ldap-security > enableSecurity.log`

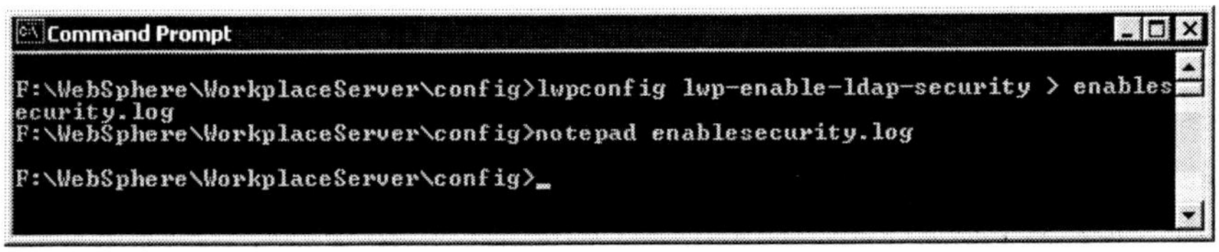

Figure F-17 Enable LDAP security.

> **NOTE**
> Ensure the LDAP server is started and you have connectivity if it is installed on a remote server.

7. Go to `C:\WebSphere\AppServer\bin` and enter the following commands (see also Figure F-18):
   - **serverStatus all -username wpsadmin -password wpsadmin**
   - Depending on which application servers are running, stop them in the following order:
     1. `stopServer LotusWorkplace_Server -username wpsadmin -password wpsadmin`
     2. `stopServer WebSphere_Portal -username wpsadmin -password wpsadmin`
     3. `stopServer server1 -username wpsadmin -password wpsadmin`
   - Now start the application servers in the following order:

     startServer server1

     startServer WebSphere_Portal

     startServer LotusWorkplace_Server

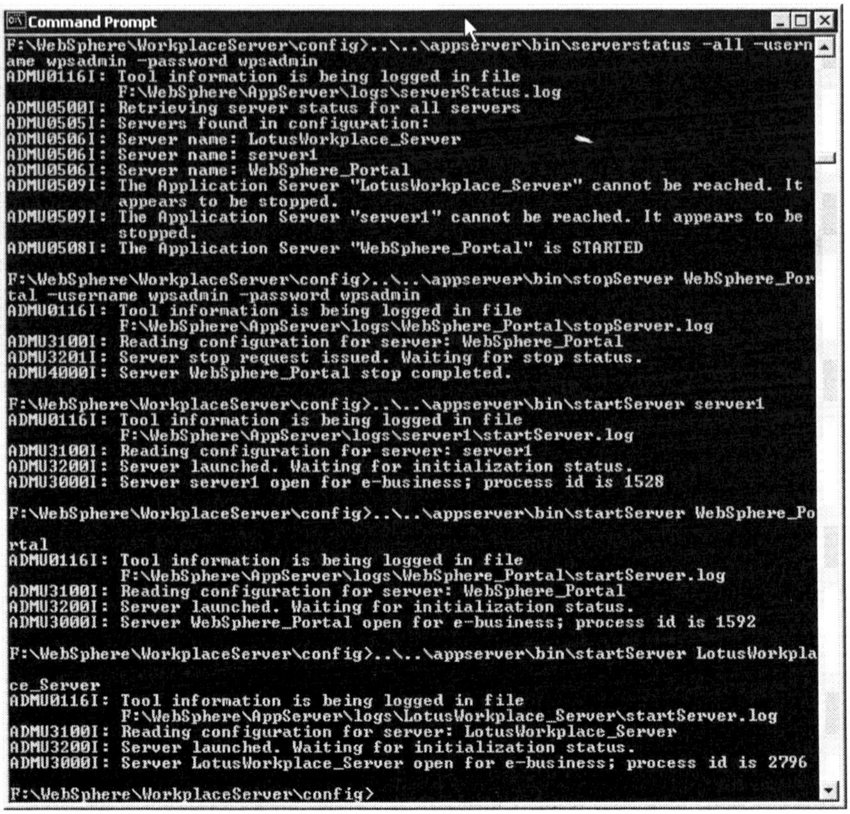

Figure F-18 StartServer commands.

- Go to Windows>Services, and stop and start the IBM HTTP server, as shown in Figure F-19.

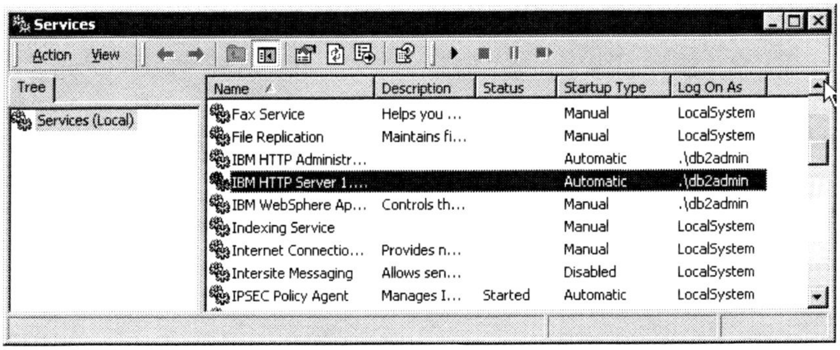

Figure F-19 Start and stop the IBM HTTP server.

> **NOTE**
> This is to allow the Updated Web Configuration to take effect on the server.

- Enter the following URL in your browser, and log in as `wpsadmin` to test and verify the completion on your IBM Lotus Workplace 1.1 installation (see Figure F-20): **`http://lwp11.techpreview.ibm.com/lwp/myworkplace`**
- Then log in as user: `wpsadmin`, password: `wpsadmin`.

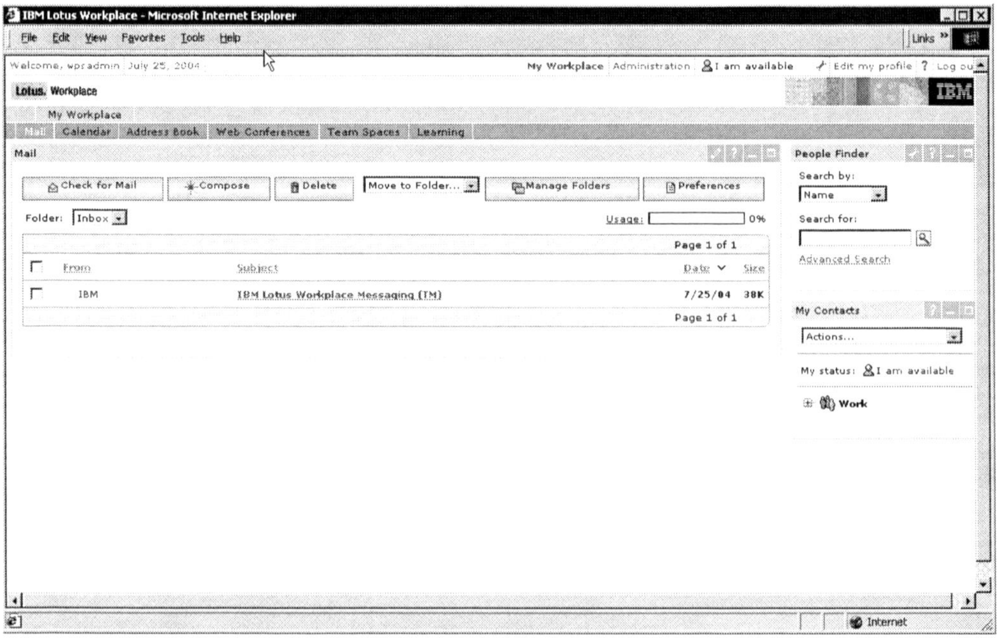

Figure F-20 Test completion of IBM Lotus Workplace 1.1 installation.

Lotus Workplace Tutorial Summary

At the end, you should have a functional install of the IBM Lotus Workplace version 1.1 environment. This proof-of-concept setup can help you understand the functionality, as well as the look and feel of this product. As a developer, you get to play with the different components of the IBM Lotus Workplace V1.1. The APIs to the different components of this product are expected to be available with the next release.

Glossary

AIX (Advanced Interactive Executive)
AIX is IBM's version of UNIX.

Applet
The Java name for a small chunk of code that is brought down over the network from a server and executed on a client.

ARPA (Advanced Research Projects Agency)
The research arm of the Department of Defense. ARPA funded the research of packet-switched networks, which led to the Internet.

Authentication
Verifying that a data transmission from a person (or a process) is indeed from that person (or process). A password is a common method used for authentication.

Awareness
This is a term used with products such as Instant Messaging and Web Conferencing. The more formal term "awareness of presence" is also used. Your Instant Messaging "buddy list" indicates who is online and hence makes you "aware" of the colleagues whom you could "chat" with online.

B2B (Business-to-Business)
Electronic commerce between companies.

Bandwidth
The amount of data that can be transmitted across a particular network. Basic Ethernet has a 10Mbps bandwidth; however, 100Mbps Ethernet and recently Gigabit Ethernet are common for corporate LANs and server farm infrastructure.

bps (bits per second)
The rate of data transmission across a network.

Browser
See **Web browser**.

Cache
A cache (pronounced CASH) is a place to store something more or less temporarily. Web pages you request are stored in your browser's cache directory on your hard disk. That way, when you return to a page you've recently looked at, the browser can get it from the cache rather than the original server, saving you time and the network the burden of some additional traffic. You can usually vary the size of your cache, depending on your particular browser.

CDN (Content Delivery Network)
A CDN consists of caching servers deployed around the world to form a network and to provide a way to offload web traffic from web servers. CDNs usually reside in co-location centers and have access to large amounts of bandwidth since they are primarily used to offload graphics and streaming media from Web servers.

Certificate
The public key and identity of an entity together with some other information, rendered unforgeable by signing the certificate with the private key of the certifying authority that issued that certificate. In the X9.59 Standard, the term "certificate" shall mean a public-key certificate.

CGI (Common Gateway Interface)
A scripting programming language used on the Internet.

Chat
A feature of most of the popular instant-messaging programs. The chat feature allows you to create your own custom chat room with friends or co-workers

Client
A computer or process that accesses the data, services, or resources of another computer or process on the network.

Collaborative Computing
A method of information sharing using computing. Collaborative computing technologies include groupware and meetingware products. These products are intended to improve personal interaction, group work, and discussion.

CORBA (Common Object Request Broker Architecture)
An architecture and specification for creating, distributing, and managing distributed program objects in a network. It allows programs at different locations and developed by different vendors

to communicate in a network through an "interface broker." CORBA was developed under the auspices of the Object Management Group (OMG) and has been sanctioned by both ISO and X/Open as the standard architecture for distributed objects (also known as components).

DARPA

Defense Advanced Research Projects Agency. See **ARPA**.

DB2 (DataBase 2)

IBM's relational database software.

DNS (Domain Name System)

The distributed name/address mechanism used in the Internet. It is through the DNS that the domain name of every computer on the Internet gets mapped to its IP address so that each computer can send information back and forth on the Internet.

Domain

Domain is a common term used in computer networks to mean: (1) that part of a network in which the data processing resources are under common control; (2) in TCP/IP, the naming system used in hierarchical networks. Domain is also a term used for a group of Domino servers with the same public Directory.

Domino

The Web-enabled Lotus Notes Server. With Release 4.5 of the Notes product, the Lotus Notes Server was renamed the Domino 4.5 Server in order to emphasize the Web technology built into that version of the server.

EAR (Enterprise Archive)

The .ear file has the same format as a regular .jar file (which is the same as ZIP, incidentally). The .ear file contains everything necessary to deploy an enterprise application on an application server. It contains both the .war (Web Archive) file containing the Web component of the application as well as the .jar file. In addition, there are some deployment descriptor files in XML.

Eclipse

The Eclipse project is an open consortium of software development tool vendors interested in creating a development environment that fosters product integration through plug-and-play, componentized tools. The core of the WebSphere Studio Application Development (WSAD) IDE platform was released as an open-source software project, Eclipse (www.eclipse.org). WebSphere Studio 5.0 is based on Eclipse 2.0.

EJB (Enterprise Java Beans)

EJB technology is the server-side component architecture for the Java 2 Platform, Enterprise Edition (J2EE) platform. EJB technology enables rapid and simplified development of distributed, transactional, secure, and portable applications based on Java technology.

GB (gigabyte)
A billion bytes of computer or hard-disk memory.

GUI (Graphical User Interface)
A pictorial way of representing to a user the capabilities of a system and the work being done on it.

HACMP (Highly Available Cluster Management Program)
HACMP is designed to allow rapid recovery from a catastrophic failure on a primary AIX machine. It allows switching of hardware resources such as external disk drives and network connections to a backup machine.

Host
In the TCP/IP sense, a computer that allows users to communicate with other host computers on a network. Individual users communicate by using application programs, such as electronic mail and FTP. Also used to refer to a large computer system, such as a mainframe.

HTML (Hypertext Markup Language)
This is the language used to write World Wide Web documents, or pages. It is a subset of ISO SGML.

HTTP (Hypertext Transfer Protocol)
This is the protocol used by the World Wide Web to transfer documents between clients and servers.

HTTPS (Hypertext Transfer Protocol, Secure)
HTTPS is a secure version of HTTP that encrypts data for transmission across the Internet.

IDE (Integrated Development Environment)
For example, the WSAD product is IBM's IDE platform for J2EE application development.

IIOP (Internet Inter-ORB Protocol)
A transport protocol used for communication between CORBA object request brokers.

IHS (IBM HTTP Server)
The IBM HTTP Server is based on the open-source Apache HTTP Server, which is the most popular server on the Web. IHS runs on AIX, HP-UX, Linux, Solaris, Windows 2000, and Windows NT.

IM (Instant messaging)
IM allows you to maintain a list of people who you want to interact with. You can send messages to any of the people in your list, often called a buddy list or contact list, as long as that person is online. Sending a message opens up a small window where you and your friend can type in messages that both of you can see.

IMAP (Internet Mail Access Protocol)
IMAP and POP3 are the Internet's standards for e-mail.

Internet
A set of connected networks. The term "Internet" refers to the large and growing public-domain internetwork developed by DARPA that uses TCP/IP. It is shared by universities, corporations, and private individuals.

InterNIC (Internet Network Information Center)
Part of the central authority for the Internet. The InterNIC (sometime called NIC for short) registers new Internet domain names and address ranges.

Intranet
A Web network that connects computers within the same company or organization over a private network. An intranet offers higher security than an Internet Web site because of the private nature of the network.

IP (Internet Protocol)
The network-layer protocol for the Internet protocol suite.

IT
Information Technology.

J2EE (Java 2 Platform, Enterprise Edition)
J2EE defines the standard for developing component-based, multi-tier enterprise applications.

JAR (Java Archive)
The JAR file format enables you to bundle multiple files into a single archive file. Typically, a JAR file will contain the class files and auxiliary resources associated with applets and applications.

Java
Sun Microsystems's object-oriented language developed for programming on the Internet. It's been stated that "Java offers the promise that the network will become the computer."

Java Beans
The applet components developed with Java.

JDK (Java Development Kit)
The JDK is the Java Development Kit distributed by JavaSoft, which provides resources for developers to develop Java applications that conform to a set of core JavaSoft APIs.

JMS (Java Message Service)
The JMS API is a messaging standard that allows application components based on the Java 2 Platform, Enterprise Edition (J2EE) to create, send, receive, and read messages. It enables distributed communication that is loosely coupled, reliable, and asynchronous.

JSP (Java Server Page)

JSP is a technology for controlling the content or appearance of Web pages through the use of servlets, small programs that are specified in the Web page and run on the Web server to modify the Web page before it is sent to the user who requested it. JSP is comparable to Microsoft's Active Server Page (ASP) technology. Whereas a Java Server Page calls a Java program that is executed by the Web server, an Active Server Page contains a script that is interpreted by a script interpreter (such as VBScript or JScript) before the page is sent to the user.

Kbps

Kilo (thousands) of bits per second.

LDAP (Lightweight Directory Access Protocol)

A set of protocols for accessing information directories. LDAP is based on the standards contained within the X.500 standard but is significantly simpler.

Linux

An Open Source implementation of UNIX initiated by Linus Torvalds, which runs on many different hardware platforms including Intel, Sparc, PowerPC, and Alpha Processors.

Lotus Notes

A workgroup application for sharing information and building business applications with integrated e-mail.

LotusScript

This is the object-oriented basic language used by Lotus Notes application developers. It is used to perform more complicated tasks than can be done using the Notes formula language.

Lotus Team Workplace

Formerly called "QuickPlace," this is a self-service Web tool for team collaboration. Also see **QuickPlace**.

Lotus Workplace

An IBM/Lotus offering that combines the collaborative capabilities of Lotus Software with the elements of the WebSphere Portal framework. Lotus Workplace is described as a portal-based collaboration platform.

LPAR (Logical Partitioning)

A system of taking a computer's total resources (processors, memory, and storage) and splitting them into smaller units that each can be run with its own instance of the operating system and applications. Logical partitioning, which requires specialized hardware circuits, is typically used to separate different functions of a system, such as Web serving, database functions, client/server actions, or systems that serve multiple time zones and/or languages.

LTPA (Lightweight Third-Party Authentication)

The LTPA mechanism provides a secure means for conveying the fact that a user has already been authenticated. Use of LTPA tokens is the typical way to set up Single Sign-On (SSO) between WebSphere and Domino.

LZ1 (Lempel-Ziv)

An adaptive data compression algorithm used by Domino 6.

MB (megabyte)

A million bytes of computer or hard-disk memory.

Netscape

A commercial browser that provides an interface to the World Wide Web.

NIC (Network Information Center)

See **InterNIC**.

NRPC

Notes Remote Procedure Call.

Packet switching

The transmission of data in small, discrete packages that are sent by different routes through a network and then reassembled at the receiving end.

PC

Personal Computer.

POP3 (Post Office Protocol 3)

POP3 and IMAP are the Internet's standards for e-mail.

Portal

A portal is an internet Web site that allows people to search for information from a number of different Internet sources, as well as the portal Web site itself. In effect, it acts as gateway to the World Wide Web.

QuickPlace

The new name for this Lotus product is "Lotus Team Workplace." This is a self-service Web tool for team collaboration. You can easily create a workspace to work in a secure environment on the Web. It provides a forum for having discussions, sharing and collaborating on documents, scheduling events, and assigning tasks.

RPC

Remote Procedure Call.

RSA
Data encryption standard that is used by the Lotus Notes/Domino products for most encryption requirements. RSA stands for Rivest-Shamir-Adleman, the three developers of the encryption technique.

Servlet
A servlet provides Web developers with a simple, consistent mechanism for extending the functionality of a Web server and for accessing existing business systems. A servlet can almost be thought of as an applet that runs on the server side—without a face. Java servlets make many Web applications possible.

SMTP (Simple Mail Transport Protocol)
This protocol utilizes TCP/IP and is a widely used e-mail protocol developed for the Internet.

SOAP (Simple Object Access protocol)
Defines a framework for passing messages between systems over the Internet. SOAP is typically used for executing remote procedure calls. The XML-based SOAP is often called a "message envelope."

SSO (Single Sign-On)
The ability to log into multiple computers or servers with a single action and the entry of a single password. Especially useful where, for example, a user on a LAN or WAN requires access to a number of different servers.

TCP/IP (Transmission Control Protocol/Internet Protocol)
The set of applications and transport protocols that uses IP (Internet protocol) to transmit data over a network. TCP/IP was developed by the Department of Defense to provide telecommunications for internetworking.

UDB (Universal DataBase)
This term is usually used with DB2, as in DB2 UDB.

UDDI (Universal Description, Discovery, and Integration)
An XML-based, platform-independent, Internet-accessible registry in which businesses, software vendors, and programmers can describe the Web services they offer and provide links on how to use them.

Unicode
Unicode provides a unique number for every character—no matter what the platform, no matter what the program, no matter what the language. The Unicode Standard has been adopted by such industry leaders as Apple, HP, IBM, JustSystem, Microsoft, Oracle, SAP, Sun, Sybase, Unisys, and many others. Unicode is required by modern standards such as XML, Java, ECMAScript (JavaScript), LDAP, CORBA 3.0, WML, and so on, and is the official way to implement ISO/IEC 10646.

UNIX

The operating system originally designed by AT&T and enhanced by the University of California at Berkeley and others. Since it was powerful and essentially available for free, it became very popular at universities. Many vendors made their own versions of UNIX available—for example, IBM's AIX, based on OSF/1. The UNIX trademark and definition have since come under the control of X/Open, which will issue a unifying specification.

URL (Universal Resource Locator)

This is the World Wide Web name for a document, file, or other resource. It describes the protocol required to access the resource, the host where it can be found, and a path to the resource on that host.

UTF-8 (Unicode Transformation Format-8)

UTF-8 is an octet (8-bit) lossless encoding of Unicode characters. It is an efficient encoding of Unicode documents that use mostly US-ASCII characters because it represents each character in the range U+0000 through U+007F as a single octet. UTF-8 is the default encoding for XML.

UTF-16 (Unicode Transformation Format-16)

A 16-bit encoding scheme for Unicode characters. UTF-16 is an efficient way of representing Asian character strings, with each character being represented by two bytes of storage. ASCII characters are usually represented by UTF-8 since one byte of storage is sufficient for ASCII.

WAN (Wide Area Network)

A long-distance network for the efficient transfer of voice, data, and/or video between local, metropolitan, campus, and site networks. WANs typically use lower transfer rates (64 Kbps) or higher-speed services such as a T3, which operates at 45 Mbps. WANs also typically use common-carrier services (communications services available to the general public) or private networking through satellite and microwave facilities.

WAR (Web Archive)

Portlet applications are packaged as Web Archive (.war) files, per the J2EE specifications.

WAS (WebSphere Application Server)

See **WebSphere** in this glossary to see how WAS fits under the WebSphere umbrella.

Web browser

An application that provides an interface to the World Wide Web.

Web conferencing

Web conferencing allows you to collaborate using the web to sharing software applications, PowerPoint presentations, or anything running on your PC with others in online meetings.

Web service

A Web-based application that can dynamically interact with other Web applications using an XML message protocol such as SOAP, XML-RPC, or XMLP. Examples of emerging standards for describing, promoting, and discovering these services are ebXML, UDDI, and WSDL, Microsoft's .NET and Sun's Sun ONE are major implementations of the concept. The goal is to enable one application to find another on the Internet that provides a needed service and to seamlessly exchange data with it. If the service is fee-based, payment processing could be included.

Webmaster
A person who manages a Web site—similar to the network administrator.

WebSphere
IBM's Web server software. A Java-based platform, designed to run on many different operating systems. IBM describes WebSphere as Internet infrastructure software, or middleware, that enables companies to develop, deploy, and integrate next-generation e-business applications. WebSphere supports business applications from simple Web publishing through enterprise-scale transaction processing. IBM identifies WebSphere products as fitting into one of three areas: foundation and tools, business portals, or business integration.

1. ***Foundation and tools.***

   The foundation and tools provide customers with the ability to build, deploy, and manage the creation of J2EE-based Web applications. The products include WebSphere Application Server, WebSphere Application Server Express, and WebSphere Studio.

2. ***Business portals.***

   Business portals let organizations extend applications to users in a single, customized environment and also extend the applications to a variety of electronic devices. The products include WebSphere Commerce, WebSphere Portal Server, WebSphere Portal Express, WebSphere Everyplace, and WebSphere Personalization.

3. ***Business integration.***

   Business Integration products provide for integration between new and existing applications in a network. The products include WebSphere MQ, WebSphere Business Integration, IBM CrossWorlds solutions, and adapters and connectors.

WebSphere portal
Software for building and deploying enterprise portals that allow users to interact in highly collaborative settings.

Wide Area Network
See **WAN**.

Workplace
See **Lotus Workplace**.

World Wide Web

The World Wide Web, or WWW, is a wide-area hypermedia information retrieval initiative aimed at giving universal access to a large universe of documents. To access the Web, you use a Web browser that provides hypertext links to jump to information on many different Internet Web servers.

WPS (WebSphere Portal Server)

See **WebSphere** in this glossary to see how WPS fits under the WebSphere umbrella.

WSAD (WebSphere Studio Application Developer)

An enterprise development tool from IBM that gleans the best features from VisualAge for Java Enterprise Edition and WebSphere Studio. It supports the full cycle of J2EE development, as well as XML and Web services development.

WSADMIN

WebSphere Administration.

WSDL (Web Services Description Language)

An XML format for describing network services as a set of endpoints operating on messages containing either document- or procedure-oriented information. WSDL allows a Web service to describe its capabilities by defining Web services interfaces and describing how to access them.

XML

Short for eXtensible Markup Language, a specification developed by the W3C (World Wide Web Consortium). XML is designed especially for Web documents; basically, it is a "light" version of SGML. It gives designers the ability to create their own customized tags to provide functionality not available with HTML. The open XML standard is emerging as the method of choice for exchanging information among computer systems.

Bibliography

The following books, articles, and web sites give additional information relating to WebSphere, Domino, and Implementing WebSphere Application Server and Lotus Domino Solutions.

Books and Articles

Conte, Paul. "Inside WAS V5." *e-Pro Magazine*, August 2002. (This article gives some good summary information on WAS 5.0.)

Dantzig, Paul. "Architecture and Design of High Volume Web Sites (A Brief History of IBM Sport and Event Web Sites)." *The Fourteenth International Conference on Software Engineering and Knowledge Engineering*, 15-19 July 2002, Ischia, Italy. (This article provided much of the information on IBM's designs for the Olympic Web Sites that appears in Chapter 15 of this book.)

Eden, Timothy. "Use WAS Clustering to Increase Application Availability." *e-Pro Magazine*, December 2003. (This article gives some good pointers on using WAS clustering based on the author's experience.)

Flanagan, David. *Java in a Nutshell, Fourth Edition*. Sebastopol: O'Reilly & Associates, Inc., 2002. (This is one of the most popular books on Java. Since it's compact and yet comprehensive, it's the kind of book you'll take with you in your knapsack so you can look up that Java command no matter where you are!)

Iyengar, Ashok, and Venkata Gadepalli. *IBM WebSphere Portal Primer*. Double Oak: IBM Press/MC Press LLC, 2003. (This book gives a good introduction to the WebSphere Portal.)

Lamb, John P., and Peter W. Lew. *Lotus Notes & Domino 5 Scalable Network Design (Web Server Network Infrastructure)*. New York: McGraw-Hill, 1999. (This book gives information on designing large-scale Domino systems and networks.)

Nielsen, Soren Peter, et al. *Domino and WebSphere Together*. Poughkeepsie: IBM Corporation, International Technical Support Organization, 2000. (This book is part of IBM's Redbook group and was written by IBM and Lotus personnel from different parts of the world.)

Rot, Dennis. "Collaboration Features in WepSphere Portal." *e-Pro Magazine*, December 2003. (This article gives a good overview of the Lotus collaboration features available with WebSphere Portal.)

Web Sites

Using WebSphere Application Server V5 for Load Balancing and Failover, <http://www.ibm.com/developerworks/library/i-wasldbal/>. (This is an excellent, short, step-by-step example that shows how to set up and test a cluster using a sample application shipped with WAS.)

WAS V5 Getting Started, <ftp://ftp.software.ibm.com/software/webserver/appserv/library/wasv5base_gs.pdf>. (This PDF manual has detailed instructions for installing the WAS base product.)

WAS ND V5 Getting Started, <ftp://ftp.software.ibm.com/software/webserver/appserv/library/wasv5nd_gs.pdf>. (This PDF manual has detailed instructions for installing the Deployment Manager and setting up a cell and cluster. Unfortunately, the explanations aren't always that clear.)

WAS V5.0 System Management and Configuration, <http://publib-b.boulder.ibm.com/Redbooks.nsf/RedbookAbstracts/sg246195.html?Open>. (Excellent IBM Redbook with installation and configuration instructions and helpful advice on choosing the right topology and configuration.)

Deploying J2EE Applications in WebSphere Application Server Network Deployment Version 5.x, <http://www7b.boulder.ibm.com/webapp/dd/transform.wss?URL=/wsdd/library/bestpractices/deploying_j2ee_applications.xml&xslURL=/wsdd/xsl/bestpractice.xsl>. (Once you have your basic cluster configured, check this article for tips on deploying applications.)

Server Clusters for High Availability in WebSphere Application Server Network Deployment Edition 5.0, <http://www.ibm.com/support/docview.wss?uid=swg27002473>. (Very broad coverage of HA topics with some helpful definitions and best practices.)

Implementing a Highly Available Infrastructure for WebSphere Application Server Network Deployment, Version 5.0 without Clustering, <http://www7b.boulder.ibm.com/webapp/dd/transform.wss?URL=/wsdd/library/techarticles/0304_alcott/alcott.xml&xslURL=/wsdd/xsl/document.xsl&format=one-column>. (If you have only a single system on which to run WAS, this article provides advice on maximizing availability using WAS ND. The main idea is protecting against extended unavailability of the Deployment Manager.)

Beaton, Wayne. Migrating from WAS V3.5 to V5, Parts 1 and 2, <http://www.e-promag.com/eparchive/index.cfm?fuseaction=viewarticle&ContentID=4209.2.http://www.e-promag.com/eparchive/index.cfm?fuseaction=viewarticle&ContentID=4330>. (Nice, practical information on migrating from V3.5 to V5.)

Summary of IBM Redbooks on WebSphere and Domino

IBM Redbooks, named for their original red covers, are how-to-books, written by very experienced IBM professionals from all over the world. They are written at the International Technical Support Organization (ITSO), which has centers close to IBM's development divisions. The following are IBM Redbooks related to implementing WebSphere and Domino together. IBM Redbooks can be ordered on the Web at `http://www.redbooks.ibm.com` using the Redbook number indicated in the following listings (the "SG" number):

Domino and WebSphere Integration on the IBM eServer iSeries Server, SG24-6223-00: Redbook, March-28-2002, last updated April-2-2002.

Portalizing Domino Applications for WebSphere Portal, SG24-7004-00: Redbook, September-22-2003.

Integrating Lotus Domino 6 and WebSphere V5 Express on the IBM eServer iSeries Server, SG24-6998-00: Redpiece, last updated October-8-2003.

Applying the Patterns for e-business to Domino and WebSphere Scenarios, SG24-6255-00: Redbook, October-19-2001.

Domino and WebSphere Together, Second Edition, SG24-5955-01: Redbook, July-3-2001.

Design and Implement Servlets, JSPs, and EJBs for IBM WebSphere Application Server, SG24-5754-00: Redbook, August-2000.

Patterns: Custom Designs for Domino & WebSphere Integration, SG24-6903-00: Redbook, April-7-2003.

Linux Web Hosting with WebSphere, DB2, and Domino, SG24-6007-00: Redbook, June-27-2000.

Domino Web Service Application Development for the IBM eServer iSeries Server, SG24-6862-00: Redbook, November-21-2002.

RDB Synchronization, Transcoding and LDAP Directory Services in IBM WebSphere Everyplace Access Version 4.1.1, SG24-6856-00: Redbook, November-22-2002, last updated November-26-2002.

Implementation and Practical Use of LDAP on the IBM eServer iSeries Server, SG24-6193-00: Redbook, April-29-2002, last updated May-21-2002.

Application Service Provider Business Model: Implementation on the iSeries Server, SG24-6053-00: Redbook, April-2-2001.

WebSphere Commerce V5.4 Handbook: Architecture and Integration Guide, SG24-6567-00: Redbook, September-30-2002, last updated October-22-2002.

IBM Host Integration in a Secure Network: A Practical Approach, SG24-5988-00: Redbook, July-26-2001.

Upgrading to Lotus Notes and Domino 6, SG24-6889-00: Redbook, December-22-2002.

IBM eServer iSeries e-business Handbook: A V5R1 Technology and Product Reference, SG24-6711-00: Redbook, October-31-2001, last updated November-1-2001.

Developing an e-Business Application Using Lotus Domino for AS/400, SG24-6052-00: Redbook, June-16-2000.

Lotus Domino for S/390 Release 5: Enterprise Integration Using Domino Connector Products, SG24-5682-00: Redbook, May-7-2001.

iSeries Access for Web V5R2 and WebSphere Host Publisher V4.0, SG24-6804-00: Redbook, September-11-2002.

IBM WebSphere V4.0 Advanced Edition Security, SG24-6520-00: Redbook, March-20-2002.

iSeries e-business Handbook: A Technology and Product Reference, SG24-5694-01: Redbook, March-20-2001.

IBM WebSphere V5.0 Security WebSphere Handbook Series, SG24-6573-00: Redbook, December-16-2002, last updated January-6-2003.

Migrating to WebSphere V5.0: An End-to-End Migration Guide, SG24-6910-01: Redbook, April-18-2003.

Index

B

C

K-L

Q-R

T

U

V

X-Z